DYNAMICS AND EVOLUTION OF MINOR BODIES WITH GALACTIC AND GEOLOGICAL IMPLICATIONS

DYNAMICS AND EVOLUTION OF MINOR BODIES WITH GALACTIC AND GEOLOGICAL IMPLICATIONS

Proceedings of the Conference held in Kyoto, Japan from October 28 to November 1, 1991

Edited by

S. V. M. CLUBE
Department of Physics, University of Oxford, U.K.

S. YABUSHITA
Department of Applied Mathematics and Physics, Kyoto University, Japan

and

J. HENRARD
Département de Mathématique, FUNDP, Namur, Belgium

Reprinted from Celestial Mechanics and Dynamical Astronomy, Volume 54, Nos. 1–3, 1992

KLUWER ACADEMIC PUBLISHERS
DORDRECHT / BOSTON / LONDON

Library of Congress Cataloging-in-Publication Data

Dynamics and evolution of minor bodies with galactic and geological implications / edited by S.V.M. Clube and S. Yabushita and J. Henrard.
p. cm.
ISBN 0-7923-1933-8
1. Comets--Congresses. 2. Asteroids--Congresses. 3. Meteors--Congresses. 4. Celestial mechanics--Congresses. I. Clube, S. V. M. II. Yabushita, S. III. Henrard, J.
QB717.D95 1992
523.6--dc20 92-23368

ISBN 0-7923-1933-8

Published by Kluwer Academic Publishers,
P.O. Box 17, 3300 AA Dordrecht, The Netherlands.

Kluwer Academic Publishers incorporates
the publishing programmes of
D. Reidel, Martinus Nijhoff, Dr W. Junk and MTP Press.

Sold and distributed in the U.S.A. and Canada
by Kluwer Academic Publishers,
101 Philip Drive, Norwell, MA 02061, U.S.A.

In all other countries, sold and distributed
by Kluwer Academic Publishers Group,
P.O. Box 322, 3300 AH Dordrecht, The Netherlands.

Printed on acid-free paper

Printed in the Netherlands

CELESTIAL MECHANICS AND DYNAMICAL ASTRONOMY

Vol. 54 Nos. 1–3 1992

SHORT CONTRIBUTIONS

PREFACE

During recent years there has been considerable discussion concerning the dynamical as well as the physical evolution of minor bodies in the Solar System, namely comets, asteroids and meteors. This, in our view, is due to three factors. First, it has come to be realized that the minor bodies, though they represent but a small fraction of the mass of the Solar System and possibly our Galaxy as a whole, may well play a fundamental role in terrestrial evolution and hence it is now important to understand their dynamics with this additional point in mind. Secondly, due to the results obtained through international collaborations leading to the observation of Comet Halley from various spacecrafts, a vast amount of new knowledge has emerged concerning the physical as well as the chemical aspects of cometary nuclei. Thirdly, along with the rapid improvement in the facilities for carrying out large scale computations, celestial mechanicians has come to realize that the Solar System is not as regular and as unchanging as a perfect clock but that it evolves and exhibits chaotic behaviour. All these factors lead one now to conclude that the minor bodies of the Solar System are far from minor in fact! We need to know what control their motion and how they evolve both dynamically and physically before intersecting the Earth.

Under these circumstances, where the climate of research opinions is itself undergoing a seachange, it seems appropriate and timely to organize a meeting on an international basis at which all these aspects of minor bodies could be considered by experts in the fields in order to gain further insights as to the problems involved and where mutual exchange of views would be expected to induce stimulating discussion. The meeting and conference thus inspired was held from October 28 through November 1, 1991 at Kyoto Sangyo University, Kyoto, Japan. The present volume is the outcome of the papers and discussions presented at this meeting. As will be seen, the volume consists of research and review papers along with abstracts of shorter contributions to the meeting.

Broadly, the volume contains investigations of the dynamics and physics of comets, asteroids and meteor streams, the rather controversial topic of periodic phenomena in the Solar System as signified by geological records and several associated developments in celestial dynamics. In fact, all these problems are interwoven, and it is hoped that the present volume will make a modest contribution toward unravelling the nature of the interactions between the Earth and its celestial environment, besides attracting the attention of newcomers to a fascinating subject.

The meeting could hardly have taken place without the generous support of the Dynic Corporation, the Murata Machinery Corporation and Kyoto Sangyo University, each of which assisted in many ways. We wish to express our cordial thanks

Celestial Mechanics and Dynamical Astronomy **54:** vii–viii, 1992.

along with those of all participants to these organizations for making the present meeting possible. In addition to the scientific sessions at Kyoto Sangyo University, a lively discussion session was also held at the Astropark Observatory, to which the participants were invited and made welcome for the final day of the meeting. Finally, we wish to thank the Local Organizing Committee (I. Hasegawa, T. Inoue and J. Yoshida) for their arrangements which made the meeting so enjoyable for all participants.

S.V.M. CLUBE
J. HENRARD
S. YABUSHITA

(*From left to right first row*) Miura, Clube, Marchal, Petrovskaya, Sabake (owner of the Dynic Corporation), Whipple, Yabushita, Henrard, Bababzhanov, Hasegawa, Yoshida, Ohnishi.
(*second row*) Yoneda, Yoshikawa, Hirayama, Yuasa, Zappalà, Froeschlé, Bailey, Kosai, Isobe, Rampino.
(*third row*) Emel'janenko, Watanabe, Inoue, Tsujii, Dvorak, Nomura, Nakamura, Rickman, Matese, Valtonen.

THE ACTIVITIES OF COMETS RELATED TO THEIR AGING AND ORIGIN

FRED L. WHIPPLE
Smithsonian Astrophysical Observatory, 60 Garden Street, Cambridge, Massachusetts, 02138, U.S.A.

Abstract. Two indices have been developed for the purpose of comparing the natures of various classes of comets. The first is the *Activity Index* (AI), measuring the inherent magnitude increase in brightness from great solar distances to maximum near perihelion. The second, or *Volatility Index* (VI), measures the variation in magnitude near perihelion. Tentative determinations of these two indices are derived from observations by Max Beyer over more than 30 years for long-period (L-P) and short-period (S-P) comets near perihelion and from other homogeneous sources. AI determinations are made for 32 long-period (L-P) comets and for 14 short-period (S-P). The range of values of AI is of the order of 3 to 10 magnitudes with a median about 6. An expected strong correlation with perihelion distance q, is found to vary as $\sim q^{-2.3}$. Residuals from a least-square solution (ΔAI) are used for comparing comets of different orbital classes, the standard deviation of a single value of ΔAI is only $\pm 1^m.1$ for L-P comets and $\pm 1^m.2$ for S-P comets.

Among the L-P comets, 19 of period P larger than 10^4 years yield $\langle \Delta \mathrm{AI} \rangle = 0^m.27 \pm 0^m.25$ compared to $0^m.39 \pm 0^m.26$ for 13 of period between 10^2 years and 10^4 years. This denies any fading with aging among the L-P comets. Also no systematic change with period occurs for the VI index, leading to the same conclusions. Weak correlations are found with the Gas/ Dust ratio of comets. No correlations are found between the two indices, nor of either index with near-perihelion magnitudes or orbital inclination.

The various data are consistent with a uniform origin for all types of comets, the nuclei being homogeneous on the large scale but quite diverse on a small scale (the order of a fraction of kilometer in extent). Small comets thus may sublimate away entirely, leaving no solid core, while huge comets may develop a less volatile core by radioactive heating and possibly become inactive like asteroids after many S-P revolutions about the Sun. When relatively new, huge comets may be quite active at great solar distances because of volatiles from the core that have refrozen in the outer layers.

Key words: Comets, aging of comets, activity of comets.

1. Introduction

In recent years the "classical" notion that short-period (S-P) comets derive from the Oort (1950) cloud of randomly moving long-period (L-P) comets is being questioned. The difficulty is purely quantitative. As Joss (1973) and others have pointed out, the calculated supply from the L-P randomly oriented orbits of the Oort cloud fails by orders of magnitude to account for the loss rate expected among the S-P comets. This failure then demands that the S-P comets derive from an inner cloud or cometary belt from which rather long-period comets in low-inclination orbits acquire Neptune-crossing orbits perhaps as deduced by Duncan, Quinn, and Tremaine (DQT) (1987, 1988). In the concept due to Hills (1981) an inner cloud within $1 - 2 \times 10^4$ AU of the Sun is much more massive than the "conventional" Oort cloud beyond. He (Hills, 1982) has postulated that the comets were formed by

Celestial Mechanics and Dynamical Astronomy **54:** 1–11, 1992.

radiation pressure from the early Sun while the protosolar nebula was collapsing, so that the inner part of the cloud is fairly symmetrical about the Sun. Most other investigators such as DQT and Fernandez (1980) rely on the accumulation of comets from a concentration of particles near the planetary plane as in the process developed by Goldreich and Ward (1973). If the protosolar nebula is assumed to have extended appreciably beyond the limits of the known planets, we return to the idea of the so-called Kuiper (1951) belt.

Fernandez brings the potential S-P comets from this roughly ecliptic-plane to the outer planetary region assuming that the largest comets attain masses of the order row 10^{24} to 10^{26} g, massive enough to disturb the orbits of the smaller comets. DQT have no detailed mechanism for the reducing the orbits of the belt comets. They note that comets of masses 10^{24} to 10^{26} g, should have been discovered, if they exist in the outskirts of the known planetary system. In any case DQT have a very convincing theory for producing the S-P comets; it requires that comets in low-inclination orbits dip into the planetary region to be further disturbed by the outer planets. On the other hand, the Hills inner cloud is too symmetrical to provide a good source for the S-P comets unless some distant relatively massive body in a low-inclination orbit still remains to be discovered.

Of course, all of this thinking is bassed on the assumption that we are now observing a fairly steady state of cometary capture and decay. If, on the other hand, the rate of new comet passages through the inner solar system varies markedly from time to time over the ages, we may be deceived by an excess or deficiency from the average rate at the present period of observation.

Because a measure of the comet rate involves impacts and crater formation on the Earth and on the solid planets and satellites, we are led into the problem of distinguishing between comets and asteroids as impactors. This further raises the thorny question as to whether some, many or no S-P comets finally become indistinguishable from asteroids at interplanetary distances. Thus this paper will deal with some of these problems of cometary origin, aging and decay utilizing the conclusions reached from two new measures of cometary activity, the *Activity* and the *Volatility Indices*, described below.

2. The Activity Indices

The *Activity Index* (AI) (Whipple, 1991) is intended to measure the increase in magnitude of the intrinsic brightness of a comet from relative inactivity at great solar distances to its maximum activity in the neighborhood of perihelion.

The *Volatility Index* (VI) (Whipple, 1992) is intended to measure the magnitude variations of a comet near the time of its maximum activity.

These indices have been approximated from certain observation sets making use of the simple formulae and definitions to follow. The classical absolute magnitude of a comet, Hy, is defined (e.g. Vsekhsvyatskii, 1964) by

$$Hy = m - 5 \log \Delta - y \log r \tag{1}$$

where m is the observed magnitude, Δ and r the geocentric and heliocentric distances, respectively, measured in AU, and $y = 2.5n$, with n being the inverse power of r in the luminosity equation.

For typical comets, $n \sim 4$ $(y \sim 10)$ when they are active near perihelion, although subject to wild deviations. Ideally n decreases to 2 $(y = 5)$ at large r when, theoretically, the comets are reflecting sunlight as inert bodies. There the absolute magnitude ideally becomes H_5. The extreme measured value of H_5 can be corrected to its opposition value by the addition of the term $2.5 \log \theta(\alpha)$ where $\theta(\alpha)$ is a function of the phase angle, α, measured at the comet between the Earth and the Sun. The phase-angle correction is yet to be well measured for comets so the correction is approximated by $-0^m.03\alpha$, where α is measured in degrees and is generally small (see e.g. Kresák, 1973). Hence the extreme absolute magnitude is defined as

$$h_0 = H_5 - 0.03\alpha, \tag{2}$$

evaluated near the maximum observed solar distance, r_0.

For comparison, the perihelion (q) value of the magnitude is taken as Hy, the absolute magnitude near perihelion later corrected to $r = q$ with $(y = 2.5n)$, the correction term being $+y \log q$, where q is measured in AU.

The Activity Index, AI, is then determined by the difference between these two magnitudes, h_0 and Hy, both evaluated at $r = q$, as

$$\mathrm{AI} = h_0 - Hy + (5 - y) \log q. \tag{3}$$

The Volatility Index requires some standard against which to measure the brightness variations as the comets' heliocentric distances changes. No such standard has universal acceptance. Nevertheless, out of necessity, I adopted H_{10}, or an inverse r^4 law, as a working standard. Consequently, the Volatility Index, VI, is defined as the standard deviation of a single cometary magnitude measurement by means of a least-squares determination of H_{10} (Equation (1)) from a series of measurement, m made near the time of the comets maximum brightness in the neighborhood of perihelion.

$$H_{10} = m - 5 \log \Delta - 10 \log r. \tag{4}$$

For VI and generally near periherion the integrated magnitude of nucleus and coma is appropriate whereas the nuclear magnitude, if observed, is appropriate for measurement of H_5 at great solar distances. In the calculation of both AI and VI, compromises from the ideal circumstances could not be avoided. In the case of magnitudes near perihelion, the measures are sensitive to the area of sky observed as well as the telescope used, introducing deviations in both indices. In the case of measures at great distances, many comets still show evidence of coma, not readily separated from the nuclear brightness (see e.g. Meech, 1991).

Hence the applications of the two indices as published must be accepted as less than ideal, although the writer used the data that seemed best suited to the two situations. Near q the extraordinarily uniform measures over more than 40 years by Max Beyer (1969, for reference summary, and 1972) provided a unique source, while for measures at great r, the major contributors were Hamilton M. Jeffers and Elizabeth Roemer with their associates, and the perennial George Van Biesbroeck. In addition, the cometary magnitudes complied by N.T. Bobrovnikoff (1941, 1942) provided values of VI for some 41 L-P and 3 S-P comets. Thus preliminary values of AI have been determined for 32 L-P and 14 S-P comets (Whipple, 1991) and VI for 109 L-P and 16 S-P comets (Whipple, 1992).

The activity index as expected, varied strongly with the inverse perihelion distance of the comet and also showed some dependence on the year of observation and the magnitude of maximum distance, h_0. Corrections for these effects were expressed in three equations, here combined as

$$\Delta \mathrm{AI} = \mathrm{AI} - (3^m.028 + 3^m.205\Delta t - 5^m.645 \log q) \tag{5}$$

where $\Delta t = (\mathrm{year} - 1900)/100$ and q is in AU.

Luminosity thus varies as $q^{-2.3}$. The average value of h_0 is $13^m.4$. The result for 32 L-P comets is $<\Delta\mathrm{AI}>= 0^m.0 \pm 0^m.2$, the standard deviation for a single comet being $\theta = \pm 1^m.06$.

2.1. Comparison of Indices with Classes of Comets

When Equation (5) was applied to the 14 S-P comets, the dependence of ΔAI with $\log q$ is shown in Figure 1 along with the values for the L-P comets.

A glance at Figure 1 shows that for the S-P comets AI exhibits a smaller inverse q dependence than the L-P comets. In fact, the least squares value for the luminosity variation is $q^{-1.4}$ instead of $q^{-2.3}$. Also the AI values for the S-P comets show no correlation with h_0 nor with the year of observation. These deviations between the S-P and L-P comets call into question the result that ΔAI for the S-P comet is less by $0^m.33 \pm 0^m.32$ as stated earlier (Whipple, 1991).

In looking for an alternative method for comparing the S-P and L-P comets, I note that the mean date of the L-P comets is only 3 years earlier than that for the S-P comets, while the mean value of $\log q$ is correspondingly only 0.05 smaller, and the mean value of h_0 is $2^m.2$ larger. Hence, with the assumed linear corrections to AI for date, $\log q$ and h_0, a direct comparison of AI for the L-P and S-P comets is perhaps nearly as significant. The result is that for 32 L-P comets $<\mathrm{AI}>= 6^m.81 \pm 0^m.40$ compared to $<\mathrm{AI}>= 6^m.39 \pm 0^m.37$ for the 14 S-P comets. The S-P comets are still systematically less active by an amount $0^m.42$ *vs* $0^m.33$, statistically not very significant. S-P comets are thus less active than L-P comets by some 35 to 47%. Note that this poorly established change can be compared to a total average

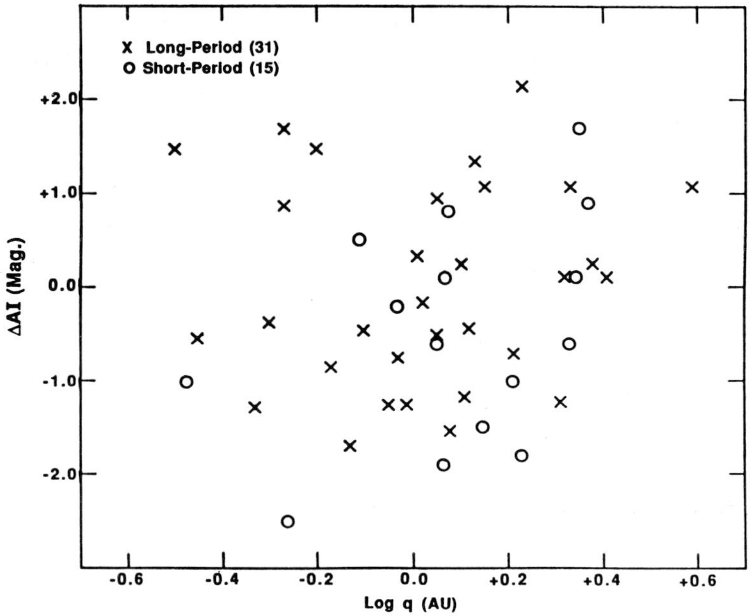

Fig. 1. Perihelion dependence of corrected activity index.

brightening of some 360 times for S-P comets and some 530 times for L-P comets, from great solar distances to perihelion.

We have seen that limited measures of the activity index, AI, for S-P comets indicates a slightly smaller range in brightness for them than for the L-P comets at a barely significant level. A similar comparison of the volatility index, VI, for 90 L-P and 14 different S-P comets results in $0^m.36 \pm 0^m.02$ vs. $0^m.35 \pm 0^m.06$, respectively. The two classes of comets vary almost identically in brightness near perihelion, with no great difference in the amount of brightening from great to small solar distance.

Although we have no orbital criteria to measure aging among S-P comets most students of comets assume that statistically, among L-P comets, those of the shorter periods have made more orbital passages and should be considered older in that sense. For these comets, Figures 2 and 3 are block diagrams showing the mean values of ΔAI and VI, respectively, versus $1/a$ (orig) and period. While Figure 3 shows no significant correlation of volatility with period, Figure 2 shows a negative correlation with ΔAI. The trend is not strikingly significant because of the paucity of the sample, the numbers within the blocks indicating the number of comets involved in each. Nevertheless, neither diagram gives any support to the concept that L-P comets become less active with increasing numbers of revolutions about the Sun. This conclusion is consistent with the relatively small, if real, reduced activity for S-P versus L-P comets.

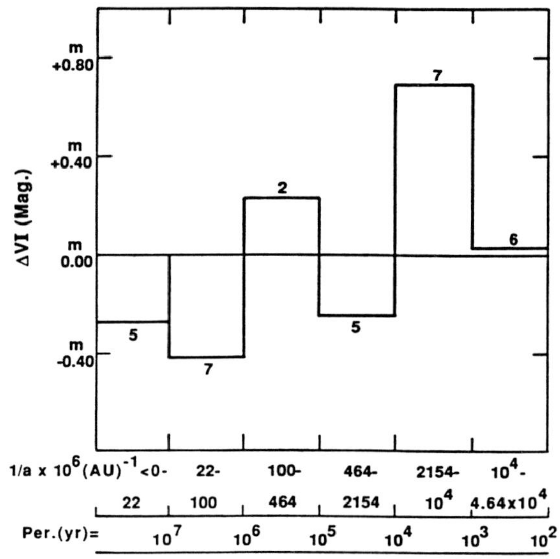

Fig. 2. Values of ΔAI for long-period comets.

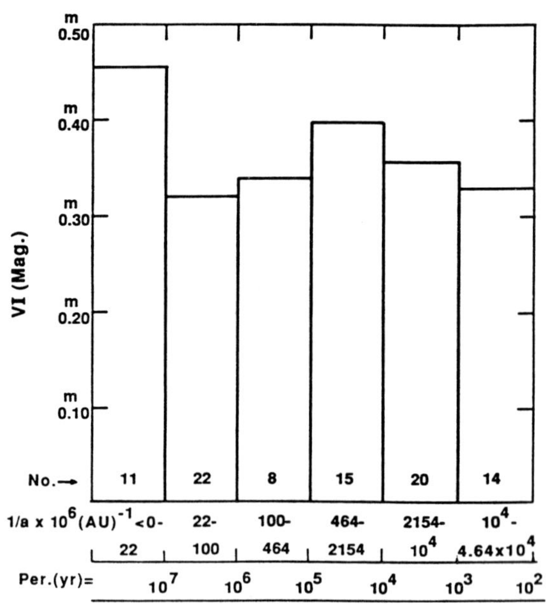

Fig. 3. Values of the Volatility Index for long-period comets.

3. Are Some Asteroids Really Old Comets?

This question, still without a definitive answer, has been with us for at least four decades (Whipple, 1950). It is not purely an academic question because its answer is critically involved in calculations of crater formation and impact rates on planets and asteroids, besides playing an important role in theories of Solar System evolution.

The asteroid 944 Hildalgo was originally the prime candidate for a dormant comet because of its large aphelion distance (9.7 AU) and large orbital inclination ($i = 43°$). Since then several likely possibilities have been suggested (see Kresák, 1979, for discussion of the dynamical aspects). The problem is thoroughly studied by Hartmann, Tholen, and Cruikshank (1987). They discuss in detail the similarities in colors, spectra and albedos of comet nuclei with the outer asteroids and, indeed, suggested asteroid 2060 Chiron as a likely comet possibility before Meech and Belton (1989) reported that they had observed a coma about the object.

Originally, I had leaned toward the idea that comet nuclei might have rocky cores, making the end product naturally an asteroid. However, the complete disappearance of the two components of Biela's comet following its apparition in 1852, subsequent to splitting in 1846 after four previous observed apparitions, was highly suggestive of complete disintegration. A similar conclusion could well have been reached by consideration of the disappearance of P/Westphal in 1913 on its way to perihelion.

Sekanina's (1982) study of split comets provided further evidence that comet interiors are not dissimilar to their exteriors, based on his derived relation between the non-gravitational accelerations of the various components and their life-times. Furthermore, this relationship was independent of the type of comet L-P or S-P.

Bortle (1991) has studied 81 L-P comets of small perihelion distance, $q < 0.51$ AU, noting their tendency to reappear or to disappear after perihelion passage. He found that of those with absolute magnitude $H_{10} > 7^m.0 + 6q(\mathrm{AU})$, 70 percent or more were never seen again after their closest approach to the Sun, even though conditions were reasonably favorable for their observation (see Figure 4). His list omitted the 13 minor sungrazing comets observed by satellites, but not visually. He found that of 23 L-P comets with H_{10} at or below the non-survival line, 16 failed to survive perihelion passage, 6 survived "unstably" and only one survived unchanged. Bortle's evidence firmly supports the theory that the small nuclei of faint L-P comets are probably active (i.e., icy) to the core and essentially dissipated when their cores are irradiated.

On the other hand, four S-P comets that violate or nearly violate Bortle's criterion have survived one or many perihelion passages. These are Bradfield 2, Brorsen-Metcalf, Encke, and Machholz, c, d, b, and a, respectively in Figure 4. Both Brorsen-Metcalf and Encke are typically volatile (VI=0.4), while P/Encke is somewhat inactive ($\Delta\mathrm{AI} = -1^m.0$). P/Honda-Mrkos-Pajdusakovà, with $q = 0.542$ AU, has been observed in 7 apparitions with one omission and is very

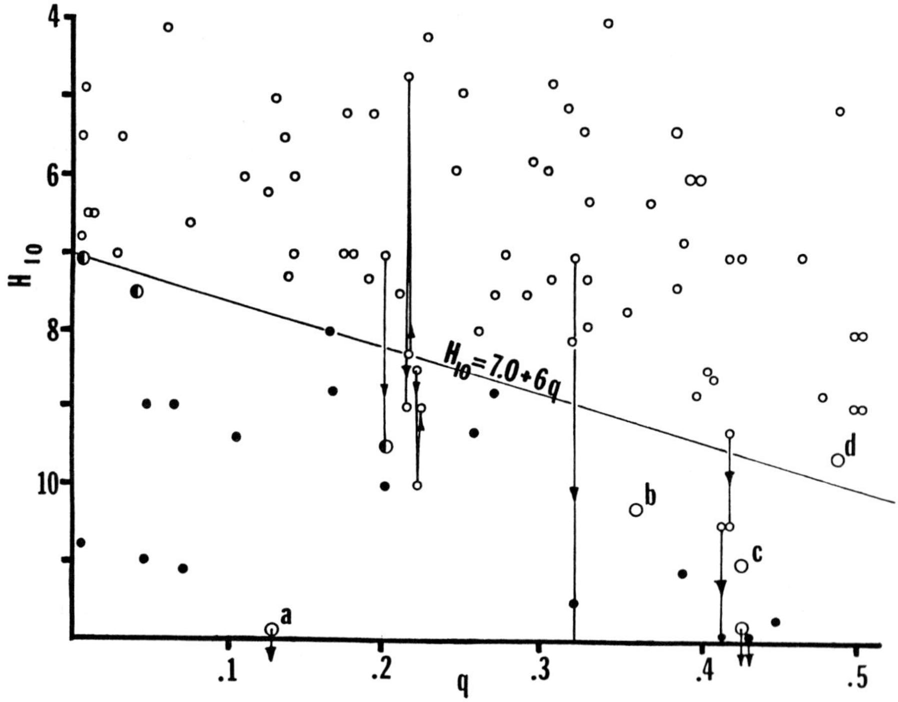

Fig. 4. Non-survival line for long-period comets in the plane 'perihelion distance' versus 'absolute magnitude'.

inactive ($\Delta AI = -1^m.5$).

Many S-P comets are still fairly active and volatile even after a great many perihelion passages. On the other hand, the huge ones such as P/Schassmann-Wachmann 1 and 2060 Chiron can be very active at extreme solar distances. But many small L-P comets disappear after one perihelion passage if q is small. These facts and the evidence from the activity index and the volatility index indicate uniformity of composition among *all* comets with just two limitations, neglecting cosmic-ray heating of the exteriors of L-P comets: (a) New very large comets may contain more volatile ices in their outer largers than is typical for smaller comets; and (b) many S-P comets are amazingly persistent.

These two limitations and the other data suggest that very large comets may have their cores heated by radioactivity as discussed by Whipple and Stefanik (1966). This is the *layered structure* of cometary nuclei as further developed by Yabushita and Wada (1988). Thus in huge comets the most volatile ices could be transferred to the outer layers while the central cores could consist mostly of the less volatile water ice or clathrates mixed with dust having some gases still absorbed. The result would be that relatively "new" huge comets could be very active at great solar distances while those that acquired short periods could persist to great ages as their core regions were more slowly sublimated. The idea of Whipple and

Stefanik that the condensation of the more volatile ices in the outer layers might be responsible for comet splitting seems unlikely to me from today's vantage point, although it might play a different role in the process (see below). Incidentally, the short life-time of ^{26}Al and the probable long accumulation times for comets probably rule out any significant heating by ^{26}Al, attested to by the decay of so many small L-P comets with small q and by P/Westphal, by P/Biela and by other split comets.

It will be interesting to see whether major break-throughs in observational techniques can soon identify old comet nuclei among the asteroids, or whether we must wait for space missions and landings to answer this important question.

4. Discussion

This concept then, that all comets were originally alike in composition and structure on a large scale of kilometers but highly heterogeneous on a small scale, seems to fit the observational data. From a study of the correlation of the production rates of C_2 and C_3 with CN in the spectra of 48 comets, Cochran (1990) notes "that only one comet, Giacobini-Zinner, deviates from this correlation". In fact it is next to the faintest inherently ($h_0 = 17^m.8$) among the S-P comets of this paper and is otherwise peculiar only in its large value of VI (0.65). In a recent paper where he discusses local activities on Halley's Comet and others, Sekanina (1991), speaks of "discrete active areas on the morphologically heterogeneous nuclei of comets", a statement consistent with the conclusions of this paper.

Following these ideas, only rather large comets with well heated cores might eventually, after long lives in short-period orbits, become inactive enough to transform into asteroids. The development of an insulating meteoric crust would not be enough to maintain for long the asteroid character of a comet with an icy core. Impacts from "debris" in the asteroid belt would break through the crust and restore the cometary activity until an extremely inactive core was exposed.

Relevant to this discussion is asteroid 3200 Phaethon, which is the only asteroid moving in the orbit of a major meteor stream. We have long known that the Geminid meteoroids are tougher and denser than those from known cometary streams. This characteristic was generally attributed to the small perihelion distance (0.14 AU) of the stream and the effects of intense solar radiations on fragile bodies – an example of "survival of the fittest". Halliday (1988), in a thorough study of the Geminid fireballs concludes "that Geminid meteoroids have too low density to associate them with meteorites or normal asteroids".

On the other hand, Green, Meadows and Davies (1985), find an IR albedo of 0.11 ± 0.02 for Phaethon and a diameter of 4.70 ± 0.5 km. They state "Both the thermal spectrum (which implies a surface of relatively high conductivity) and the reflected colors, disagree with the expected properties of an extinct cometary nucleus." But it is still "a unique object among the known Apollo asteroids". The high value of the albedo does not exceed that of P. Schwassman-Wachmann

1 but the high conductivity of the surface strongly suggests a rocky surface not consistent with the properties of the Geminid meteoroids nor with the suggestions of this paper.

A contrasting view can be gleaned from the idea by Strazzula and Johnson (1991) that the solar wind and flare ions might produce the observed density of Geminid meteoroids from lower density cometary material either on Phaethon or in space. Could the crust be thick enough to produce the small observed temperature changes on a rapidly rotating old comet nucleus? Could radioactive heating produce a similar result? If so, Phaethon may be the core of an old large comet. Similarly, the layered structures on large comets might peel off to produce the splitting phenomenon.

Whether or not we can yet certainly identify any old comets among the asteroids, the evidence discussed in this paper and the earlier one (Whipple, 1991) on the activity index, press the conclusion that comets must all or nearly all have originated near the ecliptic plane at distances beyond proto-Jupiter and to much greater distances, possibly beyond the outer planets (see Yamamoto and Kozasa, 1988). Many of the sub units in the nuclei must have been protected from the Hayashi phase of the Sun while others were not.

Acknowledgements

I am indebted to Brian G. Marsden for advise and support. This research was supported by the Planetary Geology and Geophysics Program of the U.S. National Aeronautics and Space Administration.

References

Bayer, M.: 1969, "Nachweis und Ergebnisse von Kometen-Beobachtungen aus den Jahren 1921–1968", *Astron. Nach.*, **291**, 257–264.

Beyer, M.: 1972, "Physische Beobachtungen von Kometen XVII", *Astron. Nach.*, **293**, 241–257.

Bobrovnikoff, N. T.: 1941, "Investigations of the Brightness of Comets, Part I", *Contr. of the Perkins Observatory*, No. 15.

Bobrovnikoff, N. T.: 1942, "Investigations of the Brightness of Comets, Part II", *Contr. of the Perkins Observatory*, No. 16.

Bortle, J. E. 1991, "Post-perihelion Survival of Comets with Small *q*", *Inter. Comet Quart.*, **13**, 89–91.

Cochran, A. L. 1990, "Are all Comets Created Equal ?", *Asteroids, Comets, Meteors III*, ed. C. I. Lagerkvist et al, Uppsala Univ., 281–284.

Goldreich, P. and Ward, W. R.: 1973, "The Formation of Planetesimals", *Astrophys. J.*, **183**, 1051–1061.

Duncan, M., Quinn, T., and Tremaine, S.: 1987, "The Formation and Extent of the Solar System Comet Cloud", *Astrophys. J.*, **94**, 1330–1338.

Duncan, M., Quinn, T., and Tremaine, S.: 1988, "The Origin of Short-period Comets", *Astrophys. J.*, **328**, L-69–73.

Fernandez, J. A.: 1980, "On the Existence of a Comet Belt Beyond Neptune", *Mon. Not. Roy. Astr. Soc.*, **192**, 481–491.

Green, S. F., Meadows, A. J., and Davies, J. K.: 1985, "Infrared Observations of the Extinct cometary Candidate Minor Planet (3200) 1983TB", *Mon. Not. R. A. S.*, **214**, 29p–36p.

Halliday, I.: 1988, "Geminid Fireballs and the Peculiar Asteroid 3200 Phaethon", *Icarus*, **76**, 279–294.

Hartmann, W. K., Tholen, D. J., and Cruikshank, D. P.: 1987, "The Relationship of Active Comets, 'Extinct' Comets, and Dark Asteroids", *Icarus*,**69**, 33–50.

Hills, J.G.: 1981, "Comet showers and the Steady-State Infall of Comets From the Oort Clouds", *Astron. J.*, **86**, 1730–1740.

Hills, J. G.: 1982, "The Formation of Comets by Radiation Pressure in the Outer Protosun", *Astron. J.*, **87**, 906–910.

Joss, P. C.: 1973, "On the origin of short-period Comets", *Astron. Astrophys.*, **25**, 271–273.

Kresák, L.: 1979, "Dynamical Interrelations Among Comets and Asteroids", *Asteroids*, ed. T. Gehrels, Univ. Arizona Press, 289–309.

Kuiper, G. P.: 1951, "On the Origin of the Solar System", *Astrophysics*, Ed. J. A. Hynek, McGraw-Hill, NY, 357–424.

Meech, K. J. and Belton, M. J. S.: 1989, "(2060) Chiron", *Bur. Ast. Telegrams*, IAU, No. 4770.

Meech, K. J.: 1991, "Physical Aging in Comets", Comets in The Post-Halley Era", *IAU colloquium 116*, ed. R. L. Newburn, Jr., M. Neugebuer and J. Rahe, Kluwer Acad. Pub., 625.

Oort, J.: 1950, "The Structure of the Cloud of Comets Surrounding the Solar System and a Hypothesis Concerning its Origin", *Bull. Astron. Netherlands*, **11**, 91.

Opik, E. J.: 1963, "The Stray Bodies in the Solar System. Part 1. Survival of Cometary Nuclei and the Asteroids", *Advan. Astron. Astrophys.*, **2**, 219–262.

Rickman, H.: 1991, "The Thermal History and Structure of Cometary Nuclei", *Comets in te Post-Halley Era*, Ed. R. L. Newburn, Jr., M. Neugebauer, J. Rahe, Kluwer Acad. Pub., 733–760.

Sekanina, Z.: 1982, "The problem of split Comets in Review", *Comets*, ed. L. L. Wilkening, Univ. Arizona, Tucson, 251–287.

Sekanina, Z.: 1991. "Cometary Activity, Discrete Outgassing Areas, and Dust-Jet Formation", *Comets in the Post-Halley Era*, Eds. R. L. Newburn, Jr., M. Neugebauer, J. Rahe, Kluwer Acad. Pub., 769–823.

Strazzula, G. and Johnson, R. E.: 1991, "Irradiation Effects on Comets and Cometary Debris", *Comets in the Post-Halley Era*, Ed. R. L. Newburn, Jr., M. Neugebauer, J. Rahe, Kluwer Acad. Pub., 243–275.

Vsekhsvyatskii, S. K.: 1964, *Physical Characteristics of Comets*, Israel Prog. Sci. Translations", Jerusalem, p20.

Whipple, F. L.: 1950, "A comet Model. I. The Acceleration of comet Encke", *Astrophy. J.*, **111**, 374–394.

Whipple, F. L.: 1991, "A new Activity Index for Comets", *International Conference on Asteroids, Comets, Meteors*, in press.

Whipple, F. L.: 1992, "A Volatility Index for Comets", *Icarus*, submitted.

Whipple, F. L. and Stefanik, R. P.: 1966, "On the Physics and Splitting of Cometary Nuclei", *Nature et Origine des Comètes*, Soc. Royale des Sci. de Liege, Ser. 5, Vol. 5, 33–52.

Yabushita, S. and Wada, K.: 1988, "Radioactive Heating and Layered Structure of Cometary Nuclei", *Earth, Moon, and Planets*, **40**, 303–313.

Yamamoto, T. and Kozasa, T.: 1988, "The Cometary Nucleus as an Aggregate of Planetesimals", *Icarus*, **75**, 540–551.

A MODEL OF THE GALACTIC TIDAL INTERACTION WITH THE OORT COMET CLOUD

JOHN J. MATESE and PATRICK G. WHITMAN
Department of Physics, The University of Southwestern Louisiana, Lafayette, LA 70504-4210, U.S.A.

Abstract. We consider a model of the *in situ* Oort cloud which is isotropic with a random distribution of perihelia directions and angular momenta. The energy distribution adopted has a continuous range of values appropriate for long-period (>200 yr) comets. Only the tidal torque of the Galaxy is included as a perturbation of comet orbits and it is approximated to be that due to a quasi-steady state distribution of matter with disk-like symmetry. The time evolution of *all* orbital elements can be analytically obtained for this case. In particular, the change in the perihelion distance per orbit and its dependence on other orbital elements is readily found. We further make the assumption that a comet whose perihelion distance was beyond 15 AU during its last passage through the Solar System would have orbit parameters that are essentially unchanged by planetary perturbations. Conversely, if the prior passage was inside 15 AU we assume that planetary perturbations would have removed the comet from the *in situ* energy distribution accessible by the galactic tide. Comets which had their perihelia changed from beyond 15 AU to within 5 AU in a single orbit are taken to be observable. We are able to track the evolution of 10^6 comets as they are made observable by the galactic tidal torque. Detailed results are obtained for the predicted distribution of new ($0 < 1/a < 10^{-4}$ AU^{-1}) comets. Further, correlations between orbital elements can be studied. We present predictions of observed distributions and compare them with the random *in situ* results as well as with the actual observed distributions of class I comets. The predictions are in reasonable agreement with actual observations and, in many cases, are significantly different from random when perihelia directions are separated into galactic northern and southern hemispheres. However the well-known asymmetry in the north-south populations of perihelia remains to be explained. Such an asymmetry is consistent with the dominance of tidal torques today if a major stochastic event produced it in the past since tidal torques are unable to cause the migration of perihelia across the latitude barriers $\pm 26^\circ.6$ in the disk model.

Key words: Hamiltonian mechanics, Oort comet cloud, galactic tidal torque.

1. Introduction

The Solar System is known to be surrounded by a nearly spheroidal cloud of bound comets extending over tens of thousands of AU. Oort (1950) was the first to provide evidence for this and suggested that the unusually peaked distribution in energies at the near-parabolic limit was due to random passing stars inducing small changes in the angular momentum of comets at the outer fringes of the cloud thereby reducing their perihelion distances to observable values. In the intervening decades much progress has been made in understanding the boundaries of the observable Oort cloud and the agents which make Oort cloud comets enter the inner Solar System. The reader is refered to Bailey *et al.* (1990) and Weissman

Celestial Mechanics and Dynamical Astronomy **54:** 13–35, 1992.

(1990) for more extensive reviews of the subject. For a discussion of the physical properties of comets we recommend Whipple (1989).

Bailey (1986) has concluded that giant molecular clouds dominate the energy distribution of the outer (Oort) cloud. However the parameters describing giant molecular clouds are difficult to estimate and we shall use the energy distribution of Duncan *et al.* (1987) who include the effects of stellar encounters, planetary perturbations and the galactic tide. In any case, we shall argue (in agreement with Harrington, 1985) that observations of the Oort flux do not permit a detailed determination of the *in situ* energy distribution.

Although stars and giant molecular clouds dominate the energy distribution of observed Oort comets, the galactic tidal torque dominates the distribution of angular momentum and perihelia. The importance of the interaction between the galactic tide and the Oort cloud of comets was a focus of the 1985 AAS conference *The Galaxy and the Solar System* which was motivated in part by the hypothesis of periodic showers of comets coming from the comet cloud (Rampino and Stothers, 1984a,b; Whitmire and Jackson, 1984; Davis *et al.*, 1984). Arguments were presented at this meeting which demonstrated that the galactic *disk* tidal torque dominated stellar impulses in making Oort cloud comets observable over long time scales (Heisler and Tremaine, 1986; Matese, 1985; Morris and Muller, 1986; Torbett, 1986a,b). Previous to this, Byl (1983) had considered the effects of the galactic *core* tide on comet orbits but the above authors pointed out that the disk tide is the dominant effect.

The formation and evolution of the Oort cloud is best described by Duncan *et al.* (1987). In this model unaccreted comets in the region of Uranus and Neptune are pumped by these planets into more extended orbits. When semimajor axes grow to about 5000 AU galactic tidal torques change the angular momentum, thereby moving some of the perihelia beyond the influence of the planets. Eventually passing molecular clouds and stars further diffuse and randomize the orbits to populate the inner cloud (Hills, 1981) and the outer cloud. For an alternative view of the Oort cloud as resulting from the capture of comets in the interstellar medium see Valtonen and Zheng (1990).

Although there may be infrequent intense comet showers caused by molecular clouds (Clube and Napier, 1986), the dominant quasi-steady state mechanism that makes the Oort cloud observable, but the contiguous inner cloud largely unobserved, is the galactic tidal torque (see also Byl, 1986, 1990; Matese and Whitman, 1989; Yabushita, 1988, 1989a,b; Yabushita and Tsujii, 1989, 1991). Tidal torques cause changes in the perihelion distance which increase as the seventh power of the semimajor axis, a. The *in situ* distribution of comet flux also depends strongly on a, decreasing rapidly with increasing semimajor axes. Given that a convolution of these two functions, both steeply dependent on a, determines the probability of a comet entering the inner Solar System from the Oort cloud, one can qualitatively explain the extraordinarily narrow range of semimajor axes for the observed part of the cloud.

In this model analysis we combine a prediction for the efficiency of the galactic tide in making a comet of semimajor axis a observable with a predicted *in situ* distribution of a. This permits us to model the observable distributions of all comet orbital elements. Comparisons with observed distributions are then made.

2. Model Analysis

Although galactic tides dominate stars and giant molecular clouds over long time scales in making comets observable, there is some confusion as to the degree of this dominance. Heisler and Tremaine (1986) and Torbett (1986a,b) concluded that for a typical Oort comet, tides are approximately $1.5 - 2\times$ as effective as stars would be in refilling the loss cylinder. However Matese and Whitman (1989) argued that if all stochastic impulses were ignored the Oort comet flux would still be $\approx 80\%$ of its present value for a time in excess of 10^8 yr, *i.e.*, the time scale for tidally depleting the Oort energy range via refilling the loss cylinder is the period of angular momentum oscillations. Impulses *are* needed to subsequently replenish the outer cloud from the inner cloud, to make observable those phase space regions of the outer cloud which are not efficiently torqued by the tide, or to directly make the inner cloud observable.

Assumptions made in the present analysis include the following:

- The local tidal gravitational field of the Galaxy is assumed to be axisymmetric, normal to the midplane and adiabatically changing.
- All stochastic impulses due to stars and molecular clouds are neglected.
- Orbital element evolution is described by the secularly averaged orbit equations.

The tidal field for a comet-sun separation of z at the solar location $(R_\odot(t), Z_\odot(t))$ is described by

$$\mathbf{g} = -\nabla V = -4\pi \mathrm{G}\rho(R_\odot(t), Z_\odot(t))\mathbf{z} = -4\pi \mathrm{G}\rho_\odot \mathbf{z}\,. \tag{1}$$

Our analysis proceeds by forming the adiabatic, secularly averaged Hamiltonian

$$\mathcal{H} = \mathcal{H}_{\mathrm{Kepler}} + \frac{1}{2}{\Omega_z}^2\langle z^2\rangle \tag{2}$$

where

$$\Omega_z = \frac{2\pi}{T_z} \equiv \sqrt{4\pi \mathrm{G}\rho_\odot} \approx \frac{2\pi}{62\mathrm{Myr}}\,, \tag{3}$$

using conventional estimates for the local density (Bahcall, 1986). The nominal period, T_z, would be the actual z oscillation period of the Solar System about the galactic midplane if the local density did not vary significantly in the course of a complete cycle (Innanen *et al.*, 1978).

The choice of canonical momenta and coordinates (Delaunay's action-angle variables) is somewhat arbitrary and we shall consider two sets

$$\begin{array}{llcl} \mathbf{q_k} & & \mathbf{p_k} & \\ M \equiv \dfrac{2\pi(t-\tau)}{P} & L & \equiv & \sqrt{Gm_\odot a} \\ \omega & H & \equiv & \sqrt{Gm_\odot a(1-e^2)} \\ \Omega & H_z & \equiv & \sqrt{Gm_\odot a(1-e^2)}\cos i \end{array} \tag{4}$$

or, alternatively

$$\begin{array}{llcl} M & L & \equiv & \sqrt{Gm_\odot a} \\ b & H_b & \equiv & -\sqrt{Gm_\odot a(1-e^2)}\cos\alpha \\ l & H_z & \equiv & \sqrt{Gm_\odot a(1-e^2)}\cos i \end{array} \tag{5}$$

where the two choices of angular elements are illustrated in Figures 1, 2. Although set (4) is the more common choice, for reasons to be discussed later, we shall describe the analysis in terms of both sets of variables defined by (4, 5).

In the Appendix we outline the analysis of Matese and Whitman (1989) who have shown that all orbital elements can be analytically obtained from Hamilton's equations since there exists a sufficient number of constants of the motion to reduce the problem to a single integrable equation of the motion. The angular momentum $H(t)$ is expressible in terms of a Jacobi elliptic function

$$\sqrt{H^2 - H_{\rm min}^2} = \sqrt{H_{\rm max}^2 - H_{\rm min}^2}\ \operatorname{sn}(k^2, \gamma\frac{t - t_{\rm min}}{P}) \tag{6}$$

where $H_{\rm max}, H_{\rm min}, k^2$ and γ are constants determined from the initial conditions and $t_{\rm min}$ is the time of passage of minimum angular momentum. The Appendix describes how all elements can be related to the initial conditions and $H(t)$. It also discusses the periodicity of orbital element oscillations.

There is a Newtonian analysis which, while not exact, rapidly exposes important aspects of the osculating orbit. Noting that the dominant contribution to the torque comes when the comet is near aphelion, $\mathbf{Q}$, we can say

$$\begin{aligned} \frac{d\mathbf{H}}{dt} &= \mathbf{r}\times(-\Omega_z{}^2\mathbf{z}) \\ \left\langle\frac{d\mathbf{H}}{dt}\right\rangle &= -\Omega_z{}^2\langle\mathbf{r}\times\mathbf{z}\rangle \\ &\approx -\Omega_z{}^2\langle r^2\rangle(\hat{Q}\times\hat{z})(-\sin b) \\ &\approx -\frac{5a^2}{2}\Omega_z{}^2\sin b\ (\hat{q}\times\hat{z}) \\ &\approx -\frac{5a^2}{2}\Omega_z{}^2\sin b\cos b\ \hat{H}_b. \end{aligned}$$

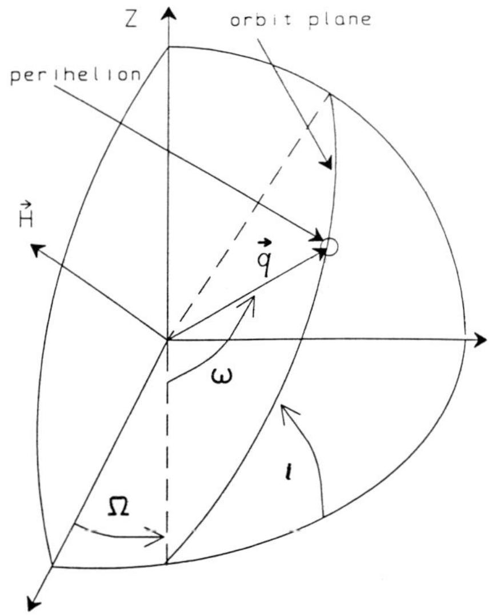

Fig. 1. The conventional orbital angles; Ω, the longitude of the ascending node, ω, the argument of perihelion, and i, the orbit plane inclination to the galactic midplane.

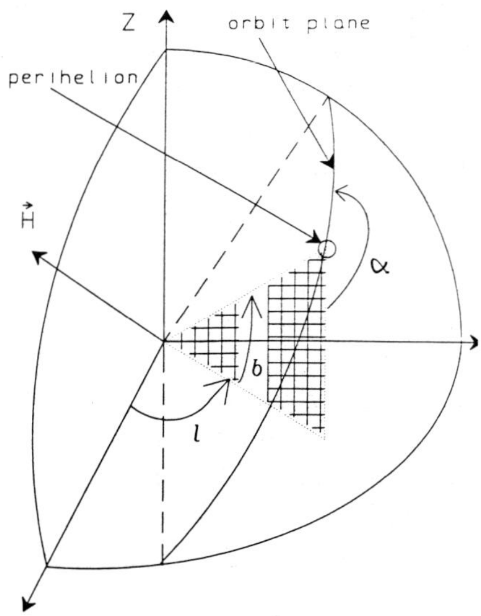

Fig. 2. Alternative orbital angles; l, the longitude of perihelion, b the latitude of perihelion, and α, the orbit plane inclination to a plane containing the galactic normal and the perihelion vector.

Defining a new angular variable F as

$$F = -\sin 2b \cos\alpha = \sin^2 i \sin 2\omega, \tag{7}$$

this yields a perihelion change in the course of a single orbit given by

$$[F\sqrt{q}\,]_{\text{previous}} - [F\sqrt{q}\,]_{\text{today}} \approx 1.6\left(\frac{a}{20000\text{AU}}\right)^{\frac{7}{2}} \sin^2 2b\ \text{AU}^{\frac{1}{2}} \tag{8}$$

which can alternatively be expressed in a form suggestive of an expansion of Equation (6) about $t \approx t_{\min}$

$$\left[\text{Sgn}(F)\sqrt{q - q_{\min}}\right]_{\text{previous}} - \left[\text{Sgn}(F)\sqrt{q - q_{\min}}\right]_{\text{today}} \approx 1.6\left(\frac{a}{20000\ \text{AU}}\right)^{\frac{7}{2}} |\sin 2b|\ \text{AU}^{\frac{1}{2}}\ . \tag{9}$$

In the Appendix it is demonstrated that b is slowly varying in the course of a single orbit in comparison to the rapid changes in F, q and α. Heisler and Tremaine (1986) argued that the secular equations of motion are invalid for large fractional changes in the angular momentum. The above analysis demonstrates that this is not always correct.

The initial conditions determine whether the minimum perihelion of the osculating ellipse would be observable. If it is, then (8, 9) indicate that during a single orbit large changes in the perihelion distance can occur for $a \geq 20000$ AU. However as $a \rightarrow 20000$ AU large changes will require the selection effects $b \rightarrow \pm 45°$ and $F_{\text{today}} \rightarrow +1$. In all cases the signature of F_{previous} necessarily is +1.

Within the context of the model, we now predict the distributions of orbital elements of observed Oort cloud comets. This requires additional assumptions to be made about the existing distribution of orbital elements in the comet cloud which we take to be:

- There is a random distribution of perihelia directions, $\hat{q}$, and angular momenta, $\mathbf{H}$ for the *in situ* Oort cloud.
- The *in situ* energy distribution of comets is taken fron Duncan, *et al.* (1987). However, every comet whose perihelion distance, q, was less than 15 AU on its last passage through the Solar System is assumed to have had an energy change which either ejected it or made it so tightly bound that it could no longer be considered a new comet.

The former assumption implies that the random probability densities

$$P_{1-e^2},\ P_l,\ P_{\sin b},\ P_\alpha,\ P_\Omega,\ P_{\cos i},\ P_\omega$$

are all *constant*. The latter assumption means that we need not consider the evolution of comets unless their previous perihelion distance was beyond the loss cylinder ($\geq$ 15AU). The Monte Carlo analysis proceeds in a manner detailed in Matese and Whitman (1989).

We now make the final model assumption:

- Every comet whose perihelion distance was less than 5 AU was observed. All comets with larger perihelion distances were unobserved.

After making a random selection of the orbital elements to determine the initial conditions for a comet one orbit previously, we analytically continue the orbit parameters to find the tidally evolved values today as described in the Appendix. We use (8, 9) to restrict the amount of phase space which we need to sample. If $q_{\rm today}$ is $\leq$ 5AU, the present orbit parameters of the observed comet are recorded. Statistics are accumulated until 10^6 comets are observed. Various choices for the model parameters $q_{\rm observable}$, $q_{\rm loss\ cylinder}$ and T_z have been considered.

In Figure 3 we show the predicted "tidal efficiency", $f_{\rho\odot}$, in overcoming the loss cylinder barrier due to Jupiter and Saturn. It is defined as

$$f_{\rho\odot}\left(\frac{1}{a}, q_{\rm observable}, q_{\rm loss\ cylinder}\right) \equiv \frac{\text{flux when } q_{\rm loss\ cylinder} \neq 0}{\text{flux when } q_{\rm loss\ cylinder} \equiv 0}. \tag{10}$$

The tidal efficiency and the assumed *in situ* flux of comets can be used to predict the flux of observable Oort cloud comets

$$\frac{d\dot{N}_{\rm obs}}{d\frac{1}{a}} = \frac{2q_{\rm obs}}{a}(1 - \frac{q_{\rm obs}}{2a})f_{\rho\odot}\left(\frac{1}{a}, q_{\rm observable}, q_{\rm loss\ cylinder}\right)\frac{d\dot{N}_{in\ situ}}{d\frac{1}{a}} \tag{11}$$

which are shown in Figure 4 for three sets of model parameters. It shows the relative insensitivity of the predictions to modest parameter changes. The predictions are qualitatively unchanged if one adopts the *in situ* flux of Bailey (1986) instead.

Figures 5–8 similarly show the arbitrarily normalized probability density for the nonrandom angular orbital elements with the canonical parameter set of $q_{\rm observable}$ = 5 AU, $q_{\rm loss\ cylinder}$ = 15 AU and T_z = 62 Myr. As mentioned, the elements Ω and l are assumed to be random. The regions of relative depletion and enhancement for these variables are entirely suggested by the predictions of (8, 9).

The variable F was discussed by several authors (Byl, 1983, 1986; Yabushita, 1988, 1989b; Matese and Whitman, 1989). We can see the significance of this variable from (8, 9). Figure 9 shows that its predicted probability distribution is uniquely distinct from the random distribution and may provide the best opportunity for distinguishing the fingerprint of the galactic tide. We now consider the comparison of these predictions with the observed distributions.

3. Comparisons with Observations

3.1. Semimajor Axis and Perihelion Distance

The data used to compare with predictions are taken from the original orbit determinations of Marsden (1989). Only those comets whose orbit determinations are deemed to be of quality class I are included. In Figure 10 we compare the tidal prediction for the canonical set of model parameters with the observations for new comets having $1/a \leq 100$ units (1 unit = 10^{-6} AU^{-1}). There are 75 comets in this data set. Included in the observed distribution are the nominally hyperbolic comets. Matese *et al.* (1991a,b) have reiterated arguments that these comets are

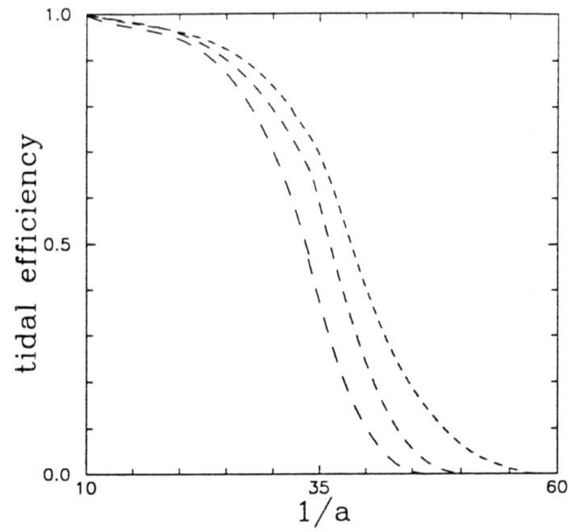

Fig. 3. The tidal efficiency of the galactic disk tide in making *in situ* Oort cloud comets observable: long dash$\Longleftrightarrow$ $q_{\text{observable}} = 5$ AU, $q_{\text{loss cylinder}} = 15$ AU and $T_z = 70$ Myr; intermediate dash$\Longleftrightarrow$ $q_{\text{observable}} = 5$ AU, $q_{\text{loss cylinder}} = 15$ AU and $T_z = 62$ Myr; short dash$\Longleftrightarrow$ $q_{\text{observable}} = 5$ AU, $q_{\text{loss cylinder}} = 10$ AU and $T_z = 62$ Myr. The units of $1/a$ are 10^{-6} AU^{-1}.

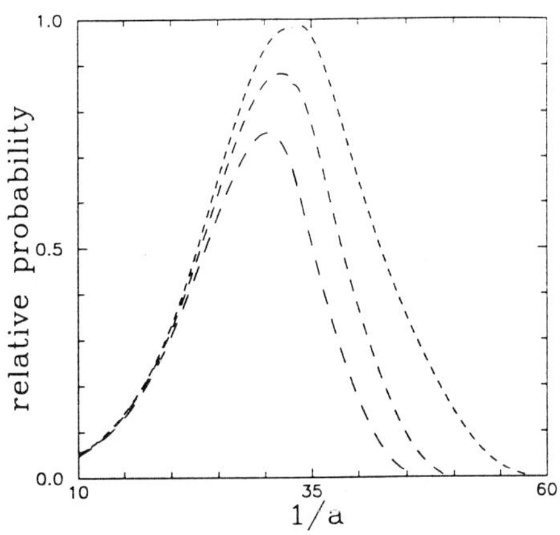

Fig. 4. The predicted flux of observable Oort cloud comets. All symbols are as indicated in Figure 3.

likely to have erroneous original energy determinations because of the omission of nongravitational forces (Marsden *et al.*, 1973, 1978). The implication is that these comets also are truly first time entrants into the Solar System from the Oort cloud. In these references it is shown that the observed weighted mean and standard deviation of reciprocal semimajor axes for new comets are $1/a \pm \sigma_{1/a} = 36 \pm 14$ units. This is to be compared to the model predictions of 32 ± 9 units for the canonical parameters. The disagreement in the mean and in the standard deviation is insignificant when one considers the inclusion in the analysis of all new comets (including hyperbolic comets) and the likely underestimation of the formal measurement error for $1/a$ (Marsden *et al.*, 1978). Although the basic agreement between observations and tidal theory is good, more work need be done to better understand the "wings" at $1/a \leq 0$ and $50 \leq 1/a$.

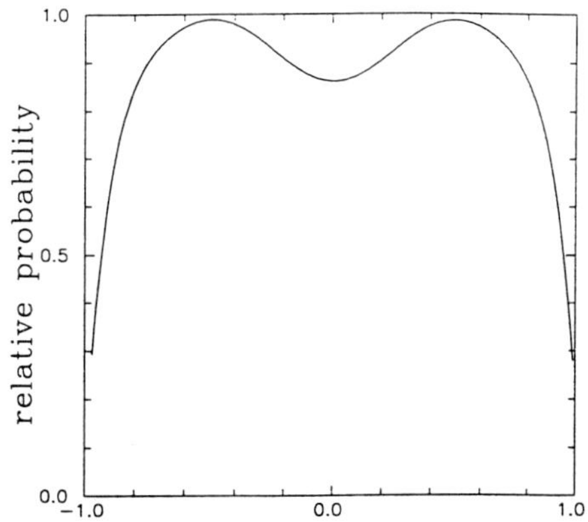

Fig. 5. The predicted probability density in $\cos i$, arbitrarily normalized, for the canonical model parameters. A random distribution of this variable is constant.

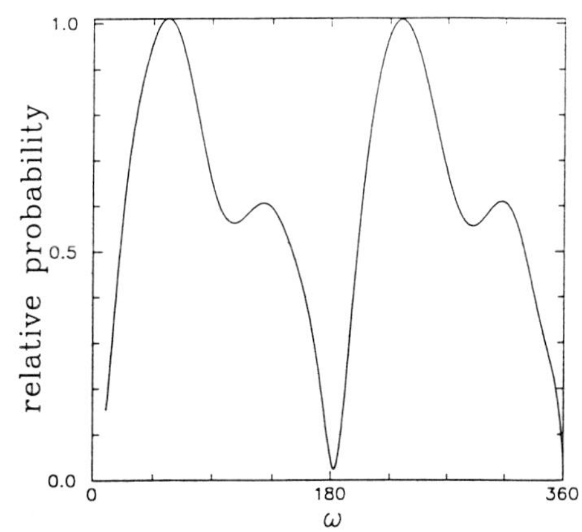

Fig. 6. The predicted probability density in ω, arbitrarily normalized, for the canonical model parameters. A random distribution of this variable is constant.

For comparison of perihelia vectors and orbit plane angles we extend the range of observed reciprocal semimajor axes to class I comets having $1/a \leq 1000$ units. The model predicts that the observed distribution should have $0 \leq 1/a \leq 50$ units, but larger values are included since these comets are expected to have passed through the inner Solar System for only a few orbits since being tidally injected from the Oort cloud and are unlikely to have suffered significant planetary impulsive changes in $\mathbf{q}$ or $\mathbf{H}$. There are a total of 113 comets in this data set. Figure 11 shows the comparison for perihelia distances. The disagreement in the observed and predicted trends is dominated by observational selection effects. In turn, the deficiency in larger values of q can be expected to slightly influence the observed distributions in orbital angles since these variables are correlated (see Equations (8–9)).

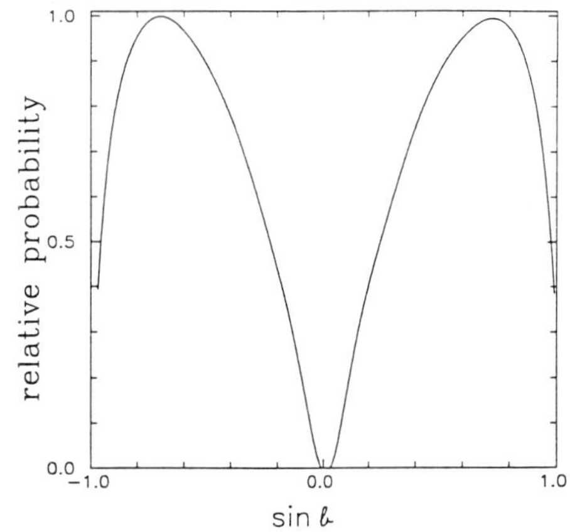

Fig. 7. The predicted probability density in $\sin b$, arbitrarily normalized, for the canonical model parameters. A random distribution of this variable is constant.

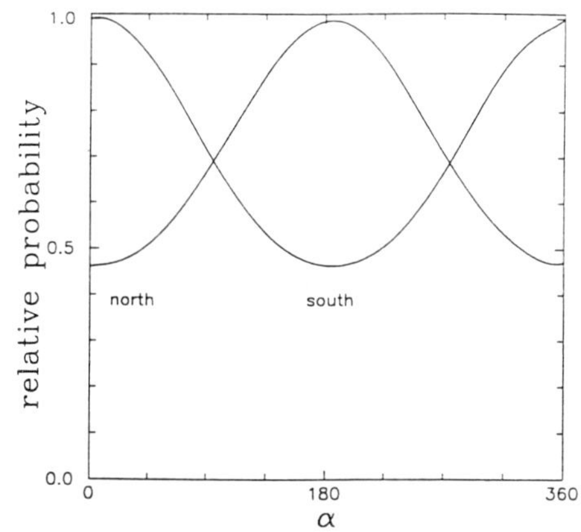

Fig. 8. The predicted probability density in α, arbitrarily normalized, for the canonical model parameters. Results are separated for perihelia in northern and southern galactic hemispheres. A random distribution of this variable is constant.

3.2. Angular Variables

To compare the angular variables, which provide the best opportunity for establishing a statistically significant correlation between theory and observations, we choose to present the data in the form of integrated (cumulative) probability densities. This avoids the problems associated with arbitrariness in bin size for procedures which bin the data for statistical comparison purposes. Relative probability densities can be recovered by comparing the slopes of the integrated curves. The results are normalized to unit total probability. As such, for a uniformly distributed random variable the integrated probability is a straight line. For comparison, the reader is referred to analyses based on two dimensional scatter plots of perihelia directions (Biermann *et al.*, 1983; Lüst, 1984; Delsemme, 1987).

The analyses are separated into two cases, perihelia latitudes $b \geq 0$ and $b \leq 0$.

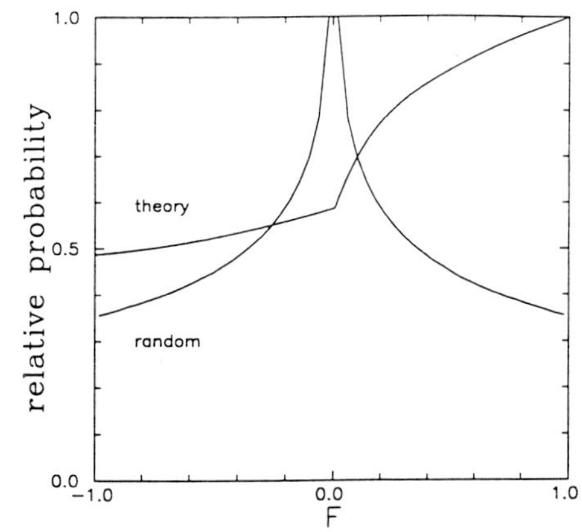

Fig. 9. The predicted probability density in F, arbitrarily normalized, for the canonical model parameters. A random distribution of this variable is also shown.

There are 70 class I northern and 43 class I southern comets. In each case we show the Kolmogorov-Smirnov acceptance limit which indicates the 95% confidence level for *rejecting* the hypothesis that the observed data set is drawn from either a random population or the galactic tidal population. The comparisons for southern comet perihelia suffer from small number statistics.

3.2.1. The conventional angles. In the first set of Figures 12–17 we show the comparisons for the conventionally defined set of orbital angles $\Omega, \cos i, \omega$. The integrated probability predictions of galactic tidal theory are represented by short dashed curves in each figure, while predictions of a random distribution are long dashed curves (for the variable Ω the tidal model assumes a random distribution). Galactic tidal theory predicts substantial deviations from a random distribution only for the argument of perihelion ω. As mentioned above, the difference in the acceptance limit for the two hemispheres is due to the different sample size.

Observations disagree with the hypothesis of a random distribution at the 95% confidence level for $\cos i$ (north). It *exceeds* this confidence level in the case ω (north). A 95% level deviation from galactic tidal theory occurs for $\cos i$ (north). This nonrandomness (for new comets) was discussed by Yabushita (1989b) and was attributed there to a disturbance by a giant molecular cloud as the Solar System crossed the galactic mid-plane a few million years ago. As the tidal theory predicts no substantial nonrandomness for the variable $\cos i$, we concur with his conclusion that a non-tidal explanation is required. Delsemme (1986) also argued for a significant nonrandomness in angular momenta which he suggested might be attributable to a solar companion. We note that in the comparisons for ω, Figures 16, 17 show that the significant deviations from random are in the same sense as predicted by tidal theory.

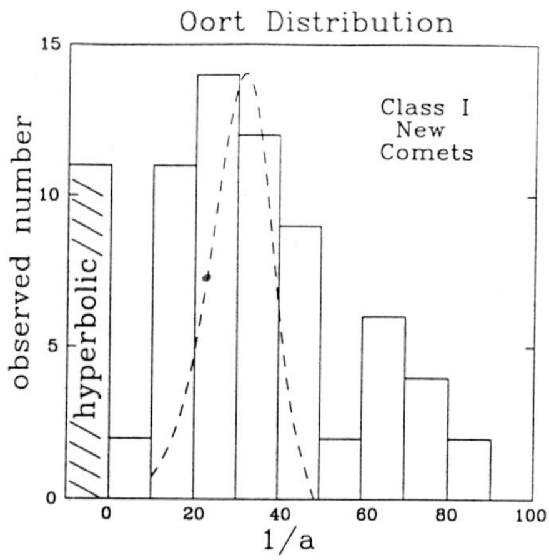

Fig. 10. The observed distribution of reciprocal semimajor axes for new class I comets. All hyperbolic values are placed in a single bin. The prediction of the galactic tidal model (arbitrarily normalized), using the canonical set of parameters, also is shown as a dashed curve. The units of reciprocal semimajor axes are 10^{-6} AU^{-1}.

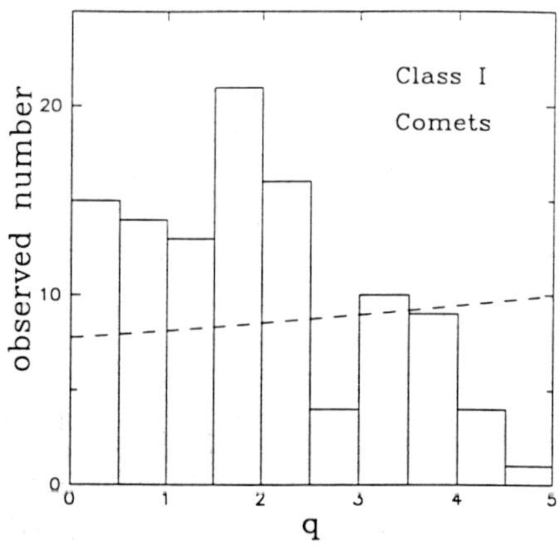

Fig. 11. The observed distribution of perihelia distances for class I comets having $1/a \leq 1000$ units. The prediction of the galactic tidal model (arbitrarily normalized), using the canonical set of parameters, also is shown as a dashed curve. The units of q are AU.

3.2.2. The alternative angles. In the set of Figures 18–23 we illustrate the comparisons between observations, galactic tidal theory and a random distribution for the alternative set of orbital angles which suit the symmetry of the problem. The format is the same as the previous set of figures. The galactic tidal model assumes that the distribution of comet perihelia is random in galactic longitude of perihelia, l, and predicts a significantly nonrandom distribution in $\sin b$ and α (when the northern and southern hemispheres are treated separately). Figure 8 demonstrates that the tidally predicted differential probability for α is 180° out of phase between the northern and southern hemispheres so that if the two hemispheres were combined

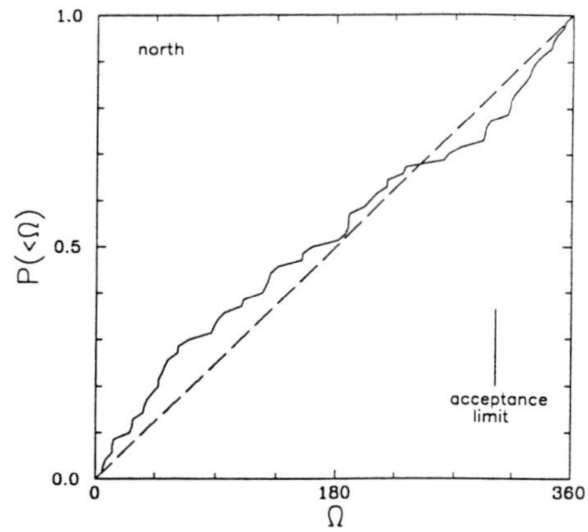

Fig. 12. Integrated probability densities for the longitude of the ascending node, Ω, for perihelia in the northern galactic hemisphere. Solid curve ⟺ observed data; dashed curve ⟺ tidal theory ≡ random. The 95% acceptance limit of the Kolmogorov-Smirnov test is illustrated.

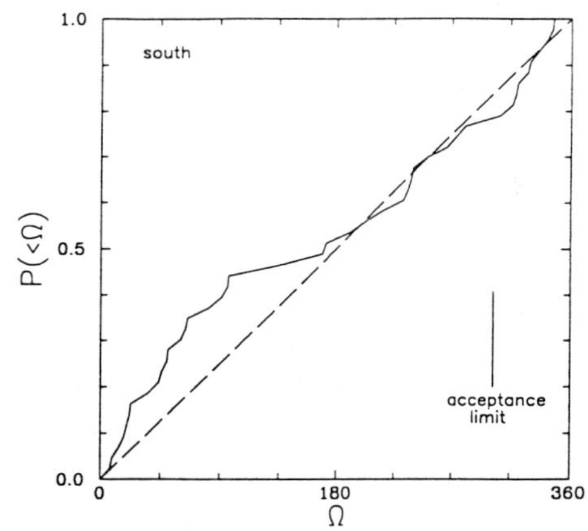

Fig. 13. Integrated probability densities for the longitude of the ascending node, Ω, for perihelia in the southern galactic hemisphere. Solid curve ⟺ observed data; dashed curve ⟺ tidal theory ≡ random.

the whole sphere distribution would be nearly random.

The trend of the data is random for the longitude of perihelion l (Figures 18, 19) even more so than for the longitude of the ascending node, Ω, for both northern and southern hemisphere comets. Examination of Figures 20, 21 shows that the observed integrated probabilities for α exhibit 95% level nonrandomness in the northern hemisphere at $\alpha \approx 180°$ where it similarly disagrees with galactic tidal theory. However we note that the curvature deviations from random are generally in the same sense as predicted by tidal theory in both hemispheres. For the latitude of perihelion, b, Figures 22, 23 illustrate that the distribution is nonrandom at a confidence level that *exceeds* 95% for the case $\sin b$ (north). A discrepancy with tidal theory at the 95% level occurs in the southern hemisphere at $\sin b \approx -0.3$

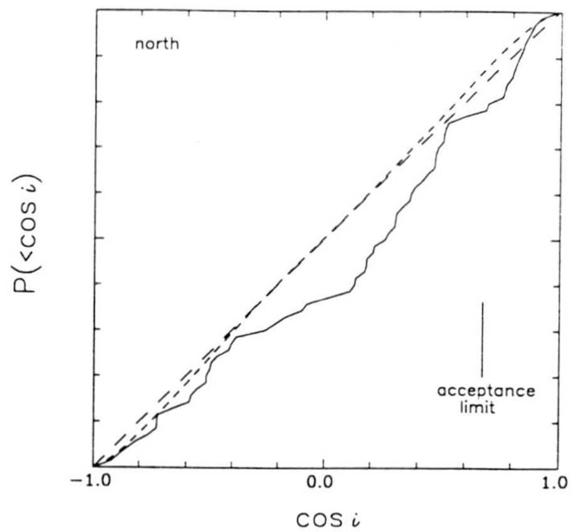

Fig. 14. Integrated probability densities for the cosine of the orbital inclination angle, i, for perihelia in the northern galactic hemisphere. Solid curve $\Longleftrightarrow$ observed data; short dashed curve $\Longleftrightarrow$ tidal theory; long dashed curve $\Longleftrightarrow$ random.

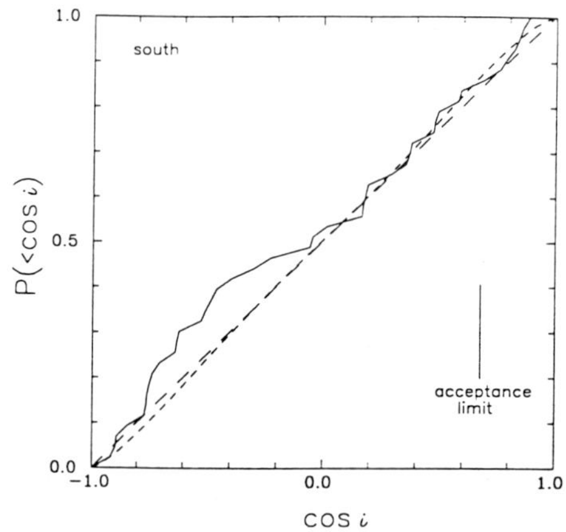

Fig. 15. Integrated probability densities for the cosine of the orbital inclination angle, i, for perihelia in the southern galactic hemisphere. Solid curve $\Longleftrightarrow$ observed data; short dashed curve $\Longleftrightarrow$ tidal theory; long dashed curve $\Longleftrightarrow$ random.

and may be due to a comet shower attributed to a stellar impulse (Biermann *et al.*, 1983; Lüst, 1984). The observed equatorial zone of avoidance for the latitude of perihelia (Byl, 1983; Biermann *et al.*, 1983; Lüst, 1984; Delsemme, 1987; Matese and Whitman, 1989) is in general agreement with galactic tidal theory for both northern and southern comet perihelia, particularly when the shower is allowed for.

3.2.3. The angular variable F. Finally we discuss the comparisons for the variable, F, whose importance is indicated in (8, 9). The trend in the observed integrated probability of F follows the predictions of galactic tidal theory in both northern and southern hemispheres as is seen in Figures 24, 25. Deviations from a random popu-

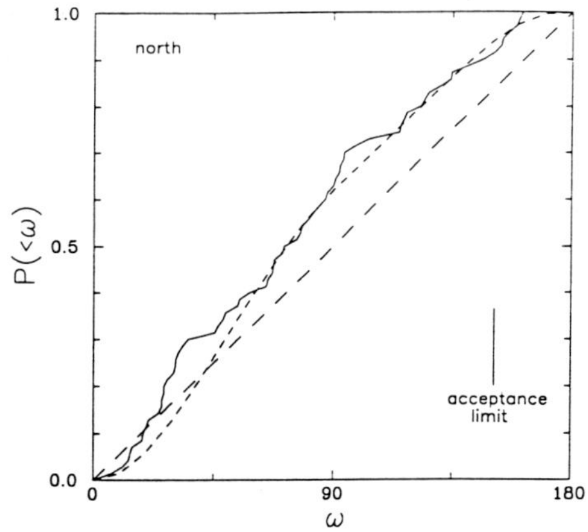

Fig. 16. Integrated probability densities for the argument of perihelion, ω, for perihelia in the northern galactic hemisphere. Solid curve $\Longleftrightarrow$ observed data; short dashed curve $\Longleftrightarrow$ tidal theory; long dashed curve $\Longleftrightarrow$ random.

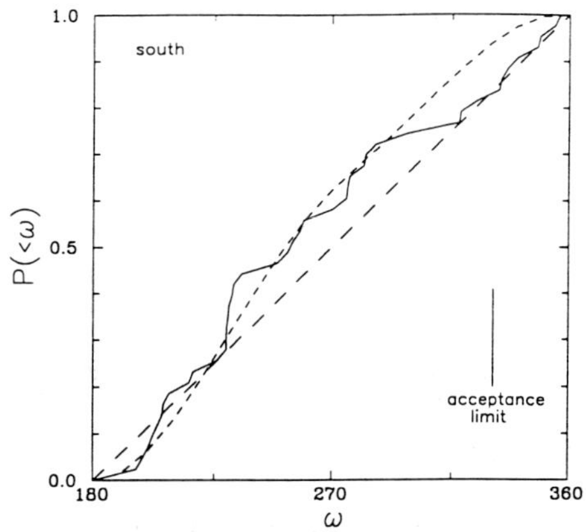

Fig. 17. Integrated probability densities for the argument of perihelion, ω, for perihelia in the southern galactic hemisphere. Solid curve $\Longleftrightarrow$ observed data; short dashed curve $\Longleftrightarrow$ tidal theory; long dashed curve $\Longleftrightarrow$ random.

lation *exceeding* the 95% confidence level occur for both hemispheres, whereas no significant deviations from galactic tidal theory occur in either hemisphere. Figure 26 shows the cumulative average of F for perihelia $q \leq q_c$ (for arbitrary q_c), indicating that a strong correlation exists in the data between F and q, in agreement with the prediction of galactic tidal theory. Uncorrelated random data would have $\langle F \rangle = 0$.

4. Summary and Conclusions

The quasi-steady state model prediction for the flux in original values of reciprocal semimajor axis of Oort cloud comets entering the Solar System for the first

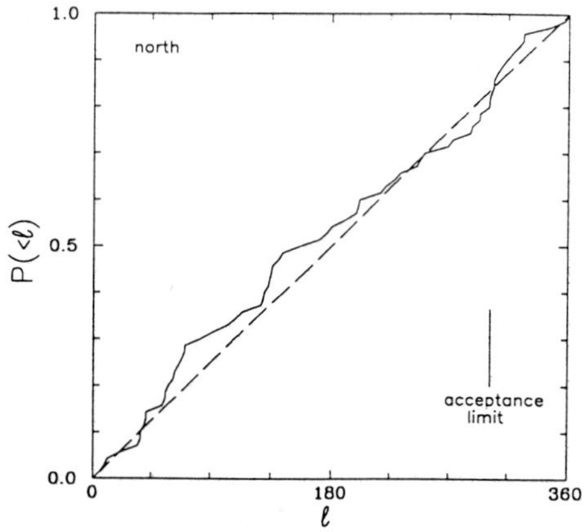

Fig. 18. Integrated probability densities for the longitude of perihelion, l, for perihelia in the northern galactic hemisphere. Solid curve $\Longleftrightarrow$ observed data; dashed curve $\Longleftrightarrow$ tidal theory $\equiv$ random.

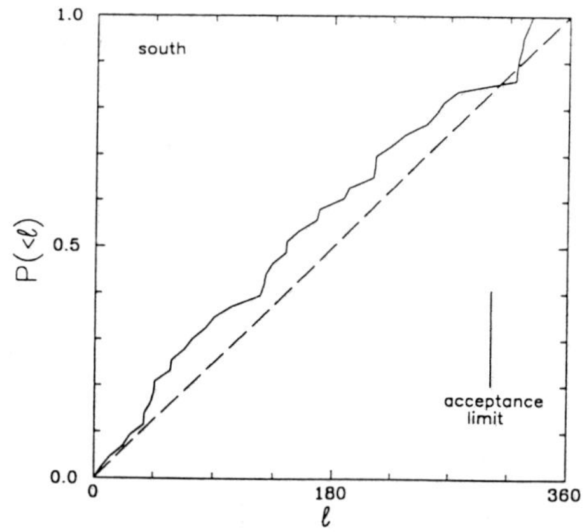

Fig. 19. Integrated probability densities for the longitude of perihelion, l, for perihelia in the southern galactic hemisphere. Solid curve $\Longleftrightarrow$ observed data; short dashed curve $\Longleftrightarrow$ tidal theory; long dashed curve $\Longleftrightarrow$ random.

time (32 ± 9 units) is in reasonable agreement with the observations (36 ± 14 units) when one considers the several factors leading to energy broadening. These include observational and calculational uncertainties, diffusion due to weak planetary perturbations on the comet's most recent passages through the outer Solar System, stochastic impulses which populate the phase space region that is tidally inaccesible and the omitted effects of nongravitational forces.

We emphasize that the majority of the loss cylinder can be refilled by the disk tidal interaction for $5 \leq 1/a \leq 35$ while a smaller fraction can be made observable for $35 \leq 1/a \leq 50$. One need invoke stellar and cloud impulses only for the complement regions of phase space that are not tidally accessible. The tidal efficiency function illustrates the dependency on reciprocal semimajor axis of the effectiveness of the tidal torque in refilling the loss cylinder. The results indicate

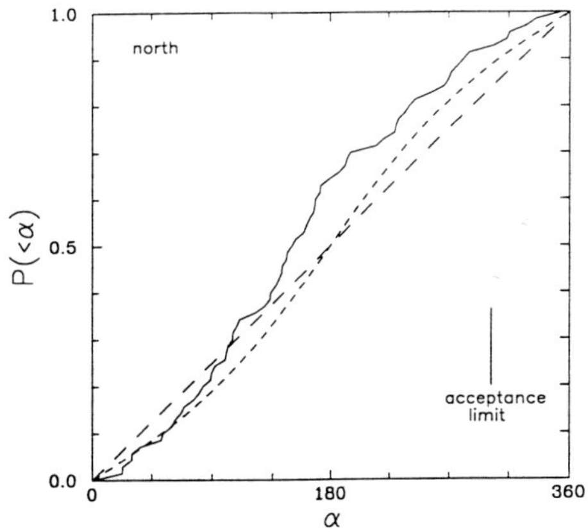

Fig. 20. Integrated probability densities for the orbit plane inclination angle, α, for perihelia in the northern galactic hemisphere. Solid curve $\Longleftrightarrow$ observed data; short dashed curve $\Longleftrightarrow$ tidal theory; long dashed curve $\Longleftrightarrow$ random.

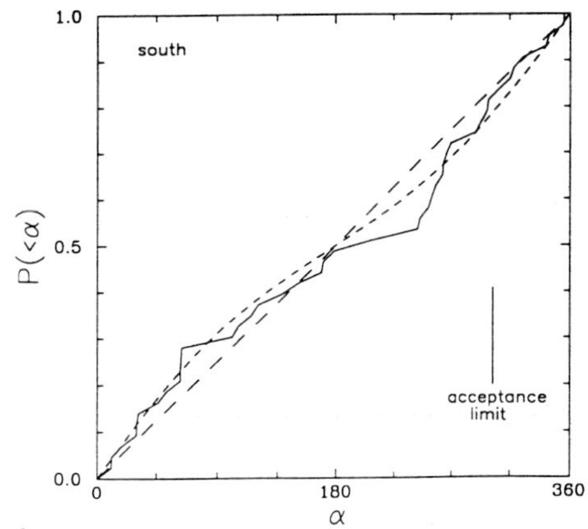

Fig. 21. Integrated probability densities for the orbit plane inclination angle, α, for perihelia in the southern galactic hemisphere. Solid curve $\Longleftrightarrow$ observed data; short dashed curve $\Longleftrightarrow$ tidal theory; long dashed curve $\Longleftrightarrow$ random.

that the majority of new class I comets have the tidal torque as a proximate cause of their becoming observable.

The importance of observational selection effects must also be recognized. There is a basic numerical imbalance in the north-south distribution of perihelia (measured relative to the galactic mid-plane) which may be due to a lack of southern geohemisphere observers. This has been discussed elsewhere (Biermann *et al.*, 1983; Lüst, 1984; Delsemme, 1987). For this reason we have separated the observations into northern and southern galactic latitudes of perihelia for comparison purposes (an additional reason is that the predicted distribution in the angle α is quite different in these cases). Yet another observational selection effect which will influence comparisons is comet observability as a function of q (Everhart, 1967).

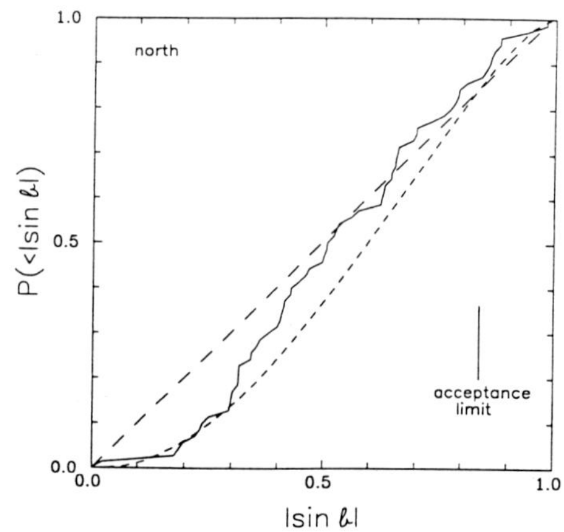

Fig. 22. Integrated probability densities for the sine of the latitude of perihelion, b, for perihelia in the northern galactic hemisphere. Solid curve $\Longleftrightarrow$ observed data; short dashed curve $\Longleftrightarrow$ tidal theory; long dashed curve $\Longleftrightarrow$ random.

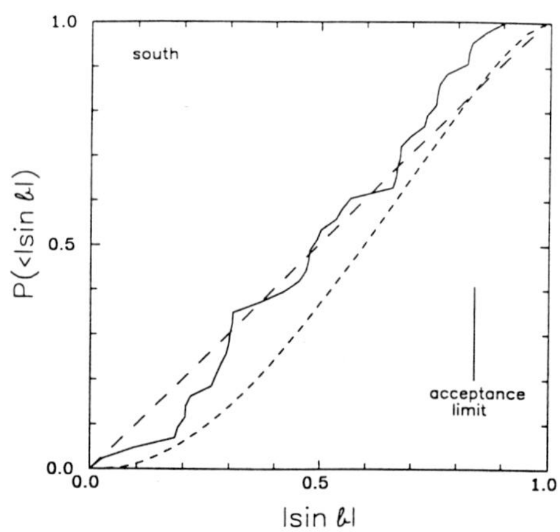

Fig. 23. Integrated probability densities for the sine of the latitude of perihelion, b, for perihelia in the southern galactic hemisphere. Solid curve $\Longleftrightarrow$ observed data; short dashed curve $\Longleftrightarrow$ tidal theory; long dashed curve $\Longleftrightarrow$ random.

We have seen that the predicted slight increase in flux as perihelion distance grows is in disagreement with observations. More work need be done to understand the role of observational selection effects in future comparisons with galactic tidal theory.

Theoretical probability densities in all relevant angular variables were presented and compared with random distributions. These results were then compared with observed data, again separated into northern and southern galactic latitudes of perihelia. We chose the Kolmogorov - Smirnov method for determining confidence levels of the null hypothesis that the observed data is *not* a representative sample of a random distribution. The data were similarly compared with the predictions of the galactic tidal model.

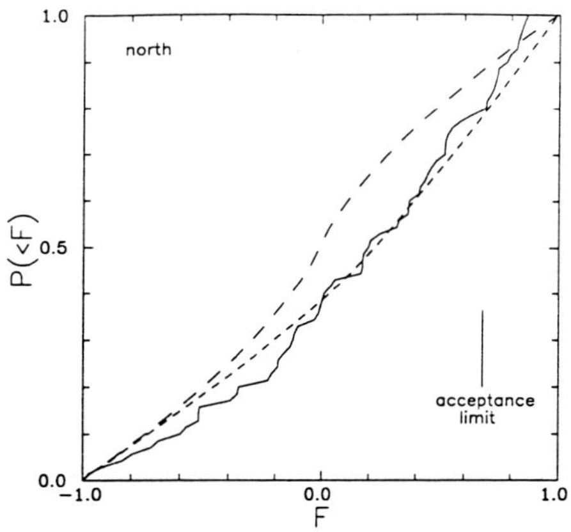

Fig. 24. Integrated probability densities for the geometrical variable $F \equiv -\sin 2b \cos \alpha$, for perihelia in the northern galactic hemisphere. Solid curve $\Longleftrightarrow$ observed data; short dashed curve $\Longleftrightarrow$ tidal theory; long dashed curve $\Longleftrightarrow$ random.

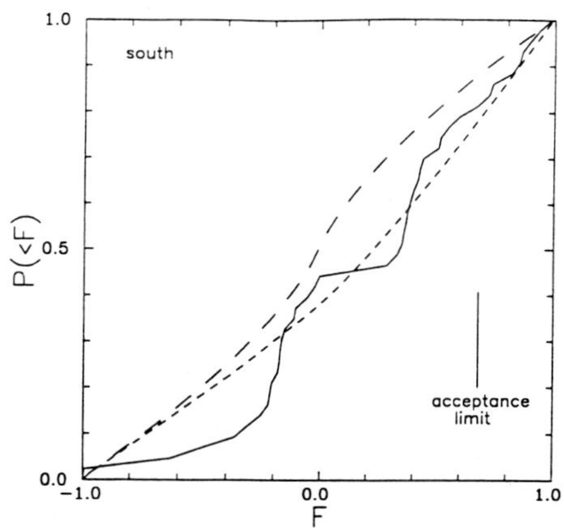

Fig. 25. Integrated probability densities for the geometrical variable $F \equiv -\sin 2b \cos \alpha$, for perihelia in the southern galactic hemisphere. Solid curve $\Longleftrightarrow$ observed data; short dashed curve $\Longleftrightarrow$ tidal theory; long dashed curve $\Longleftrightarrow$ random.

Because of the smaller sampling of perihelia in the southern hemisphere, confidence levels for rejecting the random hypothesis are small in all cases excepting the critical variable F where it exceeds 95%. That is, the observed southern hemisphere perihelia cannot be distinguished from a random distribution except for F. The comparisons are, however, more dramatic for northern hemisphere perihelia. In this case the observations are nonrandom at the 95% confidence level for the variables $\cos i$ and α. The confidence level for rejecting the random hypothesis exceeds 95% for $\sin b$, ω and F. The observations are not inconsistent with a random distribution in the variables Ω and l. If the data base is restricted to the smaller set of class I *new* comets the integrated comparisons are qualitatively unchanged, but since the acceptance limit grows as $N^{-1/2}$, the confidence levels for the null

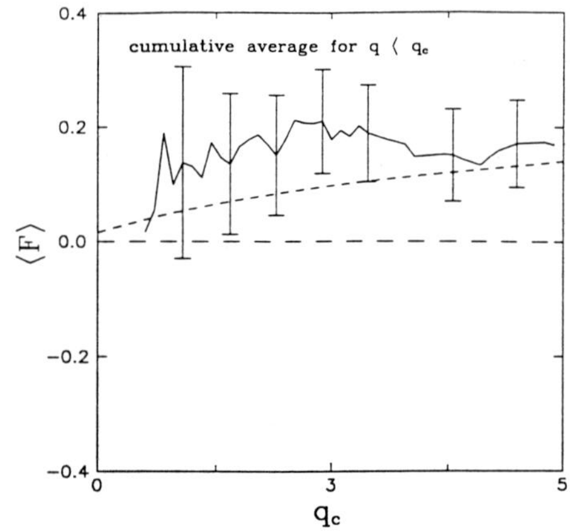

Fig. 26. The cumulative average value and standard error of the mean for the geometrical variable, F, averaged over all comet perihelia $\leq q_c$(in AU). Solid curve $\Longleftrightarrow$ observed data; short dashed curve $\Longleftrightarrow$ tidal theory; long dashed curve $\Longleftrightarrow$ random.

hypothesis are appropriately reduced and remain large only for $\cos i$, $\sin b$ and F.

When compared to the tidal model, the observations disagree at the 95% level for the variables $\cos i$ (north), $\sin b$ (south) and α (north). There was no inconsistency exceeding the 95% level in any case. It is this observation that forms the basis for the assertion that the Oort cloud is predominantly made observable by the quasi-steady state tidal torque, in contrast to the stochastic disturbances due to stars and molecular clouds.

This does not mean that individual impulses are not discernable in the observational record (indeed, the stellar impulse described by Biermann *et al.*, 1983, is likely to be real). It does mean that the number of comets involved will be relatively small in comparison to those made observable by the galactic tidal torque. In particular, it is possible that a major molecular cloud perturbation is responsible for the north/south imbalance in perihelia numbers since the tidal torque is largely unable to cause the migration of perihelia latitudes across the barriers $b = \pm 26°.6$.

Finally, the fingerprint of the galactic tidal torque may well be seen most clearly in *correlations* between variable pairs such as α - b, F - q and b - q as suggested in Equations (8, 9). Such work is in progress.

Appendix

The secularly averaged Hamiltonian (2) when alternatively expressed in terms of the two sets of Delaunay variables (4, 5) is given by

$$\mathcal{H}$$

$$= -\frac{G^2 m^2_{\odot}}{2L^2} + \left(\frac{\Omega_z L}{2Gm_{\odot}}\right)^2 \left[H_b^2\left(1 - 5\sin^2 b\right) - 4H_z^2 \tan^2 b + 5L^2 \sin^2 b\right] \quad (12)$$

$$= -\frac{G^2 m_\odot^2}{2L^2} + \left(\frac{\Omega_z L}{2Gm_\odot H}\right)^2 \left[\left(H^2 - H_z^2\right)\left(H^2 + 5\left(L^2 - H^2\right)\sin^2\omega\right)\right]. \quad (13)$$

We can then form the equations of motion for the orbital elements for each set

$$\frac{dq_k}{dt} = \frac{\partial \mathcal{H}}{\partial p_k} \quad (14)$$

$$\frac{dp_k}{dt} = -\frac{\partial \mathcal{H}}{\partial q_k}. \quad (15)$$

Observing that the Hamiltonian is independent of M and l we see that L and H_z are constants of the motion. Since $\mathcal{H}$ is also a constant one can conclude from (2) that $\langle V \rangle$ itself is a first integral. Further, the symmetry of the problem assures us that the longitude of perihelion, l, and the longitude of the ascending node, Ω, will be random variables so that their time evolution will be of no interest here.

For the remaining orbital elements of interest (q, ω, i, α and b) enough first integrals exist that we need solve only one of the equations of motion in order to obtain the analytic solutions for all. The simplest analysis is in terms of H for which we obtain

$$\frac{dH}{dt} = \frac{1}{H}\left(\frac{\Omega_z L}{Gm_\odot}\right)^2 \left[\left(H_0^2 - H^2\right)\left(H^2 - H_1^2\right)\left(H_2^2 - H^2\right)\right]^{\frac{1}{2}}. \quad (16)$$

The constants appearing in (16) are determined from the initial conditions and are defined in terms of the first integrals H_z, L and $\langle V \rangle$ as

$$H_0^2 \equiv H_z^2 + \langle V \rangle \left(\frac{2Gm_\odot}{\Omega_z L}\right)^2 \quad (17)$$

$$H_1^2 \equiv \frac{1}{8}\left[5L^2 + 5H_z^2 - H_0^2 - \sqrt{\left(5L^2 + 5H_z^2 - H_0^2\right)^2 - 80L^2 H_z^2}\right] \quad (18)$$

$$H_2^2 \equiv \frac{1}{8}\left[5L^2 + 5H_z^2 - H_0^2 + \sqrt{\left(5L^2 + 5H_z^2 - H_0^2\right)^2 - 80L^2 H_z^2}\right]. \quad (19)$$

Defining $H_{g(l)}$ as the greater(lesser) of H_0 and H_2 one can show that the following inequalities hold

$$\sqrt{5}L \geq H_g \geq L \geq H_l \geq H \geq H_1 \geq H_z \geq 0.$$

The solution to (16) is given in terms of a Jacobi elliptic function

$$H^2 = H_1^2 + \left(H_l^2 - H_1^2\right)\,\mathrm{sn}^2\left(k^2, \gamma\frac{t - t_{\min}}{P}\right) \quad (20)$$

(Gradshteyn and Rhyzhik, 1965) where

$$k^2 = \frac{H_l^2 - H_1^2}{H_g^2 - H_1^2}, \quad \gamma = 2\pi\frac{P^2}{T_z^2}\sqrt{\frac{H_g^2 - H_1^2}{L^2}} \quad (21)$$

and P is the orbital period. One finds that the angular momentum H oscillates between $H_{\max} = H_l$ and $H_{\min} = H_1$ with an oscillation period of

$$\frac{T_z^2}{2\sqrt{5}P^2} \leq \frac{P_H}{P} = \frac{2\mathbf{K}(k)}{\gamma} \leq \infty \tag{22}$$

where $\mathbf{K}(k)$ is the complete elliptic function of the first kind.

We can now determine the time dependence of the orbital elements by unfolding the defining relations of the Delaunay variables and utilizing the first integrals. They are

$$e^2 = \frac{L^2 - H^2}{L^2} \tag{23}$$

$$\sin^2 \omega = \frac{H^2\left(H_0^2 - H^2\right)}{5\left(L^2 - H^2\right)\left(H^2 - H_z^2\right)} \tag{24}$$

$$\cos^2 i = \frac{H_z^2}{H^2} \tag{25}$$

$$\sin^2 b = \frac{H_0^2 - H^2}{5\left(L^2 - H^2\right)} \tag{26}$$

$$\cos^2 \alpha = \frac{4\left(H_2^2 - H^2\right)\left(H^2 - H_1^2\right)}{H^2\left(5L^2 - 4H^2 - H_0^2\right)}. \tag{27}$$

One observes from (23–27) that of the variables q, ω, i, α and b, only the last is slowly varying as the angular momentum is tidally reduced to its minimum value.

Noting that for near-parabolic comets $L \gg (H, H_b$ and $H_z)$, we find from (12, 24) that l is also slowly varying, with a timescale (at minimum perihelion) that is longer by a factor $\mathcal{O}\left(L^2/H^2\right)$ than that for the variation of $\Omega, \omega, i, \alpha$ and q. It is for this reason that the choice of orbital elements (5) plays a non-trivial role in analyzing the secular equations and understanding the fundamental behavior of the interaction.

Matese and Whitman (1989) provide further details about the general characteristics of orbital element evolution and in particular note that the latitude of perihelion cannot migrate across the barriers $\pm 26°.6 = \pm \sin^{-1}\sqrt{1/5}$ in the tidal approximation described here.

Acknowledgements

The authors would like to acknowledge the support of a NASA/Ames University Consortium Grant and a grant from the Louisiana Education Quality Support Fund. The assistance and comments of M.E. Bailey, L.D. Lafleur and D.P. Whitmire are also appreciated.

References

Bahcall, J. N.: 1986, *The Galaxy and the Solar System* (R. Smoluchowski, J. N. Bahcall and M. S. Matthews, Eds.) Univ. of Arizona Press, Tucson, 3–12.

Bailey, M. E.: 1986, *Mon. Not. R. Astron. Soc.* **218**, 1–30.

Bailey, M. E., Clube, S. V. M., and Napier, W. M.: 1990, *The origin of comets* Pergamon Press, Oxford.

Biermann, L., Huebbner, W. F., and Lüst, R.: 1983, *Proc. Natl. Acad. Sci. USA* **80**, 5151–5155.

Byl, J.: 1983, *Moon Planets* **29**, 121–137.

Byl, J.: 1986, *Earth Moon Planets* **36**, 263–273.

Byl, J.: 1990, *Astron. J.* **99**, 1632–1635.

Clube, S. V. M., and Napier, W. M.: 1986, *The Galaxy and the Solar System* (R. Smoluchowski, J. N. Bahcall and M. S. Matthews, Eds.) Univ. of Arizona Press, Tucson, 260–285.

Davis, M., Hut, P., and Muller, R.: 1984, *Nature* **308**, 715–717.

Delsemme, A. H.: 1986, *The Galaxy and the Solar System* (R. Smoluchowski, J. N. Bahcall and M. S. Matthews, Eds.) Univ. of Arizona Press, Tucson, 173–203.

Delsemme, A.H.: 1987, *Astron. Astrophys.* **187**, 913–918.

Duncan, M., Quinn, T., and Tremaine, S.: 1987, *Astron. J.* **94**, 1330–1338.

Everhart, E.: 1967, *Astron. J.* **72**, 1002–1011.

Gradshteyn, I. S., and Rhyzhik, I. M.: 1965, *Tables of Integrals, Series and Products* (A. Jeffery, Ed.) Academic Press, New York, 904–917.

Harrington, R. S.: 1985, *Icarus* **61**, 60–62.

Heisler, J., and Tremaine, S.: 1986, *Icarus* **65**, 13–26.

Hills, J. G.: 1981, *Astron. J.* **86**, 1730–1740.

Innanen, K. A., Patrick, A. T., and Duley, W. E.: 1978, *Astrophys. Space Sci.* **57**, 511–515.

Lüst, R.: 1984, *Astron. Astrophys.* **141**, 94–100.

Marsden, B. G.: 1989, *Catalogue of Cometary Orbits 6th edn.* Smithsonian Astrophysical Observatory, Cambridge.

Marsden, B. G., Sekanina, Z., and Yeomans, D. K.: 1973, *Astron. J.* **78**, 211–225.

Marsden, B. G., Sekanina, Z., and Everhart, E.: 1978, *Astron. J.* **83**, 64–71.

Matese, J. J.: 1985, Paper presented at AAS meeting, *The Galaxy and the Solar System*, Tucson,Arizona.

Matese, J. J., and Whitman, P. G.: 1989, *Icarus* **82**, 389–401.

Matese, J. J., Whitman, P. G., and Whitmire, D. P.: 1991a, *Nature* **352**, 506–508.

Matese, J. J., Whitman, P. G., and Whitmire, D. P.: 1991b, *Asteroids, Comets, Meteors*, in press.

Morris, D. E., and Muller, R. A.: 1986, *Icarus* **65**, 1–12.

Oort, J. H.: 1950, *Bull. Astron. Inst. Neth.*, **11**, 91–110.

Rampino, M. R., and Stothers, R. B.: 1984a, *Nature* **308**, 709–712.

Rampino, M. R., and Stothers, R. B.: 1984b, *Science* **226**, 1427–1431.

Torbett, M. V.: 1986a, *The Galaxy and the Solar System* (R. Smoluchowski, J. N. Bahcall and M. S. Matthews, Eds.) Univ. of Arizona Press, Tucson, 147–162.

Torbett, M. V.: 1986b, *Mon. Not. R. Astron. Soc.* **223**, 885–895.

Valtonen, M., and Zheng, J. Q.: 1990, *Proceedings of the Nordic-Baltic Astronomy Meeting* Uppsala, 353–360.

Weissman, P. R.: 1990, *Nature* **344**, 825–830.

Whipple, F. L.: 1989, *Astrophys. J.* **341**, 1–15.

Whitmire, D.P., and Jackson, A.A.: 1984, *Nature* **308**, 713–715.

Yabushita, S.: 1988, *Mon. Not. R. Astron. Soc.* **231**, 723–733.

Yabushita, S.: 1989a, *Astron. J.* **97**, 262–264.

Yabushita, S.: 1989b, *Earth Moon Planets* **44**, 29–37.

Yabushita, S., and Tsujii, T.: 1989, *Mon. Not. R. Astron. Soc.* **241**, 59–69.

Yabushita, S., and Tsujii, T.: 1991, *Mon. Not. R. Astron. Soc.* **248**, 34–39.

ORIGIN OF OORT CLOUD COMETS IN THE INTERSTELLAR SPACE

MAURI J. VALTONEN, JIA-QING ZHENG and SEPPO MIKKOLA
Tuorla Observatory, University of Turku, SF–20520–Turku, Finland

Abstract. Comets must form a major part of the interstellar medium. The solar system provides a flux of comets into the interstellar space and there is no reason to suspect that many other stars and their surrounding cometary systems would not make a similar contribution. Occasionally interstellar comets must pass through the inner solar system, but Whipple (1975) considers it unlikely that such a comet is among the known cases of apparently hyperbolic comets. Even so the upper limit for the density of unobserved interstellar comets is relatively high.

In addition, we must consider the possibility that comets are a genuine component of interstellar medium, and that the Oort Cloud is merely a captured part of it (McCrea, 1975). Here we review various dynamical possibilities of two–way exchange of comet populations between the Solar System and the interstellar medium. We describe ways in which a traditional Oort Cloud (Oort, 1950) could be captured from the interstellar medium. However, we note that the so called Kuiper belt (Kuiper, 1951) of comets cannot arise through this process. Therefore we have to ask how necessary the concept of the yet unobserved Kuiper belt is for the theory of short period comets.

There has been considerable debate about the question whether short period comets can be understood as a captured population of the Oort Cloud of comets or whether an additional source has to be postulated. The problem is made difficult by the long integration times of comet orbits through the age of the Solar System. It would be better to have an accurate treatment of comet-planet encounters in a statistical sense, in the form of cross sections, and to carry out Monte Carlo studies. Here we describe the plan of action and initial results of the work to derive cross sections by carrying out large numbers of comet - planet encounters and by deriving approximate analytic expressions for them. Initially comets follow parabolic orbits of arbitrary inclination and perihelion distance; cross sections are derived for obtaining orbits of given energy and inclination after the encounter. The results are used in subsequent work to make evolutionary models of the comet population.

Key words: Oort cloud, comets, origin of comets, capture of comets, interstellar comets.

1. Introduction

The question of interstellar comets is a fascinating and difficult topic since we are discussing a population of objects which we have never seen, and perhaps will never see. Yet we know that interstellar comets must exist. The Solar System "leaks" comets into interstellar space due to planetary and other perturbations, and there are reasons to expect that the same process goes on also in many other planetary systems around other stars. Therefore a population of "lost" comets must exist.

In addition, it is possible that comets are truly interstellar in nature and that our Solar System has only managed to capture some of them (McCrea, 1975). The way the so called Oort Cloud (Oort, 1950) just barely hangs on to the Solar System is suggestive of a captured population. There are, of course, other mechanisms of producing such a loosely bound comet cloud, but they tend to be ineffective in that most of the comets are lost to interstellar space while populating the Oort Cloud

Celestial Mechanics and Dynamical Astronomy **54:** 37–48, 1992.

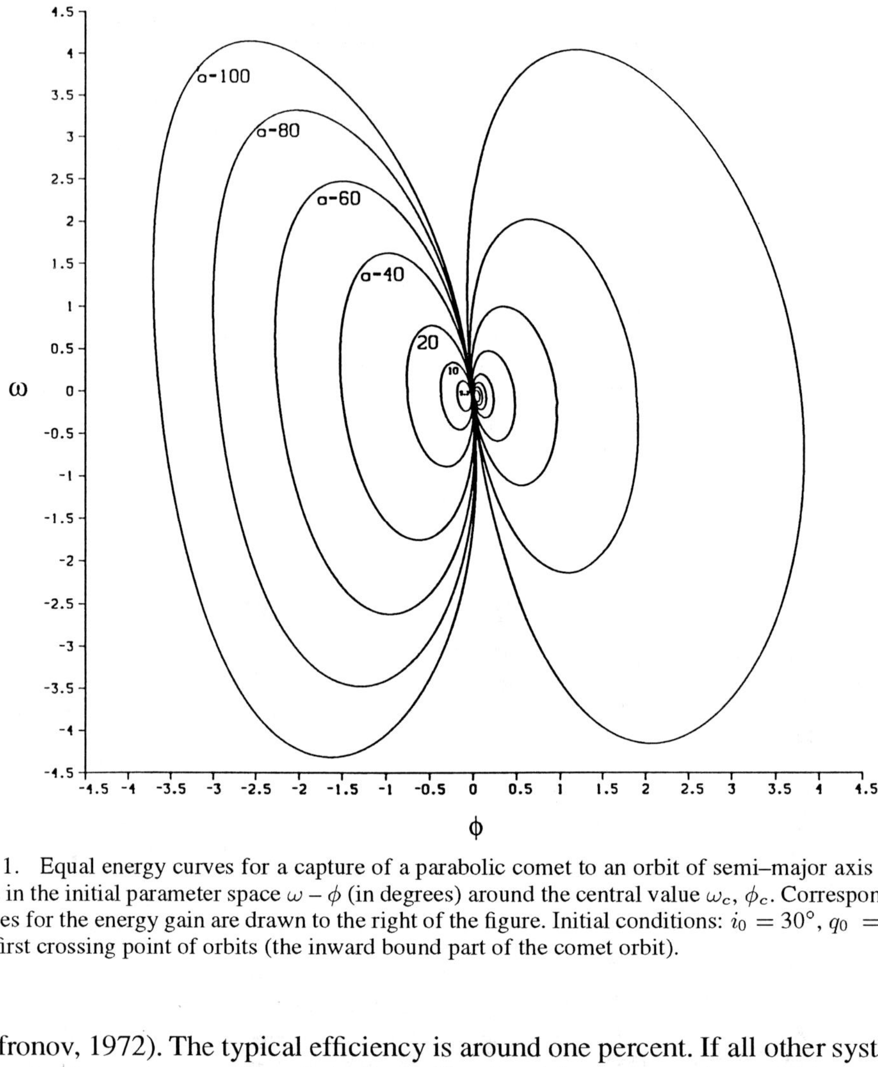

Fig. 1. Equal energy curves for a capture of a parabolic comet to an orbit of semi–major axis a (in AU) in the initial parameter space $\omega - \phi$ (in degrees) around the central value ω_c, ϕ_c. Corresponding curves for the energy gain are drawn to the right of the figure. Initial conditions: $i_0 = 30°$, $q_0 = 0.5$, the first crossing point of orbits (the inward bound part of the comet orbit).

(Safronov, 1972). The typical efficiency is around one percent. If all other systems around other stars populate their Oort Clouds with the same low efficiency, the interstellar comet density should be high and we should observe something like six extrasolar comets per century(McGlynn and Chapman, 1989). Since no clear cases of interstellar comets are known (Whipple, 1975), we are faced with a problem: where are all the interstellar comets ?

McGlynn and Chapman (1989) point to various ways out of the problem:

- nondetectability of interstellar comets. This they rule out since orbital properties of interstellar comets should not be very different from the near parabolic comets which we see frequently.
- alternative origins for the Oort Cloud. Here the main alternative to the formation of comets within the Solar Nebula is the birth of comets in interstellar medium and subsequent capture to form the Oort Cloud. We will discuss this

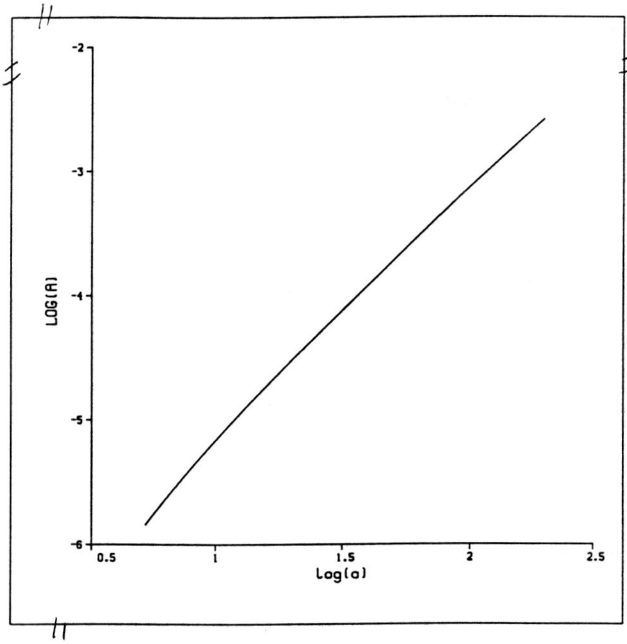

Fig. 2. The variation of the cross-section A for obtaining semi-major axis smaller than a as a function of the semi-major axis. Initial conditions: $i_0 = 30°$, $q_0 = 0.5$.

possibility below.

- Solar Systems like ours are rare. At present there is no reason to assume this to be the case.
- We are currently in a comet shower. The estimates of McGlynn and Chapman (1989) are based on the present flux of Oort Cloud comets and assuming that it represents a long-term average. If the present flux exceeds the long-term average by an order of magnitude or more, the absence of hyperbolic comets is no longer a problem.

The idea of a comet shower (Hills, 1981) at the present epoch may not be as far fetched as it may sound at first. Elsewhere in this volume it is argued by Rampino and Caldeira (1992) that there is a maximum at present time in the types of geological and paleontological events which may have a relation to a comet shower (see also the discussion by Shoemaker and Wolfe, 1986). On the other hand, Matese and Whitman (1992) point out that the maximum in Galactic tides which could cause a comet shower, also occurs close to the present epoch. Leaving aside the uncertainty about the scale of activity produced by the Galactic tide, we have a perfectly valid line of argument which supports the last mentioned solution of the problem.

As to the alternative origins for the Oort Cloud, Valtonen (1983) summarized various ways in which a cloud of comets could be captured from the interstellar medium of comets. There are primarily two alternatives:

(1) the Sun has an unknown companion, a substellar body (say, a 30 Jupiter mass object) which can act as a capture agent; or

(2) the capture of comets took place during the early stages of evolution of the Sun and the Solar System (Donn 1976).

No massive companions beyond the orbit of Pluto have been found, but the existence of certain types of substellar bodies at large distances cannot be excluded at present time (Reynolds *et al.*, 1980). The companion proposed by Valtonen (1983) differs from the hypothetical "Nemesis" (Whitmire and Jackson, 1984; Davis *et al.*, 1984) in that the orbital period of the companion is not constrained and it can have a stable orbit. Its mass could be similar to the postulated Nemesis (Delsemme, 1986). In a way it is another version of the "Planet X" hypothesis (Whitmire and Matese, 1985; Matese and Whitmire, 1986).

The attractive feature about the capture scenario is the strongly periodic nature of the capture and loss of comet clouds (Yabushita, 1983). The capture rate depends strongly on the velocity of the Sun relative to the interstellar medium of comets, as the inverse third power of the relative velocity (Valtonen and Innanen, 1982). As the solar speed varies by 10 – 40 % during the vertical oscillation about the plane of the Galaxy (Bash, 1986), the capture rate varies by 30 – 200 %. If the velocity dispersion of the cometary medium is about twice the speed of the Sun through this medium, a strong asymmetry of perihelia directions of new comets arises, much in the way found by Oja (1975) and Hasegawa (1976). While the capture of new comets would be associated with the maximum excursion of the Sun from the plane of the Galaxy, the maximum decay of the new Oort Cloud would take place subsequently (Yabushita, 1979), with a time delay of about 10 million years, a few orbital periods of the Oort Cloud. Therefore possible comet showers should occur when the Sun is already close to the midplane of the Galaxy (see Innanen *et al.*, 1978, for the motion of the Sun in the Galaxy). In fact there would seem to be up to 14 million year differences between the exact times of the Galactic plane crossings and the comet showers in the past 260 million years, if the latter are associated with mass extinctions of marine species (Rampino and Stothers, 1984).

On the other hand, if the Sun does not have a sizable companion or giant planet outside the orbit of Pluto, the present day capture rates of comets are not important (Valtonen and Innanen, 1982). However, it does not exclude the capture of the Oort Cloud of comets sometime in the past when the relative motion between the comets and the Sun was low. If the solar speed relative to the comet medium and the velocity dispersion of the cloud of comets are both below 1 km s^{-1}, the capture takes place easily with the assistance of known planets. Situation like this could have existed in the cluster of stars where the Sun presumably formed some 5 billion years ago.

The scenario of origin of the Oort Cloud in a star cluster was studied recently by Zheng *et al.* (1990) and Valtonen and Zheng (1990). It was found that comets could indeed be attracted to a single star which escapes from a star cluster. The orbits of the captured comets are loosely bound, which means that the Oort Cloud obtained in this manner is more in accordance with the ideas of Oort (1950) than the more recent ideas of the "inner Oort Cloud" (Fernandez, 1980). In this scenario most of the comets do not end up being gravitationally bound to stars but instead remain in the interstellar medium. The comet density is expected to be high but not necessarily above the observational upper limits.

One way to differentiate between the capture scenario and the scenarios where comets are born with the rest of the solar system, is to study the source of short period comets. In the interstellar capture picture the short period comets must originate from the traditional Oort Cloud which is more or less isotropic. Other scenarios have more free parameters: one may assume disk like distributions of comets, for example, and source regions at different distances from the Sun (Duncan *et al.*, 1987). Therefore we have undertaken a study of the process of making short period comets out of Oort Cloud comets through the gravitational influence of the major planets.

The source of short period comets is a long-standing puzzle (e.g. Joss, 1973). It has been variously reported that short period comets could represent the captured subpopulation of the Oort Cloud (e.g. Shoemaker and Wolfe, 1982) or that the short period comets require a yet unidentified source of their own (e.g. Duncan *et al.*, 1988). A disk-like "inner Oort Cloud" has been proposed as the source (e.g. Fernandez, 1980). The debate has concentrated on the capture process and its inclination dependence (Stagg and Bailey, 1989). It appears that further progress requires some hard data on the capture process which we accumulate via numerical scattering experiments in this paper.

2. Numerical Calculations

In the model we use a single planet of Jupiter mass which goes around the Sun. Its orbital eccentricity is zero. A comet approaches the Sun in a parabolic orbit. The K–S regularisation method (Kustaanheimo and Stiefel, 1965) is used, with the regularisation center either in the Sun or in the planet, depending on which one is deemed dominant. The results are expressed in units where the total mass = 1, the orbital radius of Jupiter $R_J = 1$ and the gravitational constant G = 1. We consider only orbits which change considerably during the passage by the planet. Thus for any given original pericenter distance q_0 and inclination i_0 we choose the argument of the perihelion ω of the comet and the phase angle ϕ of the planet in the beginning of the calculation such that a near planet–comet encounter takes place. A family of orbits is created by varying ω and ϕ near the "collision orbit" $\omega = \omega_c$ and $\phi = \phi_c$.

After the encounter has taken place we note down the new energy (= energy change) E of the comet as well as its new inclination i. With these quantities known it is possible to calculate the new pericenter distance q using the Tisserand's condition.

Figure 1 illustrates the areas in the $\phi - \omega$ plane which are required to produce

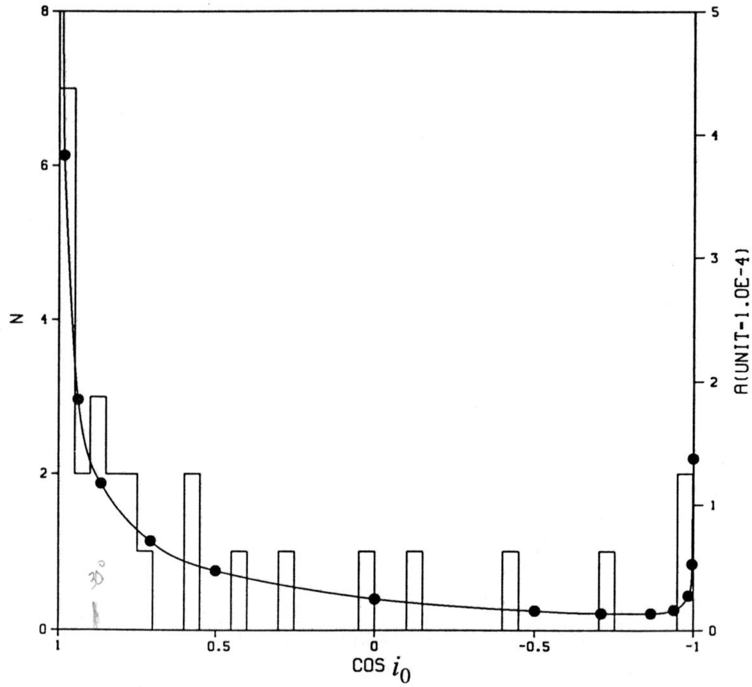

Fig. 3. The cross-section A for $E \leq -0.065$ when $q_0 = 0.5$ as a function of the initial inclination i_0 (solid line with dots). For comparison, a histogram of observed inclinations of Halley type comets.

a comet of semi–major axis a or less. The corresponding areas for a gain in comet energy lie next to the areas of energy loss. In this illustration as well as in the following ones, we study a Jupiter–comet encounter with $i_0 = 30°$ and $q_0 = 0.5$.

The change of the cross-sectional area, measured from diagrams like Figure 1, is plotted against the new semi-major axis a in Figure 2. The cross-section A gives the probability that an energy loss of magnitude $| E | \equiv 1/(2a)$ or greater takes place in a single planet–comet encounter. The cross-section is dimensionless, i.e. it is the fraction of the phase space inside the relevant equal energy curve of Figure 1. The differential cross-section is found to be well approximated by

$$dA = 10^{-6} f(i_0) E^{-3} dE, \tag{1}$$

where

$$f(i_0) = \left\{ \begin{array}{ll} 80.0, & \text{if } i_0 \leq 1° \\ 4.3 \times 10^{-3}(146° - i_0)(\sin i_0)^{-1}, & \text{if } 1° < i_0 \leq 126° \\ 1.1 \times 10^{-3}(206° - i_0)(\sin i_0)^{-1}, & \text{if } 126° \leq i_0 < 179° \\ 30.5 & \text{if } 179° \leq i_0 \leq 180° \end{array} \right\} .$$

The inclination dependence which is included in Equation (1), is shown in Figure 3 for the energy change $E \leq -0.065$, corresponding to final semi-major axis of

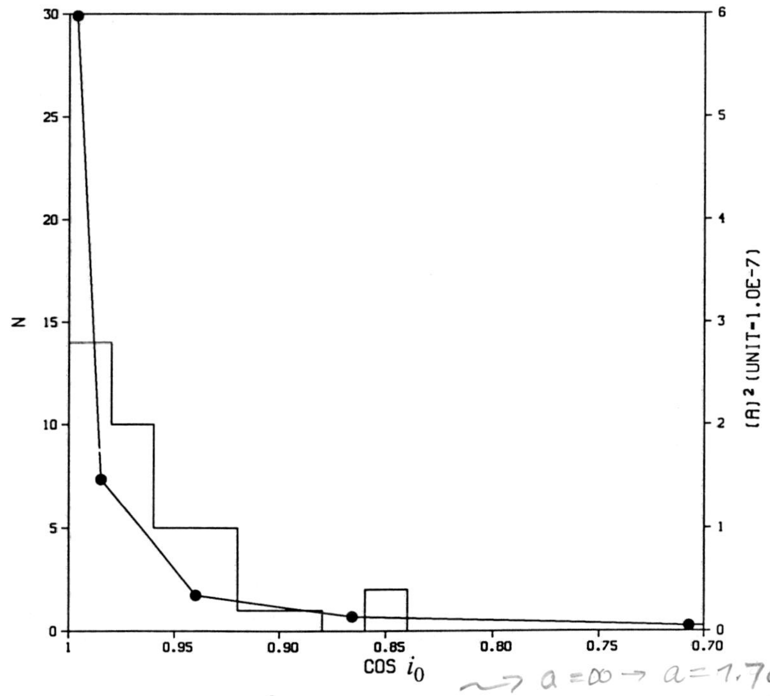

Fig. 4. The cross-section squared A^2 for $E \leq -0.065$ when $q_0 = 0.5$ as a function of the initial inclination i_0 (solid line with dots). For comparison, a histogram of observed inclinations of comets in Jupiter family.

$a \leq 40$ AU. Figure 4 is similar to Figure 3 except that we now plot the square of the cross-section. The reason for this kind of presentation becomes clear in the next section.

When the energy change of magnitude E takes place, we would like to know what happens to the inclination of the orbit. Depending on where we are in the corresponding equal energy curve of Figure 1 we obtain a value of the inclination i which is in general different from i_0. In Figure 5 we plot the new value of the inclination as a function of the angle θ in an equal energy curve, measured from the "collision center" $\omega = \omega_c$ and $\phi = \phi_c$.

In practical applications it is often enough to know what the statistical distribution of the inclinations is. Then we take a narrow strip in the $\omega - \phi$ plane, bounded by equal energy curves E and $E + \mathrm{d}E$, and calculate the number of different inclinations i which arise from this strip. The numbers are weighted according to the cross-sectional area in the $\omega - \phi$ plane.

We show an example of the inclination distributions in Figure 6. The inclination distributions are well modeled by

$$\psi(i) = 1.33E^{-1}\left\{1 + 1.42E^{-2}\left[x^2 + 0.93x^3\right]\right\}^{-2} \tag{2}$$

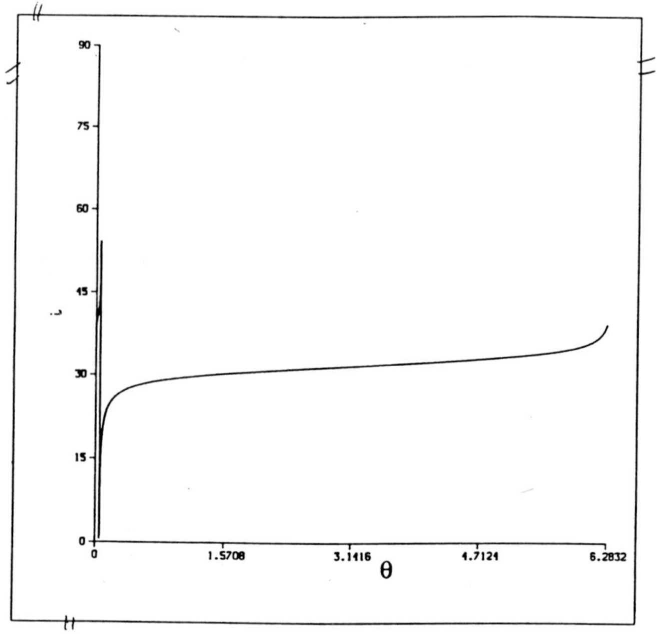

Fig. 5. The new inclination i as a function of the angle θ, measured from the collision centre (ω_c, ϕ_c), along the equal energy curve $E = -0.065$ $(i_0 = 30°, q_0 = 0.5)$.

where

$$x = i - \left(i_0 + 82.1E^{1.55}\right)\pi/180° \quad (3)$$

and the angles i and i_0 are given in degrees. The coefficients apply to $i_0 = 30°$, $q_0 = 0.5$ in a comet - Jupiter encounter. The generalization to other orbits still remains to be done. The agreement with the data between $a = 10$ AU and $a = 30$ AU is shown by the fit in Figure 7. For example, a Gaussian curve would be a much poorer fit to the data.

3. Discussion

These are merely preliminary results of a programme where the aim is to find analytic representations to the cross-sections and inclination distributions as a function of i_0 and q_0. However, as far as the present calculations are typical, we may attempt to make some general conclusions.

Supposing that the short period comets are captured from the Oort Cloud and that the orbits of Oort Cloud comets are oriented randomly, we may compare the distributions of Figures 3 and 4 with the observed inclination distributions of short period comets. These are usually classified in two groups: the Jupiter family and the Halley family.

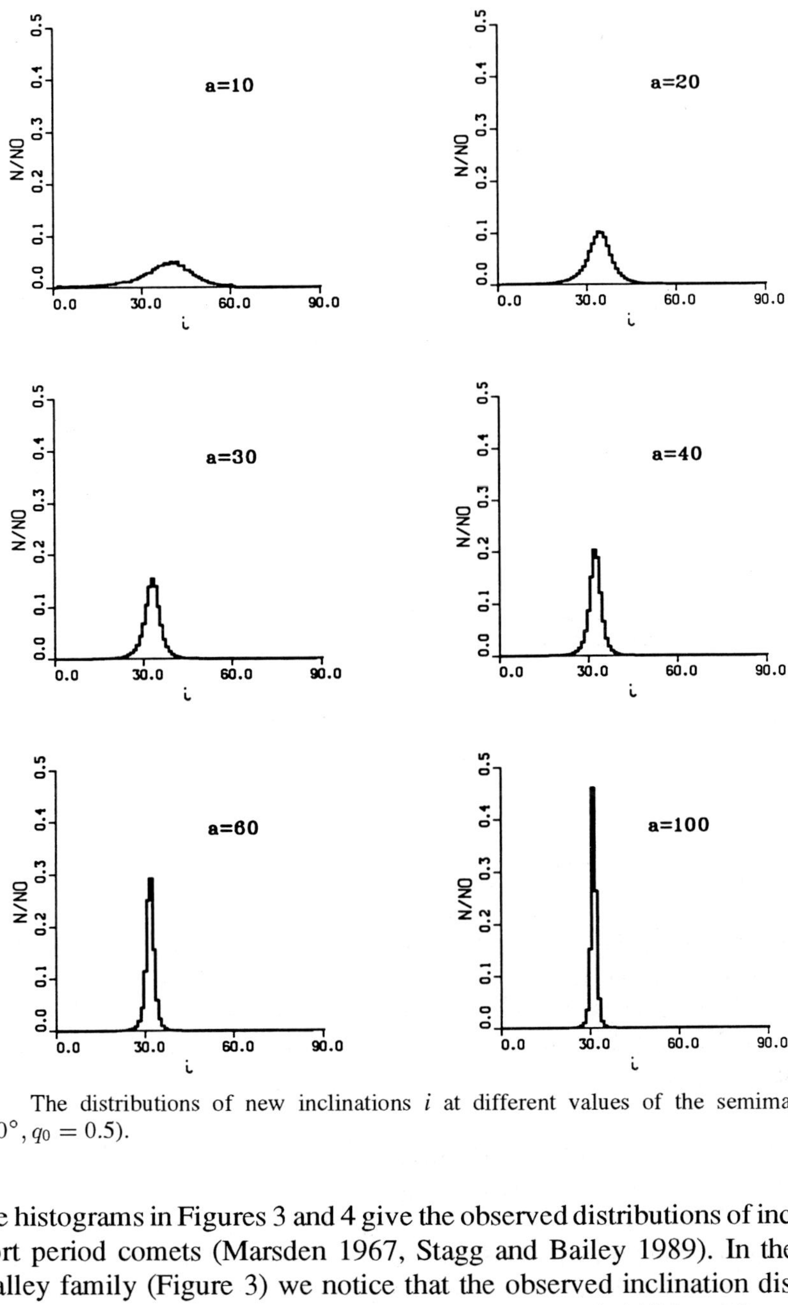

Fig. 6. The distributions of new inclinations i at different values of the semimajor axis a ($i_0 = 30°, q_0 = 0.5$).

The histograms in Figures 3 and 4 give the observed distributions of inclinations of short period comets (Marsden 1967, Stagg and Bailey 1989). In the case of the Halley family (Figure 3) we notice that the observed inclination distribution corresponds very closely to the calculated capture probabilities. Therefore one cannot rule out the possibility that the short period comets, in case of the Halley family (Figure 3), are merely the captured population of the Oort Cloud, as has been previously found out by Duncan *et al.* (1988).

On the other hand, the comets of the Jupiter family are much more tightly clustered near zero inclination than what one would expect from the capture cross-

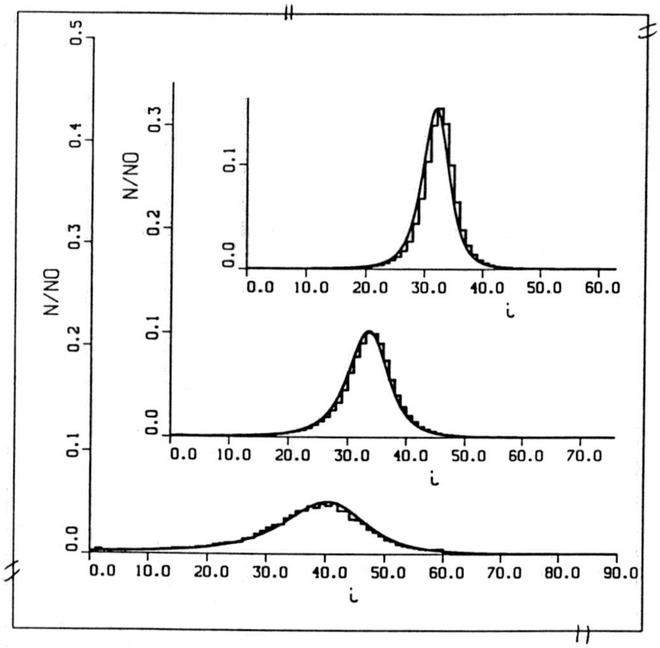

Fig. 7. The comparison of the fitting function (Equation (3)) with the data on inclination distributions. The cases of $a = 30$, 20 and 10 (from top to bottom). The initial inclination $i_0 = 30°$ and $q_0 = 0.5$.

section. This may result from the capture to the Jupiter family being a two stage process. It is easy to verify using Equation 1 that it is much more likely that a Jupiter family comet arises after two close Jupiter encounters than after a single encounter. A comet which has suffered one close encounter and has become a short period comet will sooner or later come to another close encounter. This is the reason why we plot the square of the capture cross–section in Figure 4 as being proportional to the probability of two successive encounters producing a Jupiter family comet. The agreement with the observed inclination distribution is good except that the peak at zero inclination is now more spread out in observations than in theory. But this is expected since the cross-sections in Figures 3 and 4 are plotted in terms of i_0 while the observations should really be compared with the distribution of i. The sharp features in the cross-section are spread out in the i–distribution (see Figure 6).

Acknowledgements

This work was begun after inspiring discussions with H. Rickman and M. Bailey. Their contribution to the outlining of the project is greatly appreciated.

References

Bash, F.: 1986, 'Present, past and future velocity of nearby stars: the path of the Sun in 10^8 years', in*The Galaxy and the Solar System*, R. Smoluchowski, J.N. Bahcall and M.S. Matthews (Eds.), p. 35. The Univ. Arizona Press: Tucson.

Davis, M., Hut, P. and Muller, R.A.: 1984, 'Extinction of species by periodic comet showers', *Nature*, **308**, 715.

Delsemme, A.H.: 1986, 'Cometary evidence for a solar companion', in *The Galaxy and the Solar System*, R. Smoluchowski, J.N. Bahcall and M.S. Matthews (Eds.), p. 173. The Univ. Arizona Press: Tucson.

Donn, B.: 1976, 'Comets, interstellar clouds and star clusters', in *The Study of Comets*, B. Donn, M. Mumma, W. Jackson, M. A Hearn and R. Harrington (Eds.), NASA SP–373, p. 611.

Duncan, M., Quinn, T. and Tremaine, S.: 1987, 'The formation and extent of the Solar System comet cloud', *Astron. J.*, **94**, 1330.

Duncan, M., Quinn, T. and Tremaine, S.: 1988, 'The origin of short-period comets', *Astrophys. J.*, **328**, L69.

Fernandez, J.A.: 1980, 'On the existence of a comet belt beyond Neptune', *Monthly Notices Roy. astr. Soc.*, **192**, 481.

Hasegawa, I.: 1976, 'Distribution of the aphelia of long-period comets', *Pub. Astr. Soc. Japan*, **28**, 259.

Hills, J.G.: 1981, 'Comet showers and the steady-state infall of comets from the Oort cloud', *Astron. J.*, **86**, 1730.

Innanen, K.A., Patrick, A.T. and Duley, W.W.: 1978, 'The interaction of the spiral density wave and the Sun's Galactic orbit', *Astrophys. Space Sci.*, **57**, 511.

Joss, P.C.: 1973, 'On the origin of short-period comets', *Astron. Astrophys.*, **25**, 271.

Kuiper, G.P.: 1951, 'On the origin of the solar system' in *Astrophysics*, J.A. Hynek (Ed.), p. 357, McGraw-Hill: New York.

Kustaanheimo, P. and Stiefel, E.: 1965, 'Perturbation theory of Kepler motion based on spinor regularization', *J. reine angew. Math.*, **218**, 204.

Marsden, B.G.: 1967, 'One hundred periodic comets', *Science*, **155**, 1207.

Matese, J.J. and Whitman, P.G.: 1992, 'A model of Galactic tidal interaction with the Oort comet cloud', in this volume, pp. 13–35.

Matese, J.J. and Whitmire, D.P.: 1986, 'Planet X and the origins of the shower and steady state flux of short period comets', *Icarus*, **65**, 37.

McCrea, W.H.: 1975, 'Solar System as space-probe', *Observatory*, **95**, 239.

McGlynn, T.A. and Chapman, R.D.: 1989, 'On the nondetection of extrasolar comets', *Astrophys. J.*, **346**, L105.

Oja, H.: 1975, 'Perihelion distribution of near-parabolic comets', *Astron. Astrophys.*, **43**, 317.

Oort, J.H.: 1950, 'The structure of the cloud of comets surrounding the solar system and a hypothesis concerning its origin', *Bull. Astron. Inst. Netherlands*, **11**, 91.

Rampino, M.R. and Caldeira, K.: 1992, 'Episodes of geologic activity during the past 260 million years: a quantitative approach', in this volume, pp. 143–159.

Rampino, M.R. and Stothers, R.B.: 1984, 'Terrestial mass extinctions, cometary impacts and the Sun's motion perpendicular to the galactic plane', *Nature*, **308**, 709.

Reynolds, R.T., Tarter, J.C. and Walker, R.G.: 1980, 'A proposed search of the Solar neighborhood for substellar objects', *Icarus*, **44**, 772.

Safronov, V.S.: 1972, *Evolution of the Protoplanetary Cloud and Formation of the Earth and Planets*. NASA TT-F-677. Trans. from Russian (Moscow: Nauka Press, 1969)

Shoemaker, E.M. and Wolfe, R.F.: 1982, 'Cratering time scales for the Galilean satellites', in *Satellites of Jupiter*, D. Morrison (Ed.), p. 277. The Univ. Arizona Press: Tucson.

Shoemaker, E.M. and Wolfe, R.F.: 1986, 'Mass extinctions, crater ages and comet showers' in *The Galaxy and the Solar System*, R. Smoluchowski, J.N. Bahcall and M.S. Matthews (Eds.), p. 338. The Univ. Arizona Press: Tucson.

Stagg, C.R. and Bailey, M.E.: 1989, 'Stochastic capture of short–period comets', *Monthly Notices Roy. Astron. Soc.*, **241**, 507.

Valtonen, M.J.: 1983, 'On the capture of comets into the solar system', *Observatory*, **103**, 1.

Valtonen, M.J. and Innanen, K.A.: 1982, 'The capture of interstellar comets', *Astrophys. J.*, **255**, 307.

Valtonen, M. and Zheng, J.Q.: 1990, 'On the origin of Oort's cloud of comets in a star cluster' in *Nordic-Baltic Astronomy Meeting*, C.–I. Lagerkvist, D. Kiselman and M. Lindgren (Eds.), p. 353. Reprocentralen HSC: Uppsala.

Whipple, F.L.: 1975, 'Do comets play a role in galactic chemistry and γ–ray bursts?', *Astron. J.*, **80**, 525.

Whitmire, D.P. and Jackson, A.A.: 1984, 'Are periodic mass extinctions driven by a distant solar companion?', *Nature*, **308**, 713.

Whitmire, D.P. and Matese, J.J.: 1985, 'Periodic comet showers and Planet X', *Nature*, **313**, 36.

Yabushita, S.: 1979, 'A statistical study of the evolution of the orbits of long-period comets', *Monthly Notices Roy. Astron. Soc.*, **187**, 445.

Yabushita, S.: 1983, 'Distribution of cometary binding energies based on the assumption of steady state', *Montly Notices Roy. Astron. Soc.*, **204**, 1185.

Zheng, J.-Q., Valtonen, M.J. and Valtaoja, L.: 1990, 'Capture of comets during the evolution of a star cluster and the origin of the Oort Cloud', *Celest. Mech. & Dyn. Astron.*, **49**, 265.

ORIGIN OF SHORT-PERIOD COMETS

M.E. BAILEY
Department of Astronomy, The University, Manchester M13 9PL, England, U.K.

Abstract. Reasons for interest in the origin of short-period comets and the difficulties of computing their long-term dynamical evolution are reviewed. The relative advantages of a source region in an extended 'inner core' of the Oort cloud or a compact 'comet belt' just beyond the planetary system are finely balanced, and it is premature to consider the problem solved. A complication is that some comets belonging to the Jupiter family may be part of a time-dependent system, possibly the remains of a giant comet such as Chiron which could have been part of the system $\approx 10^4$ yr ago. The origin of short-period comets plays a pivotal rôle in many areas of solar system science: planet formation, the source of water (possibly life) on the terrestrial planets, the cratering record on the terrestrial planets and satellites of the outer planets, and the environmental impact posed by massive bodies and their decay products in the Earth's near-space environment.

Key words: Short-period comets, dynamics, Oort cloud, giant comets, terrestrial catastrophism.

1. Introduction

During the past decade our understanding of comets – their origin, physical properties and dynamical evolution – has undergone something of a revolution. Ten years ago the Oort cloud was widely thought to be an extended, rather diffuse structure, stable for the age of the solar system with a total mass on the order of a few $M_\oplus$. Now it is generally agreed (though not unanimously) that its outer parts are unstable against external influences, that it has a complex 'inner' structure referred to as the dense inner core, and a total mass ranging upwards from at least 20 $M_\oplus$. Likewise, comet nuclei – which were previously regarded as rather small, insignificant members of the solar system, the remains of material which formed the planets – are now recognized as having low albedos, larger sizes and much larger masses. Their perceived importance in solar system affairs, ranging from issues of planet formation to the cratering record of the terrestrial planets and climatological effects due to accretion of cometary dust, has grown proportionately.

An active line of investigation during the past decade has concerned the origin of short-period comets, those with periods less than about 200 yr. Some 170 such comets have been discovered, most belonging to the Jupiter family having small inclinations to the ecliptic (typically less than 30°), periods $\lesssim 20$ yr and Tisserand parameters with respect to Jupiter $T_J \gtrsim 2.5$. The remainder (about 20 comets) are described as Halley-types; they have a much wider range of inclinations, and (with one or two exceptions) longer periods in the range 20–200 yr and $T_J \lesssim 2.5$.

Historically, as described in many reviews (e.g. Bailey *et al.*, 1990; Rickman, 1990), the difficulty presented by these objects has been to explain their total number: the cross-section for changing a long-period orbit into one of short period

Celestial Mechanics and Dynamical Astronomy **54:** 49–61, 1992.

by planetary perturbations is quite small, and the observed near-parabolic flux is insufficient to explain the number of short-period comets if the system is assumed to be in a steady state. This has led to the idea that the system could be time-dependent, either due to a surge in the near-parabolic flux (e.g. Bobrovnikoff, 1929), sometimes called a comet shower (Hills, 1981), or because the present short-period system is dominated by fragmentation of a single large progenitor (Alexander, 1850, 1851; Bredichin, 1889; Bobrovnikoff, 1931; Clube and Napier, 1982a, 1984, 1986a,b). A third possibility is that the distribution is indeed close to a steady state, but most short-period comets do not come from the observed near-parabolic flux.

Discussions of the origin of short-period comets during the past decade have involved ingredients of all these possibilities. The level of understanding now reached raises the significance of short-period comets in solar system affairs to a very high level, in particular as probes of the hypothetical dense inner core of the Oort cloud (Bailey, 1983a, 1990). Short-period comets provide direct constraints on theories of planet formation and on the structure and evolution of the early protoplanetary disc. At the same time they are a potentially important source of Earth-crossing asteroids (Öpik, 1963; Wetherill 1989). These bodies dominate the production of km-size craters on the Earth, while their smaller relations, representing bodies ranging in size down from that of large boulders ($\approx 100\,\mathrm{m}$) to stones and dust, and probably moving in streams analogous to those of some asteroids and meteorites (e.g. Halliday *et al.*, 1990; Drummond, 1991; Hahn, 1991), may episodically inject large amounts of extraterrestrial dust into the Earth's upper atmosphere. A full understanding of the influences affecting the Earth on timescales from as short as a few hundred years up to about 1 Myr or so, with important implications for the nature of the external stresses imposed on the planet (e.g. contributing to the onset of an ice-age; Hoyle and Wickramasinghe, 1978) hinges on the success of present studies into the origin and evolution of short-period comets.

2. Steady-State Models

The difficulty of explaining the number of short-period comets was highlighted in a calculation by Joss (1973). First, the observed near-parabolic flux corrected for incompleteness and comprising comets with total absolute magnitudes H_{10} brighter than about $H_{10} = 7$, is on the order of $\nu_{\rm new} = 0.5\,\mathrm{AU}^{-1}\,\mathrm{yr}^{-1}$. Secondly, according to Everhart (1972) the orbits of the majority of short-period comets (those comprising the Jupiter family) can be broadly understood if they originate by capture from initially parabolic orbits with inclinations i and perihelion distances q in the so-called Jupiter capture zone, with $i \lesssim 9°$ and $4 \lesssim q \lesssim 6\,\mathrm{AU}$. This range of inclinations corresponds to a fraction $(1 - \cos 9°)/2 \simeq 0.006$ of an assumed isotropic flux with inclinations uniformly distributed in $\cos i$. Thirdly, again following Everhart, only about 0.7% of such orbits end up evolving into observable Jupiter-family short-period comets. Lastly, the active lifetime $T_{\rm SP}$ of a short-period comet is probably no more than about 10^4 yr (Joss himself chose a

TABLE I
Capture probability for nearly parabolic orbits with initial perihelion distances q in the range 4–6 AU and initial inclinations i uniformly distributed in $\cos i$, to evolve to final perihelion distances $q_f < 2.5$ AU and final periods $P_f < 20$ yr and $P_f < 200$ yr respectively. The entries have been derived from Fig. 4 of Quinn *et al.* (1990).

Range of i	$P_f < 20$ yr	$P_f < 200$ yr
0–30	6.6×10^{-3}	7.9×10^{-3}
30–60	1.2×10^{-3}	3.7×10^{-3}
60–90	4.8×10^{-4}	2.0×10^{-3}
90–120	3.2×10^{-4}	1.2×10^{-3}
120–130	$< 1.4\times10^{-4}$	8.0×10^{-4}
0–180	8.6×10^{-4}	2.0×10^{-3}

value of 500 yr). Multiplying these factors together suggests that the steady-state number of short-period comets should be $\approx 10^{-1}$, three orders of magnitude less than observed. However one adjusts the individual factors in this calculation within reason, the conclusion remains that the observed short-period comets cannot come from the observed near-parabolic flux.

The accuracy of this calculation has since been greatly improved, largely due to better estimates of the capture probability $p_c(q, i)$ as a function of initial perihelion distance and inclination. In particular, Duncan *et al.* (1988) and Quinn *et al.* (1990) have shown that capture may occur from a much wider range of inclinations than found by Everhart, though still concentrated towards relatively small values. Table I shows results derived from Quinn *et al.* (1990) for the probability of comets in parabolic orbits with $4 < q < 6$ AU and inclinations i uniformly distributed in $\cos i$ to evolve to final orbits with $q_f < 2.5$ AU and $P_f < 20$ yr or 200 yr respectively.

We draw two conclusions from this Table. First, there is a significant dependence of the capture probability on initial inclination (as found by Everhart and a number of other authors, e.g. Yabushita and Tsujii 1990), although the inclination dependence is not so strong as to dominate the resulting distribution of comets captured from an isotropic source (contrary to the conclusion of Everhart). Secondly, the proportion of comets eventually captured into short-period orbits from an isotropic flux is $\approx 10^{-3}$, much greater than implied by Everhart's results. Part of the discrepancy can be attributed to different capture criteria, but despite this the capture probability obtained by Quinn *et al.* (1990) still fails to resolve the problem of the number of short-period comets. The conclusion is the same: observed short-period comets do not predominantly come directly from the observed long-period flux.

It is interesting to compare these results with those of Stagg (1991) and Wetherill (1991). The former author computed the capture probability from initial parabolic orbits with perihelion distances in the range 4–6 AU and inclinations uniformly distributed in $\cos i$ to final perihelion distances $q_f < 1.5$ AU and aphelion distances $Q_f < 10$ AU; these values being chosen (cf. Stagg and Bailey, 1989) to correspond approximately to the observed Jupiter-family. Wetherill (1991) used the same initial conditions, but computed the probability of capture to final orbits $q_f < 1.5$ AU and $Q_f < 9$ AU. However, instead of an exact numerical integration he used a Monte-Carlo method based on the Öpik scheme (cf. Carusi *et al.* 1990).

Whereas Stagg's results give a capture probability for initial inclinations 0–30° of about 1.2×10^{-3}, and an inclination-averaged result over all inclinations (0–180°) of 1.1×10^{-4}, those of Wetherill give 2.4×10^{-3} averaged over all inclinations, 20 times larger! This suggests that the Öpik scheme overestimates the number of comets captured to short-period, probably by enhancing the fraction of comets captured from inclinations greater than 30° or so. However, none of these calculations suggests that the capture probability is sufficiently high to overcome the overriding problem posed by the number of short-period comets in the Jupiter-family.

Repeating our previous schematic calculation, but substituting an inclination-averaged capture probability on the order of 10^{-3} instead of Everhart's average of $\approx 0.006 \times 0.007$ indicates that ≈ 10 short-period comets might be explained, compared to the ≈ 100 or so known. Moreover, the point emphasized by Duncan *et al.* (1988) and Quinn *et al.* (1990), that these captured comets would tend to have a broad range of inclinations, contrary to the observed Jupiter family, remains valid. Now, however, when capture of comets with $q < 4$ AU is also considered, which to order of magnitude will occur with a roughly similar capture probability (cf. Stagg and Bailey, 1989; Olsson-Steel *et al.*, 1989), certainly for Halley-types, the effect of the new results is seen to have shifted the argument. Whereas previously neither class of short-period comet could be explained by capture from the observed near-parabolic flux, Halley-types (which observationally number about 20 comets) might now be explained. The Jupiter family remains problematic, however, about an order of magnitude too large to be explained by steady-state capture, leaving aside the difficulty of their relatively low inclinations.

3. How to Increase the Jupiter Family

3.1. Time-Dependence

As already mentioned, a number of proposals have been made to increase the number of Jupiter-family comets. The first, and most obvious, considers the effect of introducing time-dependence into the system. For example, the near-parabolic flux might have been higher a capture time ago ($\approx$ 1 Myr), owing to a recent stellar or molecular cloud perturbation of the Oort cloud causing a comet shower as discussed by Hills (1981), Fernández and Ip (1987), Heisler *et al.* (1987) and Heisler (1990). Under these conditions, however, one would expect that the number of Halley-types would also be increased, and in this case (unless exceptional fine-tuning was involved, or factors introduced which depended on differential selection effects or aging between the two groups) 'solving' the problem of the Jupiter family might then lead to a new problem of making too many Halley-types.

A second way to introduce time dependence draws on the observation that comets physically decay on a much shorter timescale ($\approx 10^4$ yr) than their dynamical lifetimes, and that both long-period and short-period comets frequently split, occasionally into fragments which survive as separate comets for many revolutions. Rickman (1990) has shown that the observed rate of cometary splitting suggests that the total number of short-period comets might indeed be affected by such a process. Splitting cannot be ignored, but its effects in terms of time-dependence would be relatively small unless a very large comet were involved, as advocated by Clube and Napier (e.g. 1982a, 1984, 1986b).

The Kreutz sungrazers provide a possible example of such an effect, while orbits similar to that of P/Machholz (although not a member of the Jupiter family) indicate that the sungrazing state might be surprisingly widespread (Bailey *et al.*, 1992a,b). Long-term integrations of giant comets such as 2060 Chiron (Hahn and Bailey, 1990) and 1992 AD also show that large comets have a non-negligible probability of circulating in short-period orbits similar to those of a typical Jupiter-family comet, while evidence of a variety of short-timescale phenomena in the solar system (e.g. Dones, 1991) points in much the same way.

Further suggestions of time-dependence on a short timescale have been reviewed by Clube and Napier (1982b, 1990), Clube (1989) and Bailey *et al.* (1986, 1990). In particular, the timescale for disintegration of a giant comet in the Jupiter family is interestingly close to the 'historical' timescale ($\approx$ 5000 yr) over which both civilization and Man's ideas about the cosmos have largely developed. Current theories about the origin of short-period comets, then, may suggest that the rise of civilization since the ending of the last ice age coincided with substantial changes in both cometary activity and other celestial phenomena such as meteor storms and fireballs. Evidence for such changes, if they occurred, might reasonably be expected to be found in the form of legends or recorded knowledge. For example, it has long been a mystery why the Babylonian and megalithic civilizations were apparently so obsessed by events in the sky; and it has still not been explained why the earliest Greek philosophers taught, for example, that stars lie below the sun and the moon (Anaximander, Leucippus), that the Milky Way is the former path of the sun (Metrodorus), or that the Milky Way lies in the earth's shadow (Anaxogoras, Democritus). Such teaching makes astronomical sense only if 'stars' are interpreted to mean 'shooting stars', and if the former Milky Way is understood to refer to an earlier more intense zodiacal light. Such beliefs provide indirect support for a time-dependent system: indeed, if our remote ancestors had witnessed the break-up of a large comet, possibly in an earth-crossing orbit, with the attendant spawning of ordinary comets, bolides and dust, it is perhaps not surprising that early societies would have kept a close watch on the sky and maintained oral traditions which until now have been inexplicable. Although the probability of a body the size of Chiron breaking up in a short-period orbit several 10^4 yr ago to produce a host of 'ordinary' comets which we now recognize as a temporarily enhanced Jupiter family cannot be calculated reliably, the possibility should certainly not be ruled out.

3.2. Alternative Sources

The third class of theory to explain the excess of Jupiter-family comets, and indeed the one most amenable to quantitative study, accepts the argument that Jupiter-family comets cannot come from the observed near-parabolic flux, and looks for a solution (usually based on a steady-state model) in which they come from another source. An obvious choice is to involve the dense inner core of the Oort cloud, predicted on theoretical grounds by a number of arguments (Hills, 1981; Bailey, 1983a,b). Figure 1, taken from Bailey *et al.* (1990) illustrates the range of possible models for the inner core which have been considered, and therefore the potential range of source orbits for Jupiter-family comets.

The Figure shows that models of the dense inner core may be broadly divided into two classes: compact models, with dimensions on the order of 50–200 AU; and extended models, with diameters at least 10^3 AU possibly extending beyond 2×10^4 AU to merge into the conventional outer Oort cloud. Compact models include the suggestions (i) that primordial comets might survive for the age of the solar system in nearly circular orbits between those of the outer planets (Duncan *et al.*, 1988, 1989), although this now seems unlikely (Gladman and Duncan, 1990); (ii) that they may still exist in a belt lying beyond the present planetary system (e.g. Edgeworth, 1949; Kuiper, 1951; Whipple, 1972; Mendis, 1973; Fernández, 1980; Quinn *et al.*, 1990); and (iii) that they may move in orbits of varying eccentricity and inclination occupying a more spherical distribution about the sun (Bailey, 1983a,d).

On the other hand, the possibility of a much more extended range of cometary orbits within the Oort cloud is suggested by a number of theoretical considerations (e.g. Cameron, 1978; Biermann and Michel, 1978), and observations of stars such as β Pictoris and of extended molecular discs with dimensions on the order of 0.1 pc surrounding young stellar objects. As emphasized by Fernández and Ip (1983) and Fernández (1985), for example, the standard planetesimal theory of cometary origin also predicts an extended inner core of type (e) shown in the Figure, as demonstrated by the detailed numerical simulation of Duncan *et al.* (1987).

Among these models, several of which may be present at the same time, those

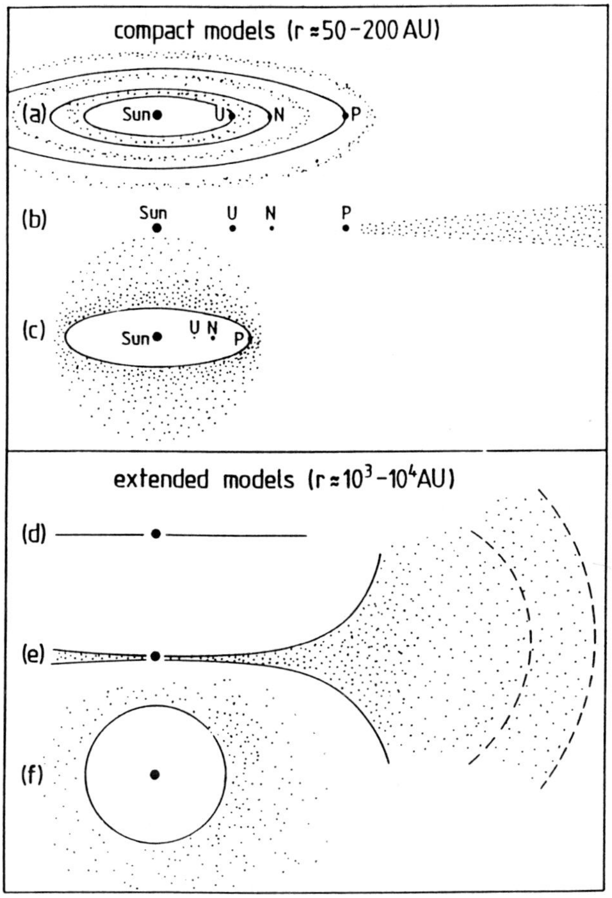

Fig. 1. Models of the dense inner core of the Oort cloud. Taken from Bailey *et al.* (1990).

most frequently discussed are (i) a compact comet-belt or 'Kuiper Belt' with dimensions on the order of 100 AU, as advocated by Duncan *et al.* (1988), Torbett (1989) and Quinn *et al.* (1990), (ii) a trans-Neptunian cometary disc ($r \approx$ 50–2000 AU) described for example by Fernández (1985) and Fernández and Ip (1983, 1991), and (iii) a more spherical cloud ($r \approx$ 5000–20000 AU) with properties similar to those predicted by the planetesimal hypothesis (Duncan *et al.* 1987) and advocated as a possible source of short-period comets by Bailey (1986a), Stagg and Bailey (1989) and Bailey and Stagg (1990). Although it should be emphasized that these models are not necessarily mutually exclusive, it is sometimes helpful to proceed as if they were, and to consider separately the respective predictions of each model in the absence of the other.

4. Compact Comet Belt Versus Extended Inner Core

The comet belt has three principal arguments in its favour: (i) comets may be the residue of the original protoplanetary disc beyond Neptune; (ii) their orbits are prograde and of low inclination; and (iii) they have initial perihelion distances close to the orbit of Neptune, making capture and subsequent handing down to the control of Jupiter possibly a more likely event. The size of the disc is on the order of 100 AU. The principal argument against the belt is that it has not yet been detected, whether by the gravitational pull exerted on the outer planets, by 'stellar occultations' due to passage of its largest members in front of distant stars (Bailey, 1976, 1983a; Dyson, 1992), or by potentially detectable far-infrared radiation (Bailey, 1983c; Bailey *et al.*, 1984; Jackson and Killen, 1988), although a recent analysis of IRAS data (Aumann and Good, 1990) suggests that an unexplained far-infrared solar system signature may be present.

Unfortunately, evidence for an extended inner core of the Oort cloud is as elusive as for the belt. Apart from a few long-period comets no direct evidence exists for any comets in this large volume of space surrounding the sun. Yet favouring the idea are the arguments that: (i) such a structure is predicted by the planetesimal theory (it may be flattened closer to the solar system, merging into the trans-Neptunian cometary disc); (ii) an inner core of substantial mass appears to be necessary to replenish the dynamically unstable outer Oort cloud over the age of the solar system when perturbations by stars and molecular clouds are included (Bailey, 1983b, 1986b; Weinberg *et al.*, 1986, 1987; Bailey, 1991); (iii) it provides a unified picture for the origin of both long-period and short-period comets (at least when based on the planetesimal picture); and (iv) such a model can easily explain the total number of short-period comets (Bailey, 1986a; Bailey and Stagg, 1990). Indeed the argument can now be turned around, and the relatively small number of comets in the Jupiter family used to provide an important constraint on the mass of any dense inner core. As emphasized by Duncan *et al.* (1988) and Quinn *et al.* (1990), however, the possibility that a spherically symmetrical dense inner core might predict too many high-inclination short-period comets could be decisive. The issue hinges on whether initially high-inclination orbits with perihelion distances in the range 15–30 AU really are captured with high efficiency.

4.1. Limitations of Capture Calculations

The problem of calculating the capture probability for nearly parabolic orbits of arbitrary inclination and perihelion distance is still beyond the capabilities of even the fastest computers. As a result, investigators have used a variety of short cuts to speed up the numerical evolution: the Monte-Carlo approaches pioneered by Everhart (1977), Froeschlé and Rickman (1980) and Nakamura (1981); increasing the planetary masses by factors ranging from 10–40 (Duncan *et al.*, 1988; Quinn *et al.*, 1990); and use of the Öpik scheme (Fernández and Ip 1991, Wetherill 1991), in which only very close encounters are considered, these being assumed to dominate the dynamics and to occur essentially randomly in time.

These techniques are beset with difficulties. The Monte-Carlo approach has particular problems in realistically accounting for the extreme asymmetric wings of the various perturbation distributions, while resonances and secular effects in the long-term evolution have not yet been properly modelled.

The validity of increasing the planetary masses has been discussed by both Stagg and Bailey (1989) and Rickman (1990); the latter makes the point that if the dynamical evolution is speeded up by a large factor, then effects due to finite cometary lifetimes should be similarly speeded up. Additional arguments against increasing the planetary masses are (i) resonances then hardly ever occur (whereas in fact they are a crucial feature in the evolution of high-inclination orbits and those with perihelia in the outer planetary system; e.g. Hahn and Bailey, 1990, 1992; Bailey and Hahn, 1992); (ii) secular effects are incorrectly modelled (Quinn *et al.*, 1990); and (iii) the asymmetric wings of the planetary perturbation function are quite wrong (Valsecchi, 1990, 1992). Insofar as the latter dominate the direction of the dynamical evolution of objects such as Chiron (Oikawa and Everhart, 1979), this is an especially severe criticism.

Finally, although the Öpik scheme does accurately describe the effects of close approaches (in particular the asymmetric wings of the energy changes; Carusi *et al.*, 1990; Valsecchi, 1992), the desired speeding up of a long-term numerical simulation requires the choice of an unrealistically small maximum impact parameter $b_{\max}$ (cf. Fernández and Ip, 1991). At the same time, the treatment of distant encounters (beyond the nominal $b_{\max}$) becomes relatively more important. Present calculations do not include effects due to resonances or secular perturbations, and they assume (along with other Monte-Carlo schemes) that captured comets in effect follow 'average' trajectories, ignoring the likelihood that one close encounter with a massive planet will probably soon be followed by another, owing to the respective orbits then being close to node-crossing. Captured comets – almost by the nature of things – end up suffering an extremely rare combination of close encounters, not randomly distributed in time, and not easily modelled by a standard Monte-Carlo scheme. A hint that the Öpik scheme may not produce reliable results is obtained by comparing the results of Wetherill (1991) with those of Stagg (mentioned above) and Quinn *et al.* (1990). In the latter case, despite the much more stringent capture criterion, Wetherill's scheme appears to capture too many comets.

In summary. it is clear that most (if not all) previous results on the capture of short-period comets, whether from a comet belt or a spherical inner core, should be treated with caution. It has *not* been shown that a spherically symmetrical, extended inner core necessarily produces too many high-inclination short-period orbits, but equally it has not been reliably demonstrated that it does not. More work is needed, and the whole range of inner core distributions suggested by the models illustrated in Figure 1 should be borne in mind and compared on their respective merits.

5. Conclusion

The principal conclusions of this study, much of which presents preliminary results from an ongoing collaboration with G. Hahn, H. Rickman, G. Tancredi and M. Valtonen, are summarized below.

(1) Halley-type comets come from the observed near-parabolic flux.

(2) Jupiter-family comets are $\approx$ 10 times too numerous to be explained by capture from the observed near-parabolic flux, or even from an extrapolated near-parabolic flux with perihelia in the neighbourhood of Jupiter's orbit. The number of short-period comets is a signature of another source, either giant-comet break-up or a separate source flux with perihelia in the outer planetary system. (These possibilities, of course, are not mutually exclusive.) An extended inner core of the Oort cloud may be required to explain the survival of the outer Oort cloud over the age of the solar system, and could explain the number of Jupiter-family comets, but it may produce too many high-inclination comets. A comet belt, on the other hand, while equally plausible on cosmogonic grounds, may in the end turn out to be unnecessary.

(3) Resonances and secular perturbations are important aspects of the dynamical evolution leading to short-period orbits, especially the production of Halley-types. Secular effects allow comets of high inclination to evolve through a large range of perihelion distances, an extreme example being comets with initial perihelion distances less than about 2 AU and inclinations near 90°. Such orbits frequently evolve to a sungrazing state (Bailey *et al.*, 1992b, Harrington 1968, Marchal 1978), and may be an important source of interplanetary matter and extinct cometary asteroids such as 1991 DA.

(4) Some short-period comets (e.g. P/Encke) evolve into orbits similar to those of Earth-crossing asteroids, while others (e.g. Halley-types) evolve into 1991 DA-types. This suggests a rather direct link between Earth-crossing asteroids and the near-parabolic flux, possibly involving time-dependence in the terrestrial cratering rate and in the flux of extraterrestrial material on to the Earth (Hahn and Bailey, 1992; Bailey and Hahn, 1992).

(5) Although short-period comets represent an unsolved problem in solar system dynamics, they also have important associations extending far beyond cometary astronomy and solar system cosmogony. The subject now covers a wide range of topics, with implications for the long-term history of the Earth, the development of Man's ideas about the cosmos (possibly influenced by events in the sky of which we now have no analogue), the impact of massive bodies on to the Earth, and the accretion of streams of dust and larger particles from comets in Earth-crossing orbits. These interrelationships give studies of 'minor bodies' in the solar system a new dimension and a heightened sense of urgency.

Acknowledgements

It is a pleasure to thank S.V.M. Clube, G. Hahn, W.M. Napier, H. Rickman, D. Steel, G.B. Valsecchi, M. Valtonen and others for discussions and correspondence during the past few years. I also thank the organizers, sponsors, and local organizing committee in Kyoto for arranging such a valuable discussion on this important subject.

References

Alexander, S., 1850. *On the classification and special points of resemblance of certain of the periodic comets; and the probability of a common origin in the case of some of them,* Astron. J., **1,** 147–150.

Alexander, S., 1851. *On the similarity of arrangement of the asteroids and the comets of short period, and the possibility of their common origin,* Astron. J., **1,** 181–184.

Aumann, H.H. & Good, J., 1990. *IRAS constraints on a cold cloud around the solar system,* Astrophys. J., **350,** 408–412.

Bailey, M.E., 1976. *Can 'invisible' bodies be observed in the solar system?* Nature, **259,** 290–291.

Bailey, M.E., 1983a. *Is there a dense primordial cloud of comets just beyond Pluto?* Asteroids Comets Meteors, 383–386, eds. Lagerkvist, C.-I. & Rickman, H. Uppsala University.

Bailey, M.E., 1983b. *The structure and evolution of the solar system comet cloud,* Mon. Not. R. Astron. Soc., **204,** 603–633.

Bailey, M.E., 1983c. *Theories of cometary origin and the brightness of the infrared sky,* Mon. Not. R. Astron. Soc., **205,** 47P–52P.

Bailey, M.E., 1983d. *Comets, Planet X, and the orbit of Neptune,* Nature, **302,** 399–400.

Bailey, M.E., 1986a. *The near-parabolic flux and the origin of short-period comets,* Nature, **324,** 350–352.

Bailey, M.E., 1986b. *The mean energy transfer rate to comets in the Oort cloud and implications for cometary origins,* Mon. Not. R. Astron. Soc., **218,** 1–30.

Bailey, M.E., 1990. *Short-period comets: probes of the inner core,* Asteroids Comets Meteors III, 221–230, eds. Lagerkvist, C.-I., Rickman, H., Lindblad B.A. & Lindgren, M. Uppsala University.

Bailey, M.E., 1991. *Comets and molecular clouds: the sink and the source,* Molecular Clouds, 273–289, eds. James, R.A. & Millar, T.J. Cambridge University Press.

Bailey, M.E. & Stagg, C.R., 1990. *The origin of short-period comets,* Icarus, **86,** 2–8.

Bailey, M.E., McBreen, B. & Ray, T.P., 1984. *Constraints on cometary origin from the isotropy of the microwave background and other measurements,* Mon. Not. R. Astron. Soc., **209,** 881–890.

Bailey, M.E., Chambers, J.E. & Hahn, G., 1992a. *Fiery fate for Comet Machholz?* Sky and Telescope, **83,** No. 1 (January), p.5.

Bailey, M.E., Chambers, J.E. & Hahn, G., 1992b. *Origin of sungrazing comets: a frequent cometary end-state,* Astron. Astrophys., in press.

Bailey, M.E., Clube, S.V.M. & Napier, W.M., 1986. *The origin of comets,* Vistas in Astron., **29,** 53–112.

Bailey, M.E., Clube, S.V.M. & Napier, W.M., 1990. *The Origin of Comets,* Pergamon Press, Oxford.

Bailey, M.E. & Hahn, G., 1992. *Orbital evolution of 1991 DA: implications for near-earth asteroids,* Periodic Comets, eds. Fernández, J.A. & Rickman, H., in press. Universidad de la Republica, Montevideo.

Biermann, L. & Michel, K.W., 1978. *The origin of cometary nuclei in the presolar nebula,* Moon & Planets, **18,** 447–464.

Bobrovnikoff, N.T., 1929. *On the disintegration of comets,* Lick Obs. Bull., **14,** 28–37.

Bobrovnikoff, N.T., 1931. *The origin of asteroids,* Publ. Astron. Soc. Pac., **43,** 324–333.

Bredichin, T., 1889. *Sur l'origine de comètes periodiques,* Ann. de l'Obs. de Moscou (Ser. 2), **2,** 1–17.

Cameron, A.G.W., 1978. *Physics of the primitive solar accretion disk,* Moon & Planets, **18,** 5–40.

Carusi, A., Valsecchi, G.B., & Greenberg, R., 1990. *Planetary close encounters: geometry of approach and post-encounter orbital parameters,* Celest. Mech., **49,** 111–131.

Clube, S.V.M., 1989. *Giant comets and catastrophism,* Catastrophes and Evolution: Astronomical Foundations, 81–105, ed. Clube, S.V.M. Cambridge University Press.

Clube, S.V.M. & Napier, W.M., 1982a. *Spiral arms, comets and terrestrial catastrophism,* Q. Jl. R. Astron. Soc., **23,** 45–66.

Clube, S.V.M. & Napier, W.M., 1982b. *The Cosmic Serpent,* Faber & Faber, London.

Clube, S.V.M. & Napier, W.M., 1984. *The microstructure of terrestrial catastrophism,* Mon. Not. R. Astron. Soc., **211,** 953–968.

Clube, S.V.M. & Napier, W.M., 1986a. *Mankind's future: an astronomical view. Comets, ice ages and catastrophes,* Interdisciplinary Sci. Rev., **11,** 236–247.

Clube, S.V.M. & Napier, W.M., 1986b. *Giant comets and the Galaxy: implications of the terrestrial record,* The Galaxy and the Solar System, eds. Smoluchowski, R., Bahcall, J.N. & Matthews, M.S., 260–285. University of Arizona Press.

Clube, S.V.M. & Napier, W.M., 1990. *The Cosmic Winter,* Basil Blackwell, Oxford.

Dones, L., 1991. *A recent cometary origin for Saturn's rings,* Icarus, **92,** 194–203.

Drummond, J.D., 1991. *Earth-approaching asteroid streams,* Icarus, **89,** 14–25.

Duncan, M., Quinn, T. & Tremaine, S., 1987. *The formation and extent of the solar system comet cloud,* Astron. J., **94,** 1330–1338.

Duncan, M., Quinn, T. & Tremaine, S., 1988. *The origin of short-period comets,* Astrophys. J. Lett., **328,** L69–L73.

Duncan, M., Quinn, T. & Tremaine, S., 1989. *The long-term evolution of orbits in the solar system: a mapping approach,* Icarus, **82,** 402–418.

Dyson, F.J., 1992. *Hunting for comets and planets,* Q. Jl. R. Astron. Soc., **33,** No. 2 (June), in press.

Edgeworth, K.E., 1949. *The origin and evolution of the solar system,* Mon. Not. R. Astron. Soc., **109,** 600–609.

Everhart, E., 1972. *The origin of short-period comets,* Astrophysical Letters, **10,** 131–135.

Everhart, E., 1977. *Evolution of comet orbits as perturbed by Uranus and Neptune,* Comets Asteroids Meteorites: Interrelations, Evolution and Origin, ed. Delsemme, A.H., IAU Coll. No. 39, 99–104. University of Toledo.

Fernández, J.A., 1980. *On the existence of a comet belt beyond Neptune,* Mon. Not. R. Astron. Soc., **192,** 481–491.

Fernández, J.A., 1985. *The formation and dynamical survival of the Oort cloud,* Dynamics of Comets: Their Origin and Evolution, IAU Coll. No. 83, 45–70, eds. Carusi, A. & Valsecchi, G.B. Reidel, Dordrecht, The Netherlands.

Fernández, J.A. & Ip, W.-H., 1983. *On the time evolution of the planetary influx in the region of the terrestrial planets,* Icarus, **54,** 377–387.

Fernández, J.A. & Ip, W.-H., 1987. *Time-dependent injection of Oort cloud comets into earth-crossing orbits,* Icarus, **71,** 46–56.

Fernández, J.A. & Ip, W.-H., 1991. *Statistical and evolutionary aspects of cometary orbits,* In: Comets in the Post-Halley Era, IAU Coll. No. 116, Vol. 1, 487–535, eds. Newburn Jr., R.L., Neugebauer, M. & Rahe, J. Kluwer, Dordrecht, The Netherlands.

Froeschlé, C. & Rickman, H., 1980. *New Monte Carlo simulations of the orbital evolution of short-period comets and comparison with observations,* Astron. & Astrophys., **82,** 183–194.

Gladman, B. & Duncan, M., 1990. *On the fates of minor bodies in the outer solar system,* Astron. J., **100,** 1680–1693.

Hahn, G., 1991. *Dynamical histories of near-earth asteroids: orbital similarities and the distribution of close encounters,* presented at San Juan Capistrano conference on Near-Earth Asteroids, July 1991.

Hahn, G. & Bailey, M.E., 1990. *Rapid dynamical evolution of giant comet Chiron,* Nature, **348,** 132–136.

Hahn, G. & Bailey, M.E., 1992. *Long-term evolution of 1991 DA: a dynamically evolved extinct Halley-type comet,* Asteroids Comets Meteors 1991, eds. Harris, A. & Bowell, E., in press. Lunar and Planetary Institute, Houston.

Halliday, I., Blackwell, A.T. & Griffin, A.A., 1990. *Evidence for the existence of groups of meteorite-*

producing asteroidal fragments, Meteoritics, **25,** 93–99.

Harrington, R.S., 1968. *Dynamical evolution of triple stars,* Astron. J., **73,** 190–194.

Heisler, J., 1990. *Monte Carlo simulations of the Oort comet cloud,* Icarus, **88,** 104–121.

Heisler, J., Tremaine, S. & Alcock, C., 1987. *The frequency and intensity of comet showers from the Oort cloud,* Icarus, **70,** 269–288.

Hills, J.G., 1981. *Comet showers and the steady-state infall of comets from the Oort cloud,* Astron. J., **86,** 1730–1740.

Hoyle, F. & Wickramasinghe, N.C., 1978. *Comets, ice ages, and ecological catastrophes,* Astrophys. Space Sci., **53,** 523–526.

Jackson, A.A. & Killen, R.M., 1988. *Infrared brightness of a comet belt beyond Neptune,* Earth, Moon & Planets, **42,** 41–47.

Joss, P.C., 1973. *On the origin of short-period comets,* Astron. Astrophys., **25,** 271–273.

Kuiper, G.P., 1951. *On the origin of the solar system,* Astrophysics, 357–424, ed. Hynek, J.A. McGraw-Hill, New York.

Marchal, C., 1978. *Collisions of stars by oscillating orbits of the second kind,* Acta Astronautica, **5,** 745–764.

Mendis, D.A., 1973. *The comet – meteor stream complex,* Astrophys. Space Sci., **20,** 165–176.

Nakamura, T., 1981. *Orbital evolution of long-period comets. I. Trivariate Monte Carlo simulation,* Icarus, **45,** 529–544.

Oikawa, S. & Everhart, E., 1979. *Past and future orbit of 1977 UB, object Chiron,* Astron. J., **84,** 134–139.

Olsson-Steel, D., Taaffe, L. & Williams, S., 1989. *Short-period comet production in close encounters with Jupiter,* Proc. Astron. Soc. Australia, **8,** 108–118.

Öpik, E.J., 1963. *The stray bodies in the solar system. Part I. Survival of cometary nuclei and the asteroids,* Advances Astron. Astrophys., **2,** 219–262.

Quinn, T., Tremaine, S. & Duncan, M., 1990. *Planetary perturbations and the origin of short-period comets,* Astrophys. J., **355,** 667–679.

Stagg, C.R., 1991. *Personal communication.*

Stagg, C.R. & Bailey, M.E., 1989. *Stochastic capture of short-period comets,* Mon. Not. R. Astron. Soc., **241,** 507–541, and Microfiche MN 241/1.

Rickman, H., 1990. *Origin and evolution of the Jupiter family,* Nordic-Baltic Astronomy Meeting, 257–273, eds. Lagerkvist, C.-I., Kiselman, D. & Lindgren, M. Uppsala University.

Torbett, M.V., 1989. *Chaotic motion in a comet disk beyond Neptune: the delivery of short-period comets,* Astron. J., **98,** 1477–1481.

Valsecchi, G.B., 1990. *Report on the workshop 'Evolution of cometary orbits',* Asteroids Comets Meteors III, 615–616, eds. Lagerkvist, C.-I., Rickman, H., Lindblad B.A. & Lindgren, M. Uppsala University.

Valsecchi, G.B., 1992. *Close encounters, planetary masses and the evolution of cometary orbits,* Periodic Comets, eds. Fernández, J.A. & Rickman, H., in press. Universidad de la Republica, Montevideo.

Weinberg, M.D., Shapiro, S.L. & Wasserman, I., 1986. *The dynamical fate of wide binaries in the solar neighborhood: encounters with stars and molecular clouds,* Icarus, **65,** 27–36.

Weinberg, M.D., Shapiro, S.L. & Wasserman, I., 1987. *The dynamical fate of wide binaries in the solar neighborhood,* Astrophys. J., **312,** 367–389.

Wetherill, G.W., 1989. *Cratering of the terrestrial planets by Apollo objects,* Meteoritics, **24,** 15–22.

Wetherill, G.W., 1991. *End products of cometary evolution: cometary origin of earth-crossing bodies of asteroidal appearance,* In: Comets in the Post-Halley Era, IAU Coll. No. 116, Vol. 1, 537–556, eds. Newburn Jr., R.L., Neugebauer, M. & Rahe, J. Kluwer, Dordrecht, The Netherlands.

Whipple, F.L., 1972. *The origin of comets,* The Motion, Evolution of Orbits, and Origin of Comets, eds. Chebotarev, G.A., Kazimirchak-Polonskaya, E.I. & Marsden, B.G., IAU Symp. No. 45, 401–408. Reidel, Dordrecht, The Netherlands.

Yabushita, S. & Tsujii, T., 1990. *On the dependence on inclination of capture probability of short-period comets,* Mon. Not. R. Astron. Soc., **244,** 663–667.

STRUCTURE AND EVOLUTION OF THE JUPITER FAMILY

H. RICKMAN
Astronomiska observatoriet, Box 515, S-751 20 Uppsala, Sweden

Abstract. Both physical and dynamical issues are important in order to judge the origin and evolution of the Jupiter family of short-period comets. The steady-state condition for maintaining this structure at its present size by captures from the classical Oort cloud is reviewed on the basis of recent results concerning the absolute number of Jupiter family comets as a function of perihelion distance as well as the coupled physical and dynamical evolutions that evidently occur. Like in most earlier investigations, a clear shortage is found in the classical Oort cloud source. The shortage seems, however, less extreme than sometimes assumed. Monte Carlo simulations are envisaged as a way to shed further light on the fate of Jupiter family comets.

Key words: Comets, Jupiter family, Oort cloud.

1. Introduction

The *Jupiter family* stands out in the distribution of short-period comets as a clustering of the majority of observed objects into low-inclination orbits with aphelia in the vicinity of Jupiter's orbit. This remarkable structure would not survive for more than a tiny fraction of the age of the solar system, unless it were replenished by captures from a distant source. The Jupiter family comets encounter Jupiter frequently and their orbits are thus strongly perturbed with ejection into long-period, eventually hyperbolic orbits as an unavoidable consequence. This dynamical ejection is associated with a typical time scale $\sim 10^6$ yrs and, in fact, it typically takes no more than $\sim 10^4$ yrs before the comets are definitively transferred into orbits with large enough perihelion distances to render most of them unobservable.

The *Oort cloud* offers one possible source, from which such captures might occur at a quasi-steady rate over the age of the solar system. The structure of this cometary reservoir has been the subject of continued debate and new ideas have led to a changeable, evolving picture (Weissman, 1986). Most importantly, it has been recognized that the cloud should be dominated by a massive *inner core*, from which new comets cannot be provided in the usual way, except following extremely close stellar encounters or passages by extremely massive objects. These would lead to cometary showers (Hills, 1981) occurring at intervals $\sim 10^8$ yrs.

We are thus confronted with a number of questions regarding cometary capture. Is the efficiency of capture from the steady flux of new comets of the classical Oort cloud high enough to provide the necessary capture rate in order to balance the Jupiter family loss rate? If not, could we be living in the aftermath of a cometary shower that still leaves its imprint in an enhanced number of Jupiter family comets? In any case, the inner core should also provide a steady rate of captures via the intermediary of all the giant planets, but is this rate significant? Moreover, since

Celestial Mechanics and Dynamical Astronomy **54:** 63–69, 1992.

the Oort cloud with its inner core should be a fairly isotropic structure, can the respective capture processes give high enough preference for low inclinations to be compatible with the observed Jupiter family? Finally, an additional source of comets may be present near the ecliptic outside Neptune (often called the *Kuiper belt*), so there may be an additional, steady rate of captures of low-inclination comets.

Many aspects of these problems still remain to be clarified and such work has important cosmogonical consequences, since the Jupiter family may be viewed as a probe of distant cometary populations, as yet inaccessible to direct observations (Bailey, 1990). This paper will review some recent progress concerning the fundamental issue of evaluating the loss rate of Jupiter family comets. For this purpose, one needs an estimate of the number of objects currently existing in orbits where the discovery completeness is reasonable, down to a certain magnitude limit. One also needs the average lifetime of such comets, which is governed by two effects: On the one hand, the *dynamical residence time* in observable orbits is limited by jovian perturbations as noted above, and on the other hand, physical evolution of the nuclei may lead to fading across the magnitude limit so that the *observable lifetime* is only a fraction of the dynamical residence time.

2. Number of Jupiter Family Comets

The distinction between the Jupiter family and the Halley-type short-period comets is to some extent arbitrary, and we shall follow Fernández *et al.* (1992; hereafter referred to as FRK) in adopting orbital periods $P < 20$ yrs as the definition of the Jupiter family. Up to 1990, there were 135 such comets on record. Their distribution of perihelion distances (q) peaks at $q \simeq 1.5$ AU, but concentrating on recent discoveries, the picture is different. While the last decades have seen a dramatic increase in the overall rate of discoveries, this increase is restricted to comets with $q > 1.5$ AU and cannot be traced for smaller perihelion distances. FRK thus deduce a high degree of discovery completeness for Earth- or Mars-crossing Jupiter family comets – a conclusion in agreement with that of Shoemaker and Wolfe (1982). Further evidence in this direction is found by Tancredi and Rickman (1992), who have computed the fraction of Jupiter family comets that experienced a decrease of q by at least 0.1 AU shortly preceding discovery as a function of time. While this fraction has remained $\sim 40\%$ for $q > 1.5$ AU since more than a century ago, it has shown a marked increase (approximately, from 15 to 30%) for $q < 1.5$ AU. As a cometary population approaches completeness, one would expect to find new discoveries restricted to objects that recently entered the population.

Taking account of the temporary nature of Jupiter family comets from dynamical or physical aspects, FRK estimate the number of objects existing on the average at any given time within different ranges of q. To go beyond absolute total magnitude $H_T = 10$ in this way proves difficult, since even at $q < 1$ AU, the level of discovery completeness for $H_T > 10$ is uncertain, and with increasing q, completeness is limited to increasingly brighter comets. One is thus forced to extrapolate the estimated numbers over significant ranges in H_T using an empirically determined distribution function. The magnitude distribution is roughly exponential (Hughes, 1987), but determination of its index is severely hampered by the need to avoid overrepresentation of bright comets by concentrating on a narrow range of q. For a cumulative number $N(H_T)$ varying as: lg $N(H_T)$ = const. + $c \cdot H_T$ with $c = 0.4 \pm 0.1$, FRK find the number of Jupiter family comets brighter than $H_T = 11$ to increase dramatically from $\simeq$ 10 for $q < 1$ AU to several hundred for $1.5 < q < 2.0$ AU and perhaps nearly a thousand for $2.0 < q < 2.5$ AU. The uncertainty caused by the error bar on c is very large, however, and the analysis cannot be extended to larger q. Restricting attention to comets with $H_T < 8$, Rickman (1992) thus states a total of $\simeq$ 40 Jupiter family comets with $q < 2.5$ AU, the expected error of this estimate being relatively modest.

3. Lifetimes of Jupiter Family Comets

Long-term orbital integrations show the dynamics of Jupiter family comets to be a mixture of quasi-regular jovian perturbations, imposing periodic variations on the orbital elements, and sudden jumps occurring as a result of close encounters with the planet. The latter component provokes chaos so that the evolutions of, e.g., perihelion distance after long enough time take on a random-walk appearance. As a consequence of this, the typical overall sojourn of a comet in the $q < 2.5$ AU range is made up of a number of temporary visits, from initial injection to the ultimate ejection. Such behaviour was also found for transitions between Jupiter family orbits and Chiron-type orbits by Hahn and Bailey (1990). For the total length of time spent on the average in Jupiter family orbits with, e.g., $q < 2.5$ AU, we have only order-of-magnitude estimates, indicating a typical duration $\sim 10^4$ yrs. Lindgren (1992) had a preliminary look at the statistics of dynamical lifetimes based on integrations of fictitious orbits, apparently supporting the above assertion.

There are two main ideas concerning physical evolution of cometary nuclei occurring during such visits, affecting the brightness and observability of the comets. On the one hand, sublimation of surface ices leads to *erosion* or ablation that in due time would consume all the volatile material if no other changes interfered. However, *dust mantling* of the surface may also occur, whereby the sublimation is choked off and outgassing eventually is limited to local vents or completely terminated. A low degree of outgassing activity, directly observed for the P/Halley nucleus and indicated by photometry for a number of other Jupiter family comets (Rickman, 1992), gives evidence that dust mantling has an important influence. Moreover, due to the possibility of mantle blow-off as a result of increased insolation or other causes, this aging process is reversible. Hence comets may undergo temporary periods of *dormancy*, after which rejuvenation occurs. Such behaviour, including more or less sporadic activity, has been inferred in a number of cases (Kresák, 1987). Assuming outgassing to be restricted to 5% of the surface area,

Rickman (1992) estimates that the mean surface erosion per revolution typically amounts to a decrease of nuclear radius by only a few dm for Jupiter family comets. Thus, over a total dynamical residence time of $\sim 10^4$ yrs, sizeable nuclei with initial radii of several km are not limited by their content of volatile material.

Modelling of the dust mantling process offers no unique prediction of the time scale of deactivation. However, based on statistics of nongravitational effects, Rickman *et al.* (1991) concluded that, following the liberation of fresh, active surface by mantle blow-off upon a large decrease of q, new mantles form over most of this surface within only 10–20 revolutions. Accordingly, the nuclear surface responds very quickly to such orbital changes that influence the amount of dust coverage. This holds in particular for large perihelion distances ($q \gtrsim 2$ AU) where the orbits tend to be particularly unstable. In the long run, we expect that decreases of q may be able to keep a reasonable fraction of the surface active until the overall minimum is reached, but eventually, as ultimate ejection is approached, dust mantling might definitively render the comet unobservable. Hence the observable lifetime should in general amount to a large fraction of the dynamical residence time and may be estimated at $\sim 10^4$ yrs.

It is difficult to find observational evidence bearing on this estimate. Kresák's (1981) statistics of *cometary disappearances* is based on few cases and necessarily refers to a time scale of $\sim 10^2$ yrs, which may be too short to be informative about the long-term behaviour. Kresák and Kresáková (1990) derived a *mean fading rate* $\sim 0^{\mathrm{m}}.01$ per revolution, which would seem to imply lifetimes $\sim 10^3$ yrs only, but as the authors themselves state, these fadings seem to be time limited and cannot be taken to be truly secular. An alternative approach is to use orbital integrations to find a *net capture rate* into, e.g., $q < 1.5$ AU (Fernández, 1985). The idea is that if physical loss occurs inside the q-limit, there must be a net capture rate of active comets to compensate for this. A sample of real Jupiter family comets should thus show an average trend for captures to dominate over ejections. Naturally, the prime feature of the variation of $\langle q \rangle$ with time is a minimum at the present, because discovery is most likely at a local minimum of q. One thus has to integrate far enough into the past and future to avoid this bias, but at the same time, the integrations become less reliable since multiple encounters with Jupiter occur. Consequently, going too far from the present, the integrations lose their affinity to the real comets and start to appear like random trajectories, for which there is no trend for captures. Tancredi and Rickman (1992) revisited this analysis using an enlarged sample of integrations. They found that the past-future asymmetry that Fernández (1985) had claimed for comets with $q < 1.5$ AU virtually disappears in the new data, indicating an upper limit to the net capture rate, which corresponds to lifetimes in excess of 1000 revolutions.

4. Loss Rate and End States

If we use 10^4 yrs as the mean observable lifetime of Jupiter family comets, arguing that this seems consistent with the observational indications at hand, and a total of 40 objects brighter than $H_T = 8$ in orbits with $q < 2.5$ AU, the loss rate is 4×10^{-3} comets per year. Estimates of the capture rate from a near-parabolic flux of Oort cloud comets with initial low-inclination orbits and perihelia near Jupiter's orbit based on capture probabilities derived from previous dynamical studies tend to yield $\sim 5 \times 10^{-5}$ comets per year, corresponding to new comets with $H_T < 6$ (Rickman 1992). Thus, if fading by two magnitudes would occur during the capture process – Hughes (1988) estimated that even larger fading is likely – the capture rate from the classical Oort cloud is no larger than 1% of that required to keep the Jupiter family in balance. How much fading actually occurs appears to be an open question, however, so let us state only that even with no fading at all, the classical Oort cloud appears to provide no more than $\sim$ 10% of the Jupiter family comets. In agreement with most other investigators, we thus find that the near-parabolic flux cannot be the most important source, but that multi-planet captures from the inner core or the Kuiper belt should dominate.

One further point concerning the balance of capture and loss rates should be noted. Implicit in the arguments described above is the assumption that physical evolution exclusively leads to loss of comets, thus restricting the lifetimes. However, one of the well-known evolutionary processes – that of *nuclear splitting* – may actually work in both ways. The ablation rate is obviously accelerated, but if the exhaustion of the volatile resources is in fact unimportant for the loss of comets, this may be of no consequence. On the other hand, although in most cases only a short-lived, minor fragment is broken off the main nucleus, there is also the case of comets P/Van Biesbroeck and P/Neujmin 3, resulting from the splitting of a common parent around 1850 (Carusi *et al.*, 1985). Both fragments are clearly comets in their own right, having existed for more than a century. Considering that the typical age of Jupiter family comets is $\sim 10^4$ yrs, hundreds of such events might have occurred during this time, if one per century can be taken as the mean rate of occurrence. Of course, the basis for this estimate is still very poor and we do not know if the resulting comets are always as enduring as assumed (cf. the case of P/Biela), but tentatively it seems wise to keep in mind that splitting may be a way of creating comets as well as destroying them.

What information do we have on the end states of cometary evolution? As dust mantling proceeds to complete dormancy, the objects take on an *asteroidal appearance*. They might thus be observed as Earth-approaching asteroids in Jupiter family orbits. Currently there are seven observed objects in this class (Rickman, 1992), and indications of cometary nature exist for a number of other, more typical Apollo asteroids with smaller aphelion distances based on associations with streams of particulate material (typically observed as meteor showers) or nongravitational effects in the orbital motions. The existence of these objects proves that some comets do evolve into asteroidal bodies, but based on observational statistics, we cannot yet say how frequent this evolution is.

The concept of a close link between the physical and dynamical aspects of cometary evolution seems a useful outcome of recent research, and such ideas may be put on a quantitative basis with the aid of Monte Carlo simulations. Rickman *et al.* (1992) describe the principles of such an investigation, where a sample of orbital integrations is used to derive jump probabilities between different ranges of q, and mantling rates and blow-off probabilities are used as model parameters. From preliminary results, it seems that most comets with sizeable nuclei that penetrate to $q_{\min} < 2$ AU become completely mantled before being dynamically ejected. However, this mantling would essentially occur on the way out when $q > 2$ AU, since in a steady state, nearly all comets with $q < 2$ AU are still active. The "meteoroidal" end state, i.e., complete ablation, seems important only for comets with $q_{\min} < 0.5$ AU, unless the nuclei are initially small (radius $\lesssim 1$ km). However, a great deal of work is still required, before the above can be regarded as safe conclusions.

References

Bailey, M.E.: 1990, 'Short-period comets: Probes of the inner core', in C.-I. Lagerkvist, H. Rickman, B.A. Lindblad and M. Lindgren, eds, *Asteroids, Comets, Meteors III*, Uppsala Univ., p. 221

Carusi, A., Kresák, L., Perozzi, E. and Valsecchi, G.B.: 1985, 'First results of the integration of motion of short-period comets over 800 years', in A. Carusi and G.B. Valsecchi, eds, *Dynamics of comets: Their origin and evolution*, D. Reidel Publ. Co., p. 319

Fernández, J.A.: 1985, 'Dynamical capture and physical decay of short-period comets', *Icarus* **64**, 308

Fernández, J.A., Rickman, H. and Kamél, L.: 1992, 'The population size and distribution of perihelion distances of the Jupiter family', in J.A. Fernández and H. Rickman, eds, Proc. of *International Workshop on Periodic Comets*, Univ. de la República, Montevideo-Uruguay, in press

Hahn, G. and Bailey, M.E.: 1990, 'Rapid dynamical evolution of giant Comet Chiron', *Nature* **348**, 132

Hills, J.G.: 1981, 'Comets showers and the steady-state infall of comets from the Oort cloud', *Astron. J.* **86**, 1730

Hughes, D.W.: 1987, 'Cometary magnitude distribution and the fading of comets', *Nature* **325**, 231

Hughes, D.W.: 1988, 'Cometary magnitude distribution and the ratio between the numbers of long- and short-period comets', *Icarus* **73**, 149

Kresák, L.: 1981, 'The lifetimes and disappearance of periodic comets', *Bull. Astron. Inst. Czechosl.* **32**, 321

Kresák, L.: 1987, 'Dormant phases in the aging of periodic comets', *Astron. Astrophys.* **187**, 906

Kresák, L. and Kresáková, M.: 1990, 'Secular brightness decrease of periodic comets', *Icarus* **86**, 82

Lindgren, M.: 1992, 'Dynamical timescales in the Jupiter family', in A. Harris and E. Bowell, eds, Proc. of *Asteroids, Comets, Meteors 1991*, Lunar Planet. Inst., Houston, in press

Rickman, H.: 1992, 'Physico-dynamical evolution of aging comets', in D. Benest and C. Froeschlé, eds, *Interrelations between the physics and dynamics of small bodies of the Solar System*, Editions Frontières, in press

Rickman, H., Bailey, M.E., Hahn, G., Tancredi, G.: 1992, 'Monte Carlo simulations of Jupiter family evolution', in J.A. Fernández and H. Rickman, eds, Proc. of *International Workshop on Periodic Comets*, Univ. de la República, Montevideo-Uruguay, in press

Rickman, H., Froeschlé, C., Kamél, L. and Festou. M.C.: 1991, 'Nongravitational effects and the aging of periodic comets', *Astron. J.* **102**, 1446

Shoemaker, E.M. and Wolfe, R.F.: 1982, 'Cratering time scales for the Galilean satellites', in D. Morrison, ed, *Satellites of Jupiter*, Univ. Arizona Press, Tucson, p. 277

Tancredi, G. and Rickman, H.: 1992, 'The evolution of Jupiter family comets over 2000 years', in S. Ferraz-Mello, ed, Proc. of IAU Symp. **152**: *Chaos, Resonance and Collective Dynamical Phenomena in the Solar System*, in press

Weissman, P.R.: 1986, 'The Oort cloud in transition', in C.-I. Lagerkvist, B.A. Lindblad, H. Lundstedt and H. Rickman, eds, *Asteroids, Comets, Meteors II*, Uppsala Univ., p. 197

COLLECTIVE RESONANT PHENOMENA ON SMALL BODIES IN THE SOLAR SYSTEM

CH. FROESCHLÉ and C. FROESCHLÉ
Observatoire de Nice – B.P. 139 06003 NICE Cedex, France

Abstract. For both asteroids and meteor streams, and also for comets, resonances play a major role for their orbital evolutions but on different time scales. For asteroids both mean motion resonances and secular resonances not only structure the phase space of regular orbits but are mainly at the origin for the inherent chaos of planet crosser objects.

For comets and their chaotic routes temporary trapping into orbital resonances is a well known phenomenon. In addition for slow diffusion through the Kuiper belt resonances are the only candidates for originating a slow chaos.

Like for asteroids, resonances with Jupiter play a major role for the orbital evolution of meteor streams. Crossing of separatrix like zones appears to be crucial for the formation of arcs and for the dissolution of streams. In particular the orbital inclination of a meteor stream appears to be a critical parameter for arc formation. Numerical results obtained in an other context show that the competition between the Poynting-Robertson drag and the gravitational interaction of grains near the 2/1 resonance might be very important in the long run for the structure of meteor streams.

Key words: Asteroid, meteor stream, comet, resonances, chaos.

1. Introduction

It is well known that the distribution of asteroid orbits is distinctly non random and many studies have been devoted to the effect of mean motion resonances.

Actually mean motion resonances are closely related both to gaps and to isolated clusters in the orbital distribution of asteroids (see Greenberg and Scholl, 1979). They also introduce interesting complexities for the definition of the so-called proper elements used to define asteroid families. These structural characteristics are related to the origin and evolution of the asteroid belt. The on-going processes that shaped the belt continue to deliver asteroidal material to the inner solar system, where some land on the Earth as meteorites.

Besides these important planetological implications the study of mean motion resonances has attracted the attention of theoriticians concerned with general issues of nonlinear processes in mathematics and physics. In particular the use of semi-computational (or semi-analytical) methods to study resonant behaviour among asteroids has led to a better understanding of various phenomena of great current interest in modern dynamics: the homoclinic point, separatrix, ordered and chaotic motions, adiabatic invariant, etc.

An earlier review by Greenberg and Scholl (1979) on resonances in the asteroid belt still provides an excellent introduction both to the resonance mechanism through Greenberg's secular perturbation theory for low eccentric, planar orbits

Celestial Mechanics and Dynamical Astronomy **54:** 71–89, 1992.

at the 2/1 resonance and to the four types of hypotheses (statistical, gravitational, collisional and cosmogonic) that have been invoked to try to explain the Kirkwood gaps. The statistical hypothesis assumes that asteroids librate around gaps and therefore are rarely seen crossing the exact resonances. The gravitational hypothesis supposes that asteroids that originally formed in the Kirkwood gaps were removed later by purely gravitational force due to Jupiter. According to the collisional hypothesis, asteroids are removed from the gaps, either through destruction or change of mean motion, by collisions with neighboring asteroids. Finally, the cosmogonic hypothesis assumes that gaps represent regions where asteroids failed to accrete or were expelled during the early history of the solar system. Actually the statistical and gravitational hypothesis have no more supporters and only two candidates remain on the line. First the collisional hypothesis but with Mars and not with other asteroids and the cosmogonic hypothesis.

However this new collisional hypothesis is the indirect consequence of the search of the gravitational believers for gravitational depletion due to Jupiter. The consequence of this work was the discovery of bursts in eccentricity leading to possible close approaches with Mars, through confined chaotic motion.

Later reviews by Froeschlé and Scholl (1983) and Froeschlé and Greenberg (1989) update that of Greenberg and Scholl (1979) and we will not enter into all the interesting developments described in these reviews, but rather concentrate on very new results concerning mainly another type of resonances: the secular resonances.

The dynamical nature of secular resonances is quite different from the one of mean-motion resonances that are related to the well-known Kirkwood gaps. In the case of mean-motion resonances, the resonance frequency is determined by the frequency of Jupiter's revolution around the Sun. In the case of secular resonances, we can imagine the planetary orbits replaced by elliptic mutually inclined rings. Gravitational attraction among these rings causes the ring to precess and causes the eccentricities and inclinations of the rings to oscillate. The major frequencies of these precessional motions and oscillations are the resonance frequencies for secular resonances. It has been shown (Scholl *et al.*, 1989; Morbidelli and Henrard, 1991a,b) that some of these resonances induce strong eccentricity or inclination variations. Of course such variations could be a clue for initiating close approaches with planets and push them into a chaotic route.

Resonant structure exists also in visual meteor orbits (Linblad, 1973). Murray (1988) found also evidence of resonant structure in some meteor streams through preliminary analysis of the Harvard Radio meteor datas. Of course because meteor stream particles have mostly orbits with high eccentricities and inclinations their long term evolution has been investigated very little in celestial mechanics since such orbits can only be approximated with difficulty by classical series expansions. Various non-gravitational mechanisms like Poynting-Robertson drag, solar wind and radiation pressure are expected to cause a mass segregation. However, for particles large enough in order not to be affected by these non gravitational mechanism mean motion resonance with Jupiter, might be responsible for the splitting of streams. The important variable which determines the splitting is the longitude of the ascending node Ω as far as high inclined orbits are concerned.

Outside a resonance region, Ω regresses almost uniformly due to Jupiter's perturbation, in particular due to orbital momentum transfer. The corresponding rate $\dot{\Omega}$ depends mainly on the semi-major axis of a particle's orbit. We would like to emphasize that outside of a resonance region, $\dot{\Omega}$ does not depend on the mean longitude of a particle with respect to Jupiter's mean longitude. In other words, the average momentum transfer to the particle's orbit which determines $\dot{\Omega}$ is basically the same for all stream particles independent of the location of a particle in its orbit. As a consequence, the whole stream precesses without splitting up.

In a resonance region, on the other hand, the situation might be different: the average angular momentum transfer might be strongly dependent on the location of a particle in the stream. Hence, the nodal lines of particle orbits might precess with very different speeds and their motion might even become prograde or librator. This is the basic mechanism for a stream to split into arcs. An arc consists of particles with about the same orbital rate $\dot{\Omega}$, which is very different from orbital rates $\dot{\Omega}$ of the other particles of the original stream.

Finally besides the well known phenomenon of temporary resonant cometary capture which might lead to peculiar chaotic streams within chaotic routes, another issue where both types of resonance are concerned is the origin of the slow chaos in the outer solar system.

In section 2 we will present briefly some general results which are relevant for understanding the peculiar astronomical problems presented in the following sections

In Section 3 we present some new results on the dynamical evolution of small solar system bodies, i.e. with low inclination.

Section 4 will be devoted to the dynamical resonant evolution of high inclined resonant orbits and some applications to meteor streams.

2. The Resonance Mechanism

2.1. Mean Motion Resonances

A mean motion resonance occurs when the period of revolution of an asteroid is a multiple $p/(p+q)$ of Jupiter's period of revolution, where p and q are small integers. Traditionally, ratios of mean motions are considered and in the following these will be called $(p+q)/p$ resonances. If a conjunction (i.e., in the planar case, alignment of the Sun, Jupiter and the asteroid) occurs at some longitude L it will be repeated at nearly the same value of L after a few revolutions of the asteroid around the Sun. For example, for the 5/2 resonance, conjunction repeats after 5 revolutions of the asteroid and 2 of Jupiter. Repetition of geometric configuration results in enhancement of perturbations. For instance, Jupiter may add orbital angular momentum to an asteroid's orbit for many periods before it later starts to take angular momentum out again. For a physical description of the resonance phenomenon, see Peale (1976) and Greenberg (1977).

Of course it is well-known that rational numbers are dense therefore a resonance arbitrary close to a given asteroid always exists. However, since the strength of a resonance n/m decreases exponentially with $\mid n \mid + \mid m \mid$ (Fourier rule) and $\mid n - m \mid$ (d'Alembert rule), only low order resonances are important.

The full equations of motion of an asteroid are nonintegrable even in the framework of the restricted three-body problem. The various approximations used represent treatment of the appropriate Hamiltonian or equivalently of the perturbing function R, in such a way to reduce the problem to one degree of freedom.

The disturbing function R can be Fourier expanded into general form

$$R = \sum \alpha(a,\, a_J,\, e,\, e_J,\, i,\, i_J) \cos(i\lambda + j\lambda_J + k\tilde{\omega} + l\tilde{\omega}_J + m\Omega + n\Omega_J) \quad (1)$$

where the coefficients α are functions of the semi-major axes, eccentricities and inclinations; $\tilde{\omega}$, Ω and λ, are respectively the longitude of pericenter, the longitude of the ascending node and the mean longitude of the asteroid; quantities with the subscript J refer to Jupiter's orbit. The summation is performed over all permissible values of the integers i, j, k, l, m, n and α is also a function of these integers. The following important mathematical properties of this expansion defines it as a d'Alembert series: α is of the order $e^{|k|}\ e_J^{|l|}\ i^{|m|}\ i_J^{|n|}$; the sum $m + n$ is even and $i + j + k + l + m + n = 0$. Therefore $\mid k \mid + \mid l \mid + \mid m \mid + \mid n \mid \geq \mid i + j \mid$. The terms with $i = 0$ and $j = 0$ are called secular since only slowly varying quantities are contained in the arguments of the cosine. However, if the ratio of the mean motion n and n_J are close to some rational number, then the quantity $i\lambda + j\lambda_J$ can also be slowly varying and will dominate the motion of the asteroid because the remaining nonsecular terms are of short period. These short-period terms are usually eliminated using the averaging principle which assumes that a particular resonant term is slowly varying, then the mean value of R and of its derivatives averaged over the resonant period only contain the secular and resonant terms which nevertheless are still given by a quadruple summation over k, l, m and n. Further truncations are of course performed and their validity depends on the location of the resonance and of the maximum eccentricity reached by the orbits since both a/a_J and $e^{|k|}$ appear in the coefficient of the cosine. The number of terms that should be retained in the d'Alembert form of the disturbing function depends crucially on the value of q. For first-order resonances of the form $(p+q)/p$, terms proportional to a single power of the eccentricity are present while for the 3/1 resonance ($q = 2$), the largest terms in the disturbing function are proportional to two powers of the eccentricities, i.e., e_J^2, e^2 or $e_j e$ and the next terms which contribute are fourth order in the eccentricities. In all cases, the problem is to retain enough terms to obtain a good representation of the motion, but when terms of high order are included, an analytical description is extremely difficult.

It is convenient to introduce the resonance variable σ which describes the location of a conjunction with respect to the asteroid's longitude of perihelion

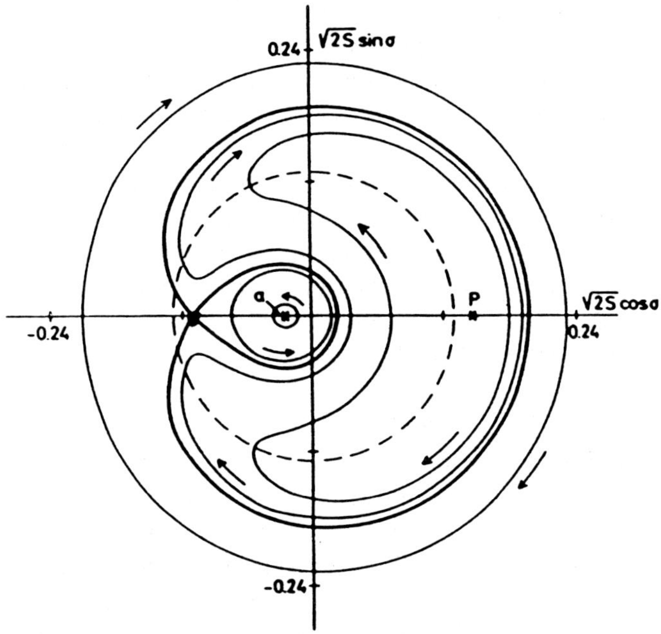

Fig. 1. Trajectories in σ *vs* $\sqrt{2S}$ space from Schubart's averaged circular model at the 2/1 commensurability for $A = 0.802$. The arrows indicate directions of motion in this space. The darker lines correspond to critical bifurcation trajectories. Paths immediatly around point a are apocentric librators; those about p are pericentric librators. The dashed circle corresponds to the exact center of the resonance. (Figure from Greenberg and Scholl, 1979).

$\tilde{\omega}$, i.e., $\sigma = L - \tilde{\omega}$, with $L = (-p\lambda + \lambda_j(p+q))/q$. For $\sigma = 0$, conjunction occurs at the asteroid's perihelion (favors stability) for $\sigma = \pi$ conjunction occurs at the asteroid's aphelion, in the case of the 2/1 resonance. The latter case could be unstable if it allows close encounters with Jupiter, a situation prevented by any resonance that maintains σ at 0. Since σ changes slowly (for resonant motion), it is natural to use σ for theoretical purpose; e.g., Poincaré (1902) used σ as a canonical variable for the 2/1 resonance in the planar problem. For the nonplanar problem, an additional canonical variable is introduced which is critical for orbital stability. Schubart (1978) used the variable τ that measures the longitude of conjunction reckoned from the longitude of the ascending node of the asteroid's orbit. For $\tau = 0$, conjunction occurs in the orbital plane of Jupiter. Motion could therefore be destabilized: τ near 90° or 270° would prevent such a possibility.

The perturbation theories use truncated power series and at large eccentricities, both analytical models and mappings break down and only numerical integration is possible.

Following an idea of Poincaré, Schubart (1968) developed a different approach. Modifying Poincaré's canonical equations for the planar-restricted Sun-Jupiter-asteroid problem, Schubart averaged the Hamiltonian over the corresponding commensurability period. Inserting the averaged Hamiltonian in the Hamiltonian equations, Schubart then integrated numerically the new set of differential equations. For the planar circular case, this set degenerates to a single implicit equation and orbits are represented by closed curves in S, σ space. Here $S = \sqrt{a/a_J}(1 - \sqrt{1 - e^2})$ is the conjugate variable of σ.

It is clear that "Poincaré à la Schubart's" topology as displayed in Figure 1 is only valid for the circular planar-averaged model as a second integral $A = \sqrt{a/a_J}(\frac{p+q}{q} - \sqrt{1 - e^2})$ exists beside this Hamiltonian. The same topology can be obtained analytically using series expansions followed by truncations (see Henrard and Lemaître, 1983), the problem being reduced to a one degree of freedom Hamiltonian. However the price to pay is that it is only valid for small eccentricity. Only in this model the problem of resonant motion is fully integrable. the critical bifurcation point is called a homoclinic point in modern dynamics (Arnold, 1978). It is well known that integrable systems are not generic; i.e., small perturbations can destroy the integrability, and the separatrix or homoclinic orbit can cause wild regions with chaotic behaviour (Arnold, 1978). This peculiar behaviour for this topology was displayed by Froeschlé and Scholl (1977) in the elliptic averaged case.

Beside ellipticity, nonaveraging as well as noncoplanarity destroys the integrability. For the case of noncoplanarity, Schubart's topology displayed in Figure 1 remains valid to some extent and can be regarded as a good starting point for understanding and describing the behaviour of resonant orbits in the three-dimensional elliptic averaged case. Schubart (1978, 1979) has extended the planar model to deal with this more general case.

For this more general case the critical arguments μ, σ and τ are such that

$$\begin{aligned} \mu &= \lambda - \lambda_J(p+q)/p \\ \sigma &= -(\tilde{\omega} + \mu\, p/q) \\ \tau &= -\Omega - \mu\, p/q \end{aligned} \tag{2}$$

with $p = q = 1$ for the 2/1 resonance. Both σ and τ vary slowly for resonant motion.

Using Schubart's model, Froeschlé and Scholl (1982) have performed a systematic exploration on the three-dimensional asteroidal motion at the 2/1 resonance. They have also shown (Froeschlé and Scholl, 1986; Scholl and Froeschlé, 1988) that the dynamical evolution of meteor stream particles in resonance appears to be affected by the same resonance mechanisms as resonant asteroids. The crossing of separatrix-like zones appears to be crucial for the formation of arcs and for the dissolution of streams. Investigating the orbital evolution of known resonant meteor streams and of model like streams, they have found examples for such a transitory arc formation. The orbital inclination of meteor streams appears to be a critical parameter for arc formation as it appears clearly in subsection (4.2).

2.2. Secular Resonances

The gravitational interactions between the planets give rise to so-called secular perturbations with periods ranging from 5 x 10^4 to 2 x 10^6 years. Not only the planetary orbits suffer secular perturbations, but they also excite secular perturbations on any body revolving about the Sun. This excitement can be particularly strong in the case of so-called secular resonances. In a first approximation, we can consider the planetary system as a system of coupled oscillators. Secular resonances are related to the eigenfrequencies of this system. A body is situated in a secular resonance when its longitude of perihelion $\tilde{\omega}$ or its ascending node Ω precesses on the mean with a frequency nearly equal to one of the eigenfrequencies g_l and s_l of the planetary system. This is the definition of secular resonant motion.

Three frequencies g_l and s_l play a major role for the asteroidal dynamics, namely g_5, g_6, s_6. They are nearly equal to the mean frequencies characterizing the secular motion of the longitude of Jupiter's perihelion $\langle\dot{\tilde{\omega}}_{\text{Jupiter}}\rangle$, of the longitude of Saturn's perihelion $\langle\dot{\tilde{\omega}}_{\text{Saturn}}\rangle$ and of the longitude of Saturn's ascending node $\langle\dot{\Omega}_{\text{Saturn}}\rangle$ respectively.

According to the theory of Brouwer and van Woerkom (1950), the rates of these three resonances frequencies are: $g_5 = 4'', 30/y$, $g_6 = 27'', 77/y$ and $s_6 = -25'', 73/y$.

The conditions for an asteroidal orbit to be located in these secular resonances can be written in the form:

$$
\begin{aligned}
g_5 &: \langle\dot{\tilde{\omega}}\rangle \simeq \langle\dot{\tilde{\omega}}_J\rangle \\
g_6 &: \langle\dot{\tilde{\omega}}\rangle \simeq \langle\dot{\tilde{\omega}}_S\rangle \\
s_6 &: \langle\dot{\Omega}\rangle \simeq \langle\dot{\Omega}_S\rangle
\end{aligned}
\tag{3}
$$

where the subscripts J and S denote Jupiter and Saturn respectively.

The positions of the secular resonances in the whole semimajor axis range between 0.1 and 55 AU have first been determined by Heppenheimer (1980), who used a purely linear, first-order analytical theory. The resonance locations were found by comparison with values of the fundamental planetary frequencies computed by the secular perturbation theory of Brouwer and van Woerkom (1950). This theory, based on low-degree truncations of the perturbatives series (dropping most perturbative terms of order two and higher in the ratio between planetary and solar mass, and including only low-degree terms in the eccentricities and inclinations) provides only approximate values for the planetary frequencies. As a consequence of this, the results of Heppenheimer – as well as those of other investigators – on the locations of the secular resonances had a limited accuracy.

Several authors in the last 20 years have determined the positions of secular resonance surfaces, although only in the region of the main asteroid belt. Their secular perturbation theories were much more refined than the linear one, but still based on some truncations and approximations. In 1969 Williams developed a mixed analytical-numerical theory based on the Gauss averaging method. which can be applied to derive proper elements. By means of this theory, Williams and Faulkner (1981) derived detailed graphical maps of the three main secular resonances $g = g_5$, $g = g_6$ and $s = s_6$ in the asteroidal belt. However, this theory, of order one in the mass ratio, breaks down near the mean motion resonances and cannot describe the topology of secular resonance surfaces in these zones. More recently, Morbidelli and Henrard (1991) revived the use of semi-numerical averaging methods. They developed a theory which avoids any expansion of the main term of the Hamiltonian (the linear one in the masses) with respect to the eccentricity or inclination of the small body, in order to achieve results valid for any values of these elements. Introducing suitable action-angle variables, they took into account the dynamics related to the motion of the argument of perihelion of the small body, which is important at high inclinations. They also computed the correction due to the quadratic term in the mass ratio, using a series expansion in eccentricity and inclination up to degree four. They then draw the surfaces of the main secular resonances $g = g_5$, $g = g_6$ and $s = s_6$ (and also some more complicated ones) in the semi-major axis interval $1.5 \leq a \leq 3.5$ AU.

In this paper we are going to discuss the possible implications on global evolutions of small bodies in the solar system of the location of secular resonances in the $a - e - i$ proper elements space both in the asteroid belt and in the outer solar system, by using a theory based on updated planetary frequencies, and yielding accurate results up to moderate values of the eccentricities and inclinations. We shall not recall the main features of this theory (for a detailed description, see (Milani and Kneževič, 1990; Kneževič *et al.*, 1991) and the papers referenced therein).

3. Dynamical Resonant Evolution of Small Solar System Bodies with Low Inclination

This chapter will report mainly of the work of Kneževič *et al.* (1991) on secular resonances from 2 to 50 AU and on the work of Farinella *et al.* (1991) on the modelisation of injecting asteroid fragments into the 3/1 and g_6 resonance.

3.1. The Potential Importance of Secular Resonances for the Dynamical Evolution of Small Solar System Bodies

Let us now discuss some implications of the positions of main secular resonances for the dynamical evolution of the small bodies orbiting the Sun. Most of these issues have to be considered only at a qualitative level, since the theory did not allow us to derive the real "strenghts" and "widths" of the resonances, i.e. their capability to induce strong variations in eccentricity and inclination at various distances from the exact resonance surfaces. On the other hand, in order to obtain quantitative results, extensive numerical experiments should be carried out for each zone, and this has not yet been done.

(a) An important application concerns the delivery of meteorites and Earth-approaching small asteroids from the inner edge of the asteroid belt. We have

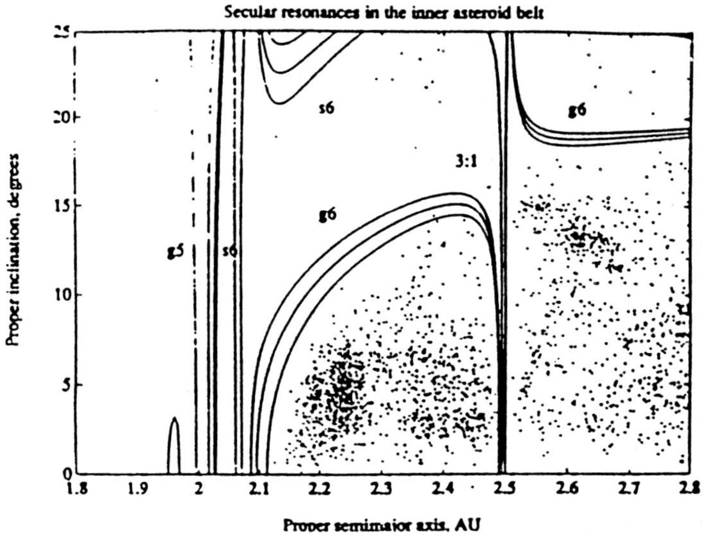

Fig. 2. An enlargement of the inner asteroid belt. The dots represent the position in proper elements space of all asteroids with proper eccentricity $e < 0.2$. The resonance lines are drawn for a constant proper eccentricity $e = 0.1$

found (see Figure 2) that the $g = g_6$ secular resonance at low inclinations occurs at a semi-major axis of about 2.10 AU, rather than at 2.05 AU as previously assumed (Williams and Faulkner, 1981; Wetherill, 1988). This difference is due in part to the inclusion of higher-order and/or higher-degree terms in the theory, and in part to the improved planetary frequency g_6 adopted in our dynamical model (we took $g_6 = 28.25$ arcsec yr^{-1} instead of 27.77 arcsec yr^{-1}, the value of Brouwer and Van Woerkom adopted by Williams and Faulkner). Although the exact location of the resonance is also affected by the inner planets (which are not taken into account by our theory), and therefore at the level of a few hundredths of AU must be considered model-dependent, there is no doubt that $g = g_6$ is closer than previously assumed to the Flora region of the inner asteroid belt, where a large number of small asteroids are crowded. We conjecture that when some of these asteroids are broken up by a collision, a significant fraction of the resulting fragments may – either directly or after close approaches with Mars – enter the secular resonance, undergo large increases in eccentricity and become Earth-crossers within a time span of the order of 1 Myr (Scholl and Froeschlé, 1991).

The mean motion resonances of order higher than two are not accounted for by our theory, so the shape of the curves representing $g = g_6$ and $s = s_6$ between 2.0 and 2.1 AU are probably not very meaningful, since at 2.065 AU we find the strong $4/1$ mean motion resonance with Jupiter. Numerical experiments (Scholl and Froeschlé, 1991) have actually shown that bodies injected in the region from 2.0 to 2.13 AU may have a complex dynamical behavior, jumping from the secular resonance $g = g_6$ to the mean motion resonance $4/1$. According to these calculations, the $4/1$ resonance broadens significantly the Earth-crossing region and decreases the transport time scales.

(b) As shown by Figure 2, several asteroids lie close (or possibly inside) the $g = g_6$ resonance between 2.3 and 2.5 AU. Of course, the presence of the $3/1$ resonance may limit the accuracy of the proper elements in the vicinity of 2.5 AU. However, recent numerical integrations (Froeschlé and Scholl, 1991) of fictitious bodies with $a = 2.3$ and 2.4 AU, starting eccentricity 0.14 and inclination $I = 14°$ and 16° show that during the whole integration time span of 2.7 Myr they were always or temporarily located in the secular resonance, suffering large variations in eccentricity (up to ~ 0.85) and becoming planet-crossers in a time scale of $\sim 10^6$ yr. Taking into account the results mentioned previously, Froeschlé and Scholl have conjectured that the secular resonance $g = g_6$ is a good candidate for producing planet-crossing objects at least up to a semi-major axis of ≤ 2.4 AU.

(c) The inner part of the asteroid belt at inclinations $I \geq 10°$ and with 2.0 AU $\leq a \leq 2.25$ AU, bordering from inside the Phocaea group, appears to be strongly depleted. As explained by Kneževič *et al.* (1991), this depletion is probably due to the fact that this region is criss-crossed by several main secular resonances ($g = g_6$, $g = g_7$, $g = g_8$, $s = s_6$) and also by other resonances like $2g = 2s$ (see Morbidelli and Henrard, 1991b, Figure 9). Due to resonance overlapping, asteroid orbits in this region suffer large perturbations in both eccentricity and inclination, possibly ending into planet-crossing orbits, at least in some cases, as confirmed by the numerical experiments by Knežević *et al.* (1991). A more extensive numerical exploration and/or a better theory (for high inclinations and eccentricities) would be needed to give a complete explanation of the observed depletion.

(d) We consider now the zones between the outer planets. Recent extensive numerical experiments (Franklin *et al.*, 1989; Weibel *et al.*, 1990; Gladman and Duncan, 1990) have shown that the majority of test particles orbiting between the giant planets are perturbed to a close approach on timescales of millions of years.

Our results are consistent with these studies. The region between Jupiter and Saturn (Knežević *et al.*, 1991, Fig. 2) is crowded with many low-order mean motion resonances which – by analogy with the Kirkwood gaps – are probably effective in causing chaotic behavior. A possible region of "relative stability" (i.e., where small bodies could escape short-term ejection), found also in the afore-mentioned numerical studies, lies between about 7 and 7.5 AU, for small inclinations; however, secular resonances are present at inclinations between 10° and 20°, and may affect the dynamical evolution of any small body injected there.

A similar situation (Kneževič *et al.*, 1991, Fig. 3) is observed between Saturn and Uranus, with a possible "stable island" at semimajor axes near 14 AU, below $i \simeq 15°$ of inclination.

A third such "island" is possibly located between Uranus and Neptune at $23.5 \leq a \leq 24.5$ AU, with a possible gap appearing at the overlapping order-two $7/5$ and $5/7$ mean motion resonances with Neptune and Uranus, respectively (Knežević *et al.*, 1991, Fig. 4).

However, all such "islands" are unlikely to contain any more surviving primordial bodies if the orbits of the giant planets underwent significant semimajor axis changes after their formation (e.g. due to ejection of cometary material into the Oort cloud), with corresponding shifts in the resonance positions.

(e) Another interesting issue concerns the possible existence of an inner "Kuiper cometary belt", acting as an effective source of low-inclination short-period comets. To confirm this hypothesis, a viable dynamical mechanism to transport the comets into planet-crossing orbits needs to be identified.

As shown in Figure 3, beyond 50 AU the secular frequencies are always too small to allow for the existence of secular resonances, which therefore cannot provide the required mechanism. Does this rule out the Kuiper belt comet source ? The problem is subtle, since it is necessary not only to show that objects formed in the Kuiper belt can become Neptune-crossers, but also to determine the time scale for this process, which must be of the order of the age of the solar system.

Numerical experiments performed by Gladman and Duncan (1990) and Torbett and Smoluchovski (1990) are not at all conclusive, as they exhibit orbits becoming Neptune-crossers in less than 10^7 yr, but for semi-major axes less than 50 AU. Only Levison (1991), who has treated the dynamical evolution of small gravitationally non-interacting objects within the solar system as a Markov process, claims to have found evidence that small objects can diffuse and leave the region between 30 and 100 AU with a lifetime of $\approx 5 \times 10^9$ yr. Our results do not contradict such a low stochasticity, which would be unlikely in a region full of strong secular resonances.

However, the problem remains on where does the stochasticity comes from. The diffusion within the Markov chain may simply arise as an artifact of the model, due for instance to very long oscillations of a quasi-periodic integrable motion. Such an artifact was harmless for the diffusion problem of the Jupiter family treated already as a Markov process by Rickman and Froeschlé (1979), since in this case the overall stochasticity comes from close approaches with Jupiter.

3.2. Injecting Asteroid Fragments into Resonances

Farinella *et al.* (1991) have recently modeled the ejection of fragments from cratering or break-up events undergone by real asteroids as a consequence of impacts. They computed the fraction of escaped fragments falling in the g_6 resonance. For some asteroids between $2.4 \leq a \leq 2.5$ AU (6 Hebe, $a = 2.42$ AU; 304 Olga, $a = 2.40$ AU; 623 Chimaera, $a = 2.46$ AU; 930 Westphalia, $a = 2.43$ AU) the percentage of ejected fragments inside the $g = g_6$ resonance is of the order of 50%. Moreover, in the region $2.55 \leq a \leq 2.8$ AU several other asteroids lie near or inside $g = g_6$ (for example, 475 Ocllo, $a = 2.59$ AU; 631 Philippina, $a = 2.79$ AU; 759 Vinifera, $a = 2.62$ AU; 907 Rhoda, $a = 2.80$ AU), and also for them the fraction of fragments ending up in the resonance is $\geq 50\%$. These preliminary results suggest that the $g = g_6$ resonance is a potential source of planet-crossing fragments also in the region beyond the $3/1$ mean motion resonance. In order to obtain quantitative results on the efficiency of the $g = g_6$ resonance in producing

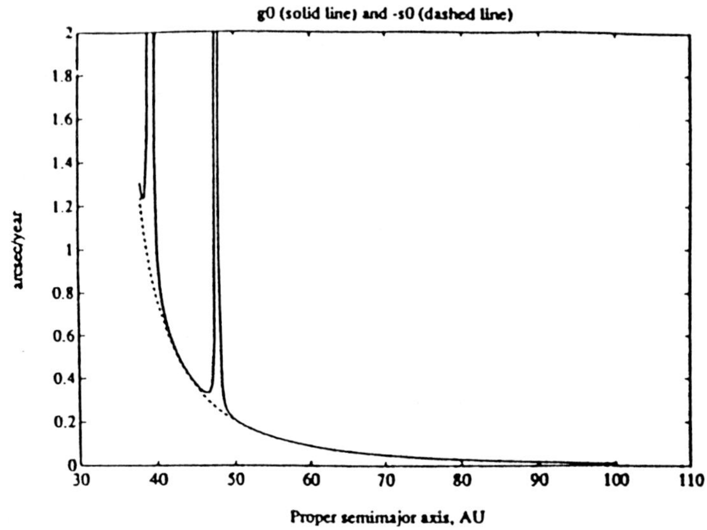

Fig. 3. Frequencies g_0 and $-s_0$ in the trans-Neptunian region up to 100 AU. Notice that for semi-major axis $a > 50$ AU both frequencies remain smaller than any planetary ones.

planet-crossing objects and also on the transport time scales, we plan to study the orbital evolution of some fictitious fragments by purely numerical computations. This will also allow us to explore the effects of the overlapping between the $g = g_6$ and $3/1$ resonances, which probably triggers chaotic behavior and may enhance the production of planet-crossers, as already found for the interaction between $g = g_6$ and $4/1$.

4. Dynamical Resonant Evolution of Small Solar System Bodies with High Inclination

This chapter deals mainly with the effect of mean motion resonances on meteor stream and on the possible relationship between comet P/Machholz and the Quadrantid meteor streams. Effect of Poynting-Robertson drag will also be discussed.

4.1. Gravitational Splitting of Meteor Streams

According to Hughes *et al.* (1979), the Quadrantid streams has the following orbital parameters with respect to ecliptic and equinox 1950: ($\omega = 170°4$, $\Omega = 292°6$, $i = 71°4$, $e = 0.681$, $a = 3.064$). Using the Schubart averaged program Froeschlé and Scholl (1986) have integrated five different streams with about the same initial conditions but situated within the resonance. Actually Babadzhov (see Froeschlé and Scholl 1987 discussion) pointed out that according to observations the semi-major axis of Quadrantids is in the range 2.79 to 3.27 AU and therefore some of the results of Froeschlé and Scholl (1986) shown here might be real. We used 12 starting values $\mu = 30°$, $60°$,..., $360°$ in order to represent a stream. We

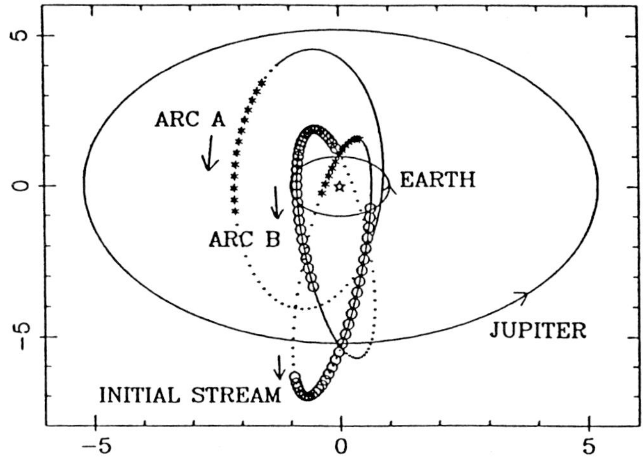

Fig. 4. The initial stream and the orbits of two resulting arc A and B after about 1000 Jupiter revolutions. Dashed parts of an orbit lie below Jupiter's orbital plane. Circle on the initial stream refer to the stream's portion with μ ranging from $-60°$ to $+90°$ (regression). Asterisks refer to the portion with μ ranging from 150° to 210° (progression). (From Froeschlé and Scholl, 1986).

have found three modes of nodal motion depending of the values of the resonance variable μ (see Equation (2)): either nodal regression with jumps of 180° or temporary libration, or progression occurs. Consequently a stream may break up into two isolated arcs.

This stream has its ascending node in the lower part of Figure 4 outside of Jupiter's orbit. The dashed part of the stream lies below Jupiter's orbital plane.

The circles on the initial stream represent the stream between $\mu = -60°$ and $\mu = 90°$. This stream portion will later form arc B. The asterisks indicate the stream portion in the range $\mu = 150° \ldots\ 210°$. A stream dispersion is possible within 10^3 years and a very slow dispersion rate of arc B is expected. On the other hand, Arc A will dissolve faster since the nodal rates of arc A orbits depend on the variable μ.

To locate the regions of the a, e, ω, i phase space where streams break up into arcs is difficult since we are concerned with a system with three degrees of freedom and therefore Figure 1 can only be a crude approximation for describing the topology of the system. Obviously there exist separatrix-like surfaces where chaotic motion is expected. We have found such orbits with temporary progression of Ω with intermittent jumps.

It is interesting to note that the jump in Ω and in eccentricity does not seem related to a crossing of the separatrix-like zone in Schubart's $S-\sigma$ plane mentioned above. They investigated the orbital evolution in a different plane, namely in the $\Psi_1 - \Psi_2$ plane, defined by $\Psi_1 = e\cos\tilde{\omega}$ and $\Psi_2 = e\sin\tilde{\omega}$.

Like in Schubart's $S - \sigma$ plane, resonant motion represented in the $\Psi_1 - \Psi_2$ plane can show three modes: Progression, regression, and libration. Alternators

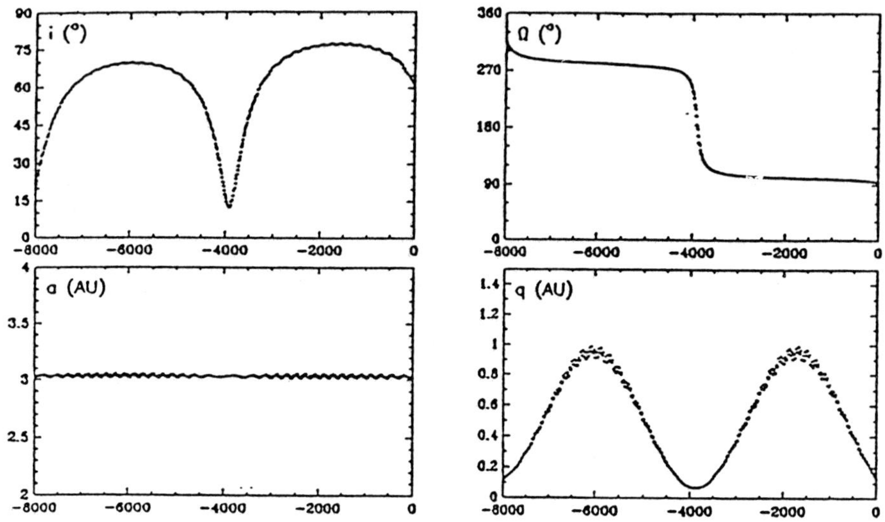

Fig. 5. Backward orbital evolution of comet P/Machholz. the four panels of this figure show the inclination i (upper left); the longitude of the ascending node Ω (upper right); the semi-major axis a (lower left); and the perihelion distance q (lower right). The abscissa of all diagrams is time in years, counted from the present.

between these three modes are also known to occur like in the $S - \sigma$ plane.

Furthermore, we have investigated the orbital evolution of seven meteor streams known to be located at mean motion resonances with Jupiter, i.e., June Bootids 211 2/1, Annual Andromedis 2/1, Librids 3/1, Lyrids 3/1, July Phoenicids 3/1, December Phoenicids 7/3.

The major goal was to find a splitting of the streams into arcs due to a resonance mechanism discovered previously by the authors for a model stream at the 2/1 resonance. None of the seven known meteor streams splits up into arcs due to this mechanism. On the other hand, close approaches to Jupiter of stream particles cause the formation of holes in some of the known meteor streams.

4.2. On the Possible Relationship between Comet P/Machholz and the Quadrantid Meteor Stream

This paragraph summarizes a recent work by Gonczi *et al.* (1991). The Quadrantid meteor shower has been recognized for more than 150 yr. The dynamics of the corresponding stream is peculiar due to the high orbital inclination and, for some particles, the closeness of the 2/1 mean motion resonance with Jupiter. It has been the subject of many investigations relating to the structure of the stream and its nodal retrogression as well as its long-term history and its likely cometary origin. Thus Hamid and Youssef (1963) found that the jovian secular perturbations lead to very large changes in the inclination and perihelion distance of typical stream particles with a period around 4000 yr. From a more extensive study by Williams *et al.* (1979) it was obvious that this period is not unique but may vary considerably between different particles. Of course related to this behaviour are the different works of Froeschlé and Scholl (1982, 1986) and Scholl and Froeschlé (1988) already described.

Babadzhanov and Obrubov (1987, 1989) studied the long-term evolution of the mean Quadrantids orbit, calculating the secular perturbations by the Halphen-Goryachev method. Interestingly, they found that the ecliptic cross-section of the stream formed by particles in all phases of the resulting q–i oscillation has four crossings with Earth's orbit. At each of these, particles can be encountered at both the ascending and descending nodes, so eight theoretical showers exist. Six of these were identified with showers listed on the basis of observed meteor statistics. Thus the Quadrantids, along with five other showers, can be argued to form part of one and the same meteor stream complex.

The parent comet of this complex was possibly not observed until quite recently. Comet P/Machholz (1986 VIII) currently has orbital elements that differ drastically from those of the Quadrantid stream, but its long-term orbital evolution has been found to involve large-scale oscillations in inclination and perihelion distance (Rickman and Froeschlé 1988). McIntosh (1990) drew attention to the fact that firstly, these variations are similar to those found for the Quadrantids by Hamid and Youssef (1963), and secondly, that the perihelion longitude of the comet is close to those of all the streams belonging to Babadzhanov and Obrubov's complex. So, conceivably, there could have been an epoch in the past, when the Quadrantid meteors were shed by comet P/Machholz into an orbit similar to that of the comet, later on to diverge from the cometary orbital evolution due to differential perturbations. Green *et al.* (1990) also commented upon this possibility as well as upon the discrepancy of about 180° in the longitude of the ascending node, which may appear embarrassing. To be precise, the meteor particles have to be ejected from the comet sufficiently far back in time that by now the nodal longitudes have taken two opposite values.

We have investigated the variation of the period of q–i oscillations between different meteor particles in relation to the hypothesis that the shedding of meteors occurred nearly 4000 yrs ago, when comet P/Machholz last had a very small perihelion distance. We have shown (Gonczi *et al.*, 1991) that this hypothesis is viable in view of the ejection velocities typically expected and the resulting spread in the period of q–i oscillations, using purely gravitational 3-body integrations. The most promising range of semimajor axes is just inside the 2/1 resonance, and a detailed study reveals many cases of chaotic behaviour due to close encounters with Jupiter. The spatial structure of the meteoroid complex originating from comet P/Machholz is thus likely to be very complicated with a broad range of semimajor axes involved and a concentration of the ascending nodes near 100° and 280° longitude. The currently observed Quadrantids belong to the second branch only, but each branch should include a wide range of perihelion distances for which the orbits are in general unobservable; cf. the above-mentioned Babadzhanov and

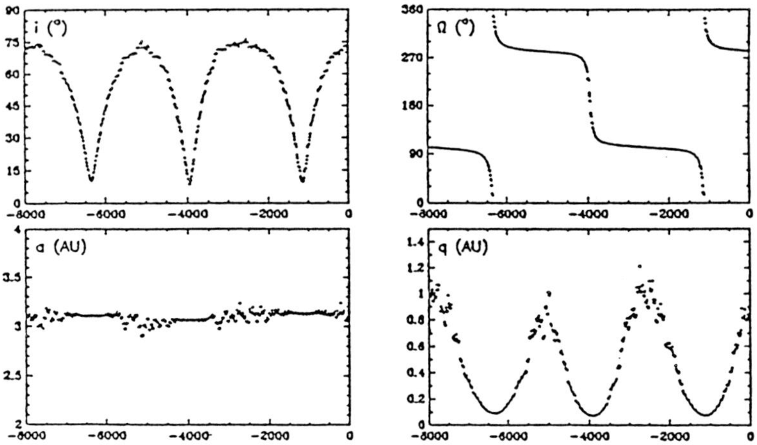

Fig. 6. Orbital evolution of a particle started in the mean Quadrantid orbit with semi-major axis 3.13 AU. The period of the $q - i$ oscillation is such as to bring the particle into virtually perfect agreement with comet P/Machholz at the time of the previous minimum perihelion distance of the comet.

Obrubov meteor stream complex. Babadzhanov and Obrubov (1991) argued that the complex can be understood on the basis of ejection from comet P/Machholz around 4500 B.C. – an epoch partly chosen for maximum orbital similarity between the comet and the Quadrantid meteors. Our results indicate that the same goal can be achieved even for an ejection epoch 2500 yr later. However, investigating the detailed structure and evolution of the complex requires a more complete dynamical model than the one used here. Over several millennia the Poynting-Robertson effect will induce important reductions of orbital eccentricity and semimajor axis, and this is bound to have an important effect on the structure of the meteoroid complex. We are continuing our work along these lines.

4.3. Evolution of Resonant Orbits in Presence of Poynting-Robertson Drag

Gonczi *et al.* (1982, 1983) have studied numerically the competition between the Poynting-Robertson drag and the gravitational interaction of grains with Jupiter near the 2/1 resonance. Computations are based on the three-dimensional restricted elliptic three-body problem. A numerical exploration of numerous orbits has been performed. While in the two-dimensional problem (orbit inclination $i = 0$) they found only two kinds of orbits now a third behaviour appears: a grain coming from the outer side of the resonance as a circulator can be temporarily captured inside. The orbits either always cross the resonance region without any oscillation (see Figure 7, 2b) or in the special case where the grains were initially inside the resonance they remain inside the resonance librators which become circulators after ejection of the particle from the resonance. A third case (Figure 7, 2c) corresponds

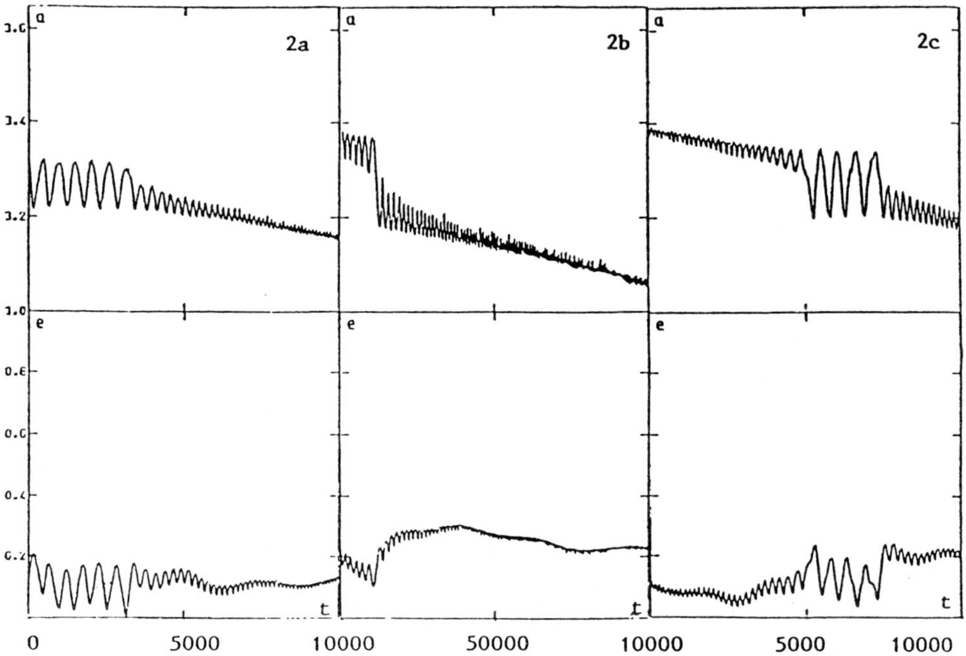

Fig. 7. Semi-major axis and eccentricity of three kinds of orbits found in the three-dimensional problem. (a) librator; $a_0 = 3.33$, $e_0 = 0.12$, $i_0 = 15°.4$, $\Omega_0 = -84°.4$, $\omega_0 = 43°.7$, $M_0 = 54°.0$; (b) circulator; $a_0 = 3.34$, $e_0 = 0.22$, $i_0 = 7°.9$, $\Omega_0 = 64°.5$, $\omega_0 = 105°.1$, $M_0 = 112°.1$; (c) capture $a_0 = 3.44$, $e_0 = 0.12$, $i_0 = 15°.3$, $\Omega_0 = 71°.8$, $\omega_0 = -114°.4$, $M_0 = -110°.3$. ($M_0 =$ inital mean anomaly). From Gonczi *et al.* (1983).

to capture into the resonance where the orbit remains for a while as librator. A systematic study of the trapping time shows that for some range of initial inclination i, the orbits are more stable at the edges than at the center of the resonance.

These computations were carried out in order to explore a cosmological mechanism of depletion or more precisely of non formation of asteroids through snowballing in the Kirkwood gaps. New computations are now performed in the framework of the dynamical evolution of meteor streams.

Acknowledgements

It is a pleasure to thank both members of the Scientific Organizing Committee, Drs. S.V.M. Clube, J. Henrard and S. Yabushita and the members of the Local Organizing Comittee, Drs. I. Hasegawa, T. Inoue and J. Yoshida, for their invitation to present this lecture. The Scientific program of the Conference, and the friendly atmosphere made the stay a memorable one.

References

Arnold, V.: 1978, *Mathematical Methods of Classical Mechanics*, Springer Verlag, New York

Babadzhanov, P.B. and Obrubov, Yu.V.: 1987, 'Evolution of Meteor Streams', in Z. Ceplecha and P. Pecina (eds.) *Interplanetary Matter*, Publ. Astron. Inst. Czechosl. Acad. Sci. No. **67**, 141–150

Babadzhanov, P.B. and Obrubov, Yu.V.: 1989, 'Dynamics and Spatial Shape of Short-Period Meteoroid Streams', in D. McNally (ed.), *Highlights of Astronomy*, Kluwer Acad. Publ., vol. **8**, 287–293.

Babadzhanov, P.B. and Obrubov, Yu.V.: 1991, 'P/Machholz 1986 VIII and Quadrantid Meteorid Stream. Orbital Evolution and Interrelation', to appear in *Asteroids, Comets, Meteors 1991*, Flagstaff, USA

Brouwer, D., van Woerkom, A.J.J.: 1950, 'The secular variation of the orbital elements of the principal planets', *Astron. Papers U.S. Naval Obs.*, **13**, 85–107.

Farinella, P., Gonczi, R., Froeschlé, Ch., Froeschlé, C.: 1991, 'Injecting asteroids fragments into resonances', submitted for *Asteroids, Comets, Meteors 1991*, Flagstaff, USA

Franklin, F., Lecar, M. and Soper, P.: 1989, 'On the original distribution of the asteroids. Do stable orbits exist between Jupiter and Saturn ?', *Icarus* **79**, 223–227

Froeschlé, C., and Scholl, H.: 1977, 'A qualitative comparison between the circular and elliptic model for resonant asteroidal orbits', *Astron. astrophys.* **57**, 33–39

Froeschlé, C. and Scholl, H.: 1982, 'A systematic exploration of three-dimensional asteroidal motion at the 2/1 resonance', *Astron. Astrophys* **111**, 346–356

Froeschlé, C., and Scholl, H.: 1983, 'The dynamical structure of the Asteroidal Belt', in C.I. Lagerkvist and H. Rickman (eds.) *Asteroids, Comets, Meteors*, Uppsala Univ., pp. 115-125.

Froeschlé, C. and Scholl, H.: 1986, 'Gravitational splitting of quadrantid-like Meteor Streams in resonance with Jupiter', *Astron. Astrophys.* **158**, 259–265

Froeschlé, C., Scholl, H.: 1987, 'Resonance Intermittence causes the Gravitational Splitting of Meteor Streams', *Proceedings of the tenth European Regional Astronomical meeting of the IAU*, Prague, 151–155

Froeschlé, C., and Greenberg, R.: 1989, 'Mean Motion Resonances', in Binzel R.P. Gehrels and Mattews M.S. (eds.) *Asteroids*, Tucson Univ. of Arizona Press, 827–844

Froeschlé, Ch., Scholl, H.: 1991, 'The Effect of Secular Resonances in the Asteroid Region between 2.1 and 2.4 AU', submitted for *Asteroids, Comets, Meteors 1991*, Flagstaff, USA

Gladman, B., and Duncan, M: 1990, 'On the fates of minor bodies in the outer Solar System', *Astron. J.* **100**, 1680–1693

Gonczi, R., Froeschlé, Ch., Froeschlé, C.: 1982, 'Poynting-Robertson Drag and Orbital Resonance', *Icarus* **51**, 633–654.

Gonczi, R., Froeschlé, Ch., Froeschlé, C.: 1983, 'Evolution of Three Dimensional Resonant Orbits in Presence of Poynting-Robertson Drag', in C.I. Lagerkvist and H. Rickman (eds.) *Asteroids, Comets, Meteors*, Uppsala Univ., 137–143

Gonczi, R., Rickman, H. and Froeschlé, C.: 1991, 'The Connection between comet P/Machholz and the Quadrantid meteor streams', *M. N. R. A. S*, in press

Greenberg, R: 1977, 'Orbit-orbit resonances in the Solar System: Varieties and similarities', *Vistas in Astron.* **21**, 209–239

Greenberg, R. and Scholl, H.: 1979, 'Resonances in the asteroidal belt', in T. Gehrels (ed.) *Asteroids*, Univ. of Arizona Press, 310–333

Green, D.W.E., Rickman, H., Porter, A.C. and Meech, K.J.: 1990, 'The Strange Periodic Comet Machholz', *Science* **247**, 1063–1067

Hamid, S.E. and Youssef, M.N.: 1963, 'A short Note on the Origin and Age of the Quadrantids', *Smithsonian Contr. Astrophys.* **7**, 309–311

Heppenheimer, T.A.: 1980, 'Secular resonances and the origin of eccentricities of Mars and the Asteroids', *Icarus* **41**, 76-88

Hugues, D.W., Williams, I.P., Murray, C.D.: 1979, 'The orbital evolution of the Quadrantid Meteor Stream between AD 1830 and 2030', *M.N.R.A.S.* **189**, 493–500

Knežević, Z., Milani, A., Farinella, P., Froeschlé, Ch. and Froeschlé, C.: 1991, 'Secular resonances from 2 to 50 AU', *Icarus* **93**, 316–330

Levinson, H.F.: 1991, 'The long-term dynamical behaviour of small bodies in the Kuiper belt', *Astron.*

J. **102**, 787–794

Lindblad, B.A.: 1973, 'The distribution of 1/a in photographic meteor orbits.', in Hemenway, C.L., Millman, P.M. and Cook, A.F. (eds.) *Evolutionary and physical properties of meteoroids*, NASA Special Publication, Washington, D.C., **319**, 175–181

McIntosh, B.A.: 1990, 'Comet P/Machholz and the Quadrantid Meteor Stream', *Icarus* **86**, 299–304

Milani, A. and Knežević, Z.: 1990, 'Secular perturbation theory and computation of asteroid proper elements', *Celest. Mech.* **49**, 347–411

Morbidelli, A. and Henrard, J.: 1991a, 'Secular resonances in the asteroidal belt. Theoritical perturbation approach and the problem of their location', *Celest. Mech.* **51**, 131–168

Morbidelli. A., and Henrard, J.: 1991b, 'The main secular resonances ν_5 ν_6 and ν_{16} in the asteroidal belt', *Celest. Mech.* **51**, 169–198

Peale, S.J.: 1976, 'Orbital resonances in the Solar System', *Astron. Astrophys.* **14**, 215–215.

Poincaré, H.: 1902, 'Sur les planètes du type d'Hécube', *Bull. Astron.* **19**, 289–315

Rickman, H. and Froeschlé, C.: 1979, 'Orbital evolution of short period comet treated as a Markov process', *Astron. J.* **84**, 1910–1917

Rickman, H. and Froeschlé, C.: 1988, 'Cometary dynamics', *Celest. Mech.* **43**, 243–263

Scholl, H., and Froeschlé, C.: 1988, 'Gravitational breaking of meteor streams in resonance with Jupiter', *Astron. Astrophys.* **195**, 345–349

Scholl, H., Froeschlé, Ch., Kinoshita, H., Yoshikawa, M. and Williams, J.G.: 1989, 'Secular resonances', in Binzel R.P. Gehrels and Mattews M.S. (eds.) *Asteroids*, Tucson Univ. of Arizona Press, 845–861.

Scholl, H. and Froeschlé, Ch.: 1991, 'The ν_6 Secular Resonance Region near 2.AU: a Possible Source of Meteorites', *Astron. Astrophys.* **245**, 316–321

Schubart, J.: 1968, 'Long-period effects in the motion of Hilda-type planets', *Astron. J.* **73**, 99–103

Schubart, J.: 1978, 'New results on the commensurability cases of the problem Sun-Jupiter-Asteroids, in V. Szebehely (ed.) *Dynamics of Planets of Satellites and Theories of Their Motion*, IAU Coll N° 48, D. Reidel: Dordrecht, 137–143

Schubart, J.: 1979, 'Asteroidal motion at commensurabilities treated in three dimensions', in R.L. Duncombe (ed.) *Dynamics of the Solar System*, D. Reidel: Dordrecht, 205–215

Torbett, M.V., and Smoluchowski, R.: 1990, 'Chaotic motion in a primordial comet disk beyond Neptune and comet influx to the Solar System', *Nature* **345**, 49–51

Weibel, N.M., Kaula, W.M. and Newmann, W.I.: 1990, 'A computer search for stable orbits between Jupiter and Saturn', *Icarus* **83**, 382–392

Wetherill, G.W.: 1988, 'Where do the Apollo objects come from ?', *Icarus* **76**, 1–18

Williams, J.G.: 1969, ' Secular Perturbations in the Solar System', Ph. D. Thesis , Univ. of California, Los Angeles.

Williams, I.P., Murray, C.D. and Hughes, D.W.: 1979, 'The Long-Term Orbital Evolution of the Quadrantid Meteor Stream', *M. N. R. A. S.* **189**, 483–492

Williams, J.G., and Faulker, J.: 1981, 'The positions of secular resonance surfaces', *Icarus* **46**, 390–399

DYNAMICS OF PERIODIC COMETS AND METEOR STREAMS

V.V. EMEL'YANENKO
Department of Mathematics, Technical University, Chelyabinsk, 454080, U.S.S.R.

Abstract. The methods for an investigation of the evolution of orbits with large eccentricities are presented. The role of resonances in the motion of comets is considered. The expansion of the disturbing function in the restricted three-body problem is obtained for cometary orbits. The method is based on the proximity of cometary eccentricities to 1. The Hamiltonian describing the basic perturbations in the motion of long-period comets is constructed. It leads to mappings for the description of cometary dynamics.

The long-term evolution of short-period comets is considered. The importance of temporary librations near commensurabilities with Jupiter is emphasized. The analytical dependence of the diffusion rate upon the orbital elements of nearly-parabolic comets is found. The results of calculations of the diffusion rate taking into account perturbations from the outer planets are presented. A numerical study of cometary dynamics using mappings is carried out. A distribution like the Oort cloud is a typical stage of the evolution of comets in nearly-parabolic orbits under planetary perturbations.

A change of the partial density of meteor streams is studied. For the case of librations the dispersion speed of meteor streams under the action of planetary perturbations is substantially less than that for the case of typical chaos. A similarity between the observed structure of the meteor streams and resonant properties of the motion is discussed.

Key words: Resonance, libration, chaos, diffusion.

1. Introduction

The main features of the motion of comets and meteor streams are caused by the large eccentricities of their orbits. As the cometary orbits are situated close to the planetary orbits they are capable of being greatly perturbed. Moreover, it is impossible to apply classical methods of celestial mechanics, based on small eccentricities. Therefore, a calculation of the long-term evolution of cometary orbits requires special methods.

First, we consider the possibility of an application of resonant perturbation theory, obtain an analytical expansion of the disturbing function and construct mappings to describe the motion of cometary orbits. Later, we present the numerical results of the investigation of short-period comets, long-period comets and meteor streams, based on those methods.

2. The Role of Resonances in the Motion of Comets

Let n and n_p be the mean motions of a comet and a planet. We shall suppose that n/n_p is close to a rational number j'/j . Let us denote in the astronomical system of units $u_1 = \mu_1\sqrt{a}$, $u_2 = \mu_1\sqrt{a(1-e^2)}$, $u_3 = \mu_1\sqrt{a(1-e^2)}\cos I$, $v_1 = M$, $v_2 = \omega$, $v_3 = \Omega$ (the canonical Delaunay's elements), $\mu_1 = \mu$ in this

Celestial Mechanics and Dynamical Astronomy **54:** 91–110, 1992.

section, μ is the gravitational constant, a the semi-major axis, e the eccentricity, I the inclination, M the mean anomaly, ω the argument of the perihelion, Ω the longitude of the ascending node. The orbital plane of the perturbing planet is the plane of reference for ω, Ω and I, and the origin of longitudes is situated at the perihelion of the planet. Elements of the planet carry the subscript P. Furthermore, we use the heliocentric coordinate system in this section.

Let us take only the secular part $R_{\text{sec},P}$ and the resonant part $R_{\text{res},P}$ of the disturbing function R_P of the restricted three-body problem (the Sun-planet-comet). Under certain conditions, there is a canonical transformation, which leads to the separation of variables (Giacalgia, 1972; Charlier 1927; Emel'yanenko, 1988b): two pairs of variables describe the secular motion but the action of the resonant perturbations may be reflected by the equations.

$$\frac{dx_1}{dt} = c\sin(y_1 - \bar{y}_1) \quad , \quad \frac{dy_1}{dt} = b - \frac{3\mu^4}{j^2\bar{x}_1^4}(x_1 - \bar{x}_1) \; , \tag{1}$$

where $x_1 = u_1/j$, $y_1 = jv_1 - j'M_P + s_1v_2 + s_1'v_3$ is the critical argument, $\bar{x}_1$ is the value of x calculated exactly at the mean motion commensurability, $y_1 = y_1^0 + ht$ and c , b , s_1 , s_1' , y_1^0 , h are constants. The Equations (1) lead to the equation of motion of a simple pendulum with the well-known solution: the phase plane is divided by a separatrix into a libration and two circulation regions. The general properties of nonintegrable systems which are close to (1) (typical resonant systems) have been well studied (Chirikov, 1979; Lichtenberg and Liberman, 1983).

An analytical determination of the conditions for obtaining the Equations (1) is very difficult for cometary orbits. Therefore we applied numerical methods for the detailed study of resonant motions. We used the method proposed by Schubart (1968) for the investigation of resonant asteroids. The disturbing function $R_P(u_1, u_2, u_3, v_1, v_2, v_3, y_1)$ is averaged over segment $[0, 2\pi j']$ of v_1 :

$$\bar{R}_p = \frac{1}{2\pi j'}\int_0^{2\pi j'} R_p dv_1 \; . \tag{2}$$

This procedure retains the secular and the resonant terms in the expansion of the disturbing function. The numerical integrations of averaged Newton's equations for the orbital elements have shown that the motion of comets near the commensurabilities with Jupiter is close to the typical resonant motion (Emel'yanenko, 1988b).

It is important to mention that librations have a particular stability with respect to gravitational and nongravitational perturbations. We stress three basic points (Emel'yanenko, 1988b):

(1) The rate of separation of neighbouring trajectories undergoing libration is substantially less than that for typical chaos.

(2) For libration, the semi-major axis oscillates within small limits near the resonant value. This allows the investigation of the long-term evolution of orbits

taking into account the secular perturbations only. For typical chaotic motion, the deviation of the semi-major axis is unlimited. Therefore secular perturbations reflect inaccurately the evolution of orbits for this case.

(3) The action of nongravitational forces need not break the libration. As the comet moves outside the libration region however, the action of nongravitational forces results in slow secular variation of the semi-major axis.

3. An Expansion of the Disturbing Function for Cometary Orbits

The basis of an analytic theory of cometary motion is the expansion of the quantity $1/\Delta_P$, where Δ_P is the distance between the comet and the perturbing planet:

$$\Delta_P = \sqrt{r^2 + r_P^2 - 2rr_P \cos h_P}$$

(r and r_P are the position vectors of the comet and the planet, h_P is the angle between r and r_P).

Let us denote

$$\eta = \frac{2rr_P(1 + \cos h_P)}{(r + r_P)^2} ,$$

then Δ_P^{-1} can be written in the form

$$\frac{1}{\Delta_P} = \frac{1}{r + r_P} \sum_{\alpha=0}^{\infty} \frac{\left(\frac{1}{2}\right)_\alpha}{(1)_\alpha} \eta^\alpha \tag{3}$$

where $(W)_\alpha = W(W+1)\cdots(W+\alpha-1)$ and $(W)_0 = 1$.

The power series (3) converges everywhere, with the exception of collision points. This expansion was considered in earlier papers (Boda, 1931; Petrovskaya, 1970).

Taking into account that

$$(1 + \cos h_p)^\alpha = 2^\alpha \sum_{k=0}^{\alpha} \frac{(\alpha - k + 1)_k}{(\alpha + 1)_{k+1}} (2k+1) P_k(\cos h_P) ,$$

where $P_k(\cos h_P)$ are Legendre polynomials, and

$$\cos h_P = \cos^2 \frac{I}{2} \cos(\nu - \nu_P + \omega + \Omega) + \sin^2 \frac{I}{2} \cos(\nu + \nu_P + \omega - \Omega) ,$$

where ν is the true anomaly, we get for Δ_P^{-1} the expression:

$$\frac{1}{\Delta_P} = \sum_{\alpha=0}^{\infty} \sum_{k=0}^{\alpha} \sum_{s} \sum_{s'} \frac{(\alpha+1)_\alpha (\alpha-k+1)_k (2k+1)}{(1)_{\alpha+k+1}} \times$$
$$\times \; C^k_{s',s} \frac{r^\alpha r_P^\alpha}{(r+r_P)^{2\alpha+1}} \exp[i(\beta\nu + \beta'\nu_P + \beta\omega - \beta'\Omega)] \,,$$

where $\beta = s - s'$, $\beta' = s + s'$, and (Krasinsky 1973):

$$C^k_{s',s} = \frac{\left(\frac{1}{2}\right)_{k-\bar{m}} \left(\frac{1}{2}+|s|\right)_{\bar{m}}}{(1)_{|s|}(1)_{|s'|}(1+|s'|)_{\bar{m}}(1)_{\bar{m}}} \left(\cos\frac{I}{2}\right)^{2|s'|} \left(\sin\frac{I}{2}\right)^{2|s|} \times$$
$$\times \; F\left(-2m, 2k - 2\bar{m} + 1, 1 + 2|s|; \sin^2\frac{I}{2}\right)$$

where F is the hypergeometric function, s and s' are such integer numbers that the quantity $k - |s| - |s'| = 2\bar{m}$, $\bar{m}$ being a nonnegative integer. Using the relation

$$\frac{1}{(r+r_P)^{2\alpha+1}} = \sum_{\ell=0}^{\infty} (-1)^\ell \frac{(2\alpha+1)_\ell (r_P - a_P)^\ell}{(1)_\ell (r+a_P)^{2\alpha+1+\ell}} \,,$$

we can write

$$\frac{1}{\Delta_P} = \sum_{\alpha=0}^{\infty} \sum_{\ell=0}^{\infty} \sum_{k=0}^{\alpha} \sum_{s} \sum_{s'} (-1)^\ell \frac{(\alpha+1)_\alpha (\alpha-k+1)_k (2\alpha+1)_\ell (2k+1)}{(1)_\ell (1)_{\alpha+k+1}} \times$$
$$\times C^k_{s',s} \, E^{\alpha,\beta,\ell} E_p^{\alpha,\beta',\ell} \exp[i(\beta\omega - \beta'\Omega)] \,, \tag{4}$$

where

$$E^{\alpha,\beta,\ell} = \frac{r^\alpha}{(r+\alpha_P)^{2\alpha+1+\ell}} \exp(i\beta\nu), \quad E_p^{\alpha,\beta',\ell} = r_P^\alpha (r_P - a_P)^\ell \exp(i\beta'\nu_p) \,.$$

The quantities $E^{\alpha,\beta,\ell}$ and $E_P^{\alpha,\beta',\ell}$ may be presented as Fourier series with respect to M and M_p, respectively.

The calculation of Fourier coefficients for the $E_p^{\alpha,\beta',\ell}$ with accuracy to any power of the small planetary eccentricity e_P is a standard problem in the planetary theory of motion. Note that $E_P^{\alpha,\beta',\ell} = 0(e_P^\ell)$ for $e_P \to 0$.

Substantially more complicated is the calculation of the coefficients $U_j^{\alpha,\beta,\ell}$ in the expression

$$E^{\alpha,\beta,\ell} = \sum_{j=-\infty}^{\infty} U_j^{\alpha,\beta,\ell} \exp(ijM) \,.$$

The classical methods of representation of $U_j^{\alpha,\beta,\ell}$ in the form of a power series with respect to e are not valid because the eccentricities of most cometary orbits exceed the well-known Laplace's limit.

We expand $E^{\alpha,\beta,\ell}$ in power series of the quantity $\varepsilon = (1-e)/2$, developing a suggestion of Elenevskaya (1962). The basis of this approach consists in the transition from mean anomaly M to the true anomaly ν_0 of the parabolic motion with the same perihelion distance q and the same moment of perihelion passage as the given elliptic motion.

For this purpose, the following formulae may be used:

$$\begin{aligned} M &= \sqrt{\frac{2q^3}{a^3}}\left(\tau + \frac{\tau^3}{3}\right), \\ \tau &= \tan\frac{\nu_0}{2}, \\ r &= q(1+e\sigma^2), \\ \exp(i\nu) &= \frac{1-\sigma^2 + i\sigma\sqrt{2(1+e)(1-\sigma^2\varepsilon)}}{1+e\sigma^2}, \end{aligned} \tag{5}$$

where

$$\sigma\cos g + \sigma^3 G(g) - \tau - \frac{\tau^3}{3} = 0, \tag{6}$$

$$G(g) = \frac{2g - \sin 2g}{4\sin^3 g}, \quad \sin g = \sigma\sqrt{\varepsilon}.$$

Let us transform (6) to

$$\sigma - \tau - \varepsilon f(\sigma, \varepsilon, \tau) = 0,$$

where

$$f(\sigma, \varepsilon, \tau) = \frac{\varphi(\sigma, \varepsilon)}{1 + (\sigma^2 + \sigma\tau + \tau^2)/3}, \tag{7}$$

Then the function $\varphi(\sigma, \varepsilon)$ has an analytical continuation into the complex domain $|\sigma\sqrt{\varepsilon}| < 1$ and may be written in the form of a series

$$\varphi(\sigma, \varepsilon) = \frac{1}{2}\sum_{x=0}^{\infty} \frac{\left(\frac{1}{2}\right)_x}{(1)_{x+1}} \sigma^{2x+3}\left(1 - \sigma^2\frac{2x+1}{2x+5}\right)\varepsilon^x. \tag{8}$$

In an earlier paper (Emel'yanenko, 1991) we obtained the following representation of the quantity $E^{\alpha,\beta,\ell}$ using the theory of analytical functions

$$
\begin{aligned}
E^{\alpha,\beta,\ell} &= \sum_{m_1=0}^{\infty} \frac{\varepsilon^{m_1}}{(1)_{m_1}} \frac{\partial E^{\alpha,\beta,\ell}}{\partial \varepsilon^{m_1}} \Big|_{\sigma=\tau}^{\varepsilon=0} + \\
&+ \sum_{m_2=1}^{\infty} \sum_{m_1=0}^{\infty} \frac{\varepsilon^{m_1+m_2}}{(1)_{m_1}(1)_{m_2}} \frac{\partial^{m_1+m_2-1}}{\partial \sigma^{m_2-1} \partial \varepsilon^{m_1}} \left[\frac{\partial E^{\alpha,\beta,\ell}}{\partial \sigma} f^{m_2}(\sigma, \varepsilon, \tau) \right]_{|\substack{\varepsilon=0 \\ \sigma=\tau}} .
\end{aligned}
\tag{9}
$$

The limits of the convergence region are given in Table I. Here e^* is the least value of the eccentricity for which the series (9) converges for $\tau = \tau^*$, and $r^* = r(q, e^*, \tau^*)$.

The Fourier coefficients are calculated from the formula

$$
U_j^{\alpha,\beta,\ell} = \frac{1}{2\pi} \int_{-\pi}^{\pi} E^{\alpha,\beta,\ell} \exp(-ijM) dM . \tag{10}
$$

Let us replace M in the integral (10) by τ using the first relation from (5). The series (9) converges uniformly for any ε inside the convergence circle and may be integrated on the interval $[-\tau^*, \tau^*]$. For all the elliptic orbits the convergence interval $[-\tau^*, \tau^*]$ is less than the interval corresponding to the variation of M from $-\pi$ to π. But the convergence region of this series reaches to large distances, where planetary perturbations can be neglected (Table I).

Thus, we can write for small ε

$$
U_j^{\alpha,\beta,\ell} = \frac{x}{2\pi j} \int_{-\tau^*}^{\tau^*} E^{\alpha,\beta,\ell} (1 + \tau^2) \exp \left[-ix \left(\tau + \frac{\tau^3}{3} \right) \right] d\tau , \tag{11}
$$

where $x = j\sqrt{2q^3/a^3}$.

The formulae (7), (8), (9), (11) show that the calculation of $U_j^{\alpha,\beta,\ell}$ leads to the integrals

$$
\int_{-\tau^*}^{\tau^*} \frac{\cos\left[x\left(\tau + \tau^3/3\right)\right] d\tau}{(1 + y + \tau^2)^{\nu-1}} , \int_{-\tau^*}^{\tau^*} \frac{\tau \sin\left[x\left(\tau + \tau^3/3\right)\right] d\tau}{(1 + y + \tau^2)^{\nu-1}} \tag{12}
$$

where $y \geq 0$ and ν is an integer number.

Hence the secular terms of the disturbing function are very easily found, and for $x = 0$ the integrand is a readily integrated rational function.

In order to find the resonant terms for the case of long-period comets we have to compute the integrals (12) for $|j| \gg 1$. In these integrals $y = a_p/q$ or $y = 0$. The calculation of the integrals (12) has been considered elsewhere (Emel'yanenko, 1991) in detail. It has been shown that the expressions for the Fourier coefficients have a connection with the Macdonald functions. The recurrence relations for the successive calculations of integrals (12) have been found and problems associated with numerical errors have been discussed.

TABLE I
The convergence region of the series (9)

τ^*	e^*	r^*/q
2.0	0.61	3.4
3.0	0.87	8.3
4.0	0.93	15
5.0	0.96	23
6.0	0.970	32
7.0	0.978	44
8.0	0.984	58
9.0	0.987	73
10.0	0.990	90
11.0	0.992	109
12.0	0.993	129
13.0	0.994	152
14.0	0.995	176
15.0	0.9954	201
20.0	0.9974	357
30.0	0.9989	807
50.0	0.9996	2239

4. The Construction of Mappings for Cometary Dynamics

Let us consider the motion of a long-period comet under the gravitational attraction of the Sun and N_P planets of masses m_P. Let us introduce the barycentric coordinate system. The disturbing function of this problem is

$$R = \mu^2 \sum_{P=1}^{N_P} m_P \left(\frac{1}{\Delta_p} - \frac{1}{r} \right) + \mu^2 \left(\frac{1}{\Delta_s} - \frac{1}{r} \right) , \tag{13}$$

where

$$\Delta_s = \sqrt{r^2 + r_s^2 - 2rr_s \cos h_s} \ ,$$

(r_s is the position vector of the Sun, h_s is the angle between r_s and r).

The second term in (13) can be expanded in a power series of the ratio r_s/r. We shall omit second and higher-order terms with respect to r_s/r because $r_s \ll r$ (excluding the case of very small q). Then

$$R = \mu^2 \sum_{P=1}^{N_P} m_P \left(\frac{1}{\Delta_P} - \frac{1}{r} - \frac{r_P \cos h_P}{r^2} \right) = \sum_{P=1}^{N_P} \left(R_P - \frac{m_P}{r} \right) . \tag{14}$$

The form (14) allows one to represent the disturbing function as a sum of disturbing functions for each planet. Let us assume for simplicity that the planets move in circular orbits in the same plane.

Let us consider the Hamiltonian

$$\bar{H} = -\frac{\mu_1^4}{2u_1^2} - R_{\text{sec}} - \sum_{P=1}^{N_P} \sum_{j=-\infty}^{\infty} \sum_{j'=1}^{\infty} R_{\text{res},P}^{(j,j')} , \tag{15}$$

where R_{sec} is the secular part of the disturbing function R, $R_{\text{res},P}^{(j,j')}$ is the resonant part of the function R_p at the exact commensurability with the corresponding planet (i.e., at $n = j'n_p/j$), and

$$\mu_1 = \mu\sqrt{1 + \sum_{P=1}^{N_P} m_P} .$$

The Hamiltonian (15) describes very exactly the secular and resonant perturbations for a chaotic motion close to the $j' : j$ resonance and reflects with a sufficient accuracy the main periodic perturbations from neighbouring resonances. We use for the calculation of $R_{\text{res},P}^{(j,j')}$ the relations (4), (9), (11) if $q < r_P$. For the case $q > r_P$ we use the standard expansion of $R_{\text{res},P}^{(j,j')}$ in power series of the ratio r/r_P. The orbits of most of the long-period comets have eccentricities which do not differ from 1 practically. Since here τ^* is very large we set infinite limits to the integration in (11). Also we keep only the main term in (9) admitting an error $0(\varepsilon)$ for $\varepsilon \to 0$.

In such an approximation for $q < r_P$

$$\begin{aligned} R_{\text{res},P}^{(j,j')} &= \frac{\mu^2 m_P x}{\pi q j} \sum_{\alpha=j'}^{\infty} \sum_{k=j'}^{\alpha} \sum_{S} \frac{(\alpha+1)_\alpha (\alpha-k+1)_k (2k+1)}{(1)_{\alpha+k+1} z^\alpha} \times \\ &\times C_{s+j',s}^k V_{\alpha,2s+j} \cos[jM - j'\lambda_p + (2s+j')\omega + j'\Omega] \\ &- \frac{\mu^2 m_P x}{\pi q j z} \Big[C_{1,0}^1 S_{1,1} \cos(jM - \lambda_p + \omega + \Omega) \\ &+ C_{0,1}^1 S_{1,-1} \cos(jM - \lambda_p - \omega + \Omega) \Big] , \end{aligned} \tag{16}$$

where λ_p is the mean longitude of the planet,

$$k - |s| - |s + j'| = 2\bar{m} ,$$

$$V_{\alpha,\beta} = \int_{-\infty}^{\infty} \frac{(1+\tau^2)^{\alpha+1}}{(1+y+\tau^2)^{2\alpha+1}} \exp(i\beta\nu_0) \exp\left[-ix\left(\tau + \tau^3/3\right)\right] d\tau ,$$

$$S_{k,\beta} = \int_{-\infty}^{\infty} \frac{1}{(1+\tau^2)^k} \exp(i\beta\nu_0) \exp\left[-ix\left(\tau + \tau^3/3\right)\right] d\tau ,$$

$$z = \frac{q}{r_p}\,, \qquad x = \frac{j' n_p \sqrt{2q^3}}{\mu_1}\,.$$

We shall not discuss here the methods of calculation for $V_{\alpha,\beta}$ and $S_{k;\beta}$. Evidently the integrals $V_{\alpha,\beta}$ and $S_{k,\beta}$ are connected with the integrals (12) in which $\tau^* \to \infty$.

Substituting (16) into the Hamiltonian (15) gives

$$\begin{aligned} \bar{H} &= -\frac{\mu_1^4}{2u_1^2} - R_{\rm sec} \\ &+u(\nu_1) \sum_{P=1}^{N_P} \sum_{j'=1}^{\infty} \sum_{\alpha=j'}^{\infty} \sum_{k=j'}^{\alpha} \sum_{s} G_{\alpha,k,s}^{(j',P)} \sin[(2s+j')v_2 + j'v_3 - j'\lambda_p]\,, \end{aligned} \tag{17}$$

where $u(v_1)$ is such periodic function with period 2π that

$$u(v_1) = \begin{cases} \pi - v_1\,, & \text{if } 0 < v_1 < 2\pi\,, \\ 0\,, & \text{if } v_1 = 0\,, \end{cases}$$

$$\begin{aligned} G_{1,1,s}^{(1,P)} &= \frac{\mu^2 m_P x}{\pi q z} C_{s+1,s}^{1} (V_{1,2s+1} - S_{1,2s+1})\,, \\ G_{\alpha,k,s}^{(j',P)} &= \frac{\mu^2 m_P x (\alpha+1)_\alpha (\alpha-k+1)_k (2k+1)}{\pi q z^\alpha (1)_{\alpha+k+1}} C_{s+j',s}^{k} V_{\alpha,2s+j'} \text{ for } j' \geq 2\,. \end{aligned}$$

With an error of $0(\varepsilon)$ the coefficients $G_{\alpha,k,s}^{(j',P)}$ depend on u_2, u_3 only.

The Hamiltonian $\bar{H}$ leads to mappings for the description of motion (Emel'yanenko, 1990b). In particular, the equations for u_1, v_1 are

$$\begin{aligned} \frac{du_1}{dt} &= [1 - 2\pi\delta_{2\pi}(v_1)] \times \\ &\times \sum_{P=1}^{N_p} \sum_{j'=1}^{\infty} \sum_{\alpha=j'}^{\infty} \sum_{k=j'}^{\alpha} \sum_{s} G_{\alpha,k,s}^{(j',P)} \sin[(2s+j')v_2 + j'v_3 - j'\lambda_p]\,, \\ \frac{dv_1}{dt} &= \frac{\mu_1^4}{u_1^3} - \frac{\partial R_{\rm sec}}{\partial u_1}\,. \end{aligned}$$

where $\delta_{2\pi}$ is a periodic delta function with period 2π;

The delta function acts at the perihelion ($v_1 = 0$). Ignoring the secular part $R_{\rm sec}$ and the small deviations between perihelions leads to the mapping

$$
\begin{aligned}
w^{(m+1)} &= w^{(m)} \qquad\qquad (18)\\
&+ \frac{4\pi}{\mu_1^2} \sum_{P=1}^{N_P} \sum_{j'=1}^{\infty} \sum_{\alpha=j'}^{\infty} \sum_{k=j'}^{\alpha} \sum_{s} G_{\alpha,k,s}^{(j',P)} \sin[(2s+j')w + j'\Omega - j'\lambda_P^{(m)}] , \\
\lambda_P^{(m+1)} &= \lambda_P^{(m)} + \frac{2\pi n_P}{\mu_1 \left(w^{(m+1)}\right)^{3/2}} , \qquad (q < r_P)
\end{aligned}
$$

where $w^{(m+1)} - w^{(m)}$ is the change in $w = 1/a$ during the m-th comet perihelion passage, $\lambda_P^{(m)}$ is the value of λ_P at this comet perihelion time.

For the case $q > r_P$ we have the analogous mapping

$$
\begin{aligned}
w^{(m+1)} &= w^{(m)} \qquad\qquad (19)\\
&+ \frac{4\pi}{\mu_1^2} \sum_{P=1}^{N_P} \sum_{j'=1}^{\infty} \sum_{k=2}^{\infty} \sum_{s} G_{0,k,s}^{(j',P)} \sin[(2s+j')\omega + j'\Omega - j'\lambda_p^{(m)}] , \\
\lambda_P^{(m+1)} &= \lambda_P^{(m)} + \frac{2\pi n_P}{\mu_1 \left(w^{(m+1)}\right)^{3/2}} , \qquad (q > r_P)
\end{aligned}
$$

where

$$
G_{0,k,s}^{(j',P)} = \frac{\mu^2 m_P x}{\Pi q z^k} C_{s+j',s}^k S_{k,2s+j'}
$$

The mappings (18), (19) have the same form as the ones used by Petrosky (1986) and Chirikov, Vecheslavov (1986). But here the mappings have been written analytically. We would like to emphasize that the Hamiltonian (17) permits one to construct more general mappings for the detailed consideration of motion in the near-parabolic orbits. In particular, expressions for the changes in u_2, u_3, v_2, v_3 can be written too.

5. The Motion of Short-Period Comets

There are many works on the motion of short-period comets. Here we consider briefly the features of a calculation of the long-term evolution for the observed short-period comets.

Short-period comets can approach the outer planets during an evolution time (the only exception is P/Encke). Therefore we have to apply numerical methods in the investigation of cometary motion. This investigation has been based on an integration of the equations of cometary motion including perturbations by all the planets (Emel'yanenko, 1987). Everhart's method (1973) of integration has been

TABLE II
Comets which librate during ±50 revolutions

Name	Commensurability	d_J	d_S	d_U	d_N
P/Brorsen-Metcalf	1:6	0.5	1.7	3.8	7.6
P/Pons-Brooks	1:6	2.0	0.8	1.1	5.3
P/Neuimin 1	2:3	1.0	0.8	6.2	18.0
P/du Toit	4:5	0.9	2.6	9.3	20.0
P/Peters-Hartly	3:2	0.5	2.9	13.0	23.5
P/Dubiago	1:5	0.8	2.2	6.7	10.5
P/Giacobini	9:5	0.5	3.8	12.8	24.6
P.Pigott	2:1	0.3	4.9	13.7	25.3
P/Tritton	2:1	0.2	4.7	13.7	24.8

TABLE III
Planet avoiding comets during ± 50 revolutions

Name	d_J	d_S	d_U	d_N
P/Olbers	0.7	3.4	9.1	16.4
P/Tuttle	0.8	1.8	10.1	20.6
P/Arend	0.6	4.3	12.7	24.6
P/Borrelly	0.5	4.2	13.1	24.5
P/Johnson	0.5	4.2	13.5	25.2
P/Clarl	0.6	4.4	14.3	25.2
P/Tempel 2	0.6	4.4	13.8	25.3
P/Mellish	1.5	3.0	7.9	11.9
P/Vaisala 2	1.6	0.9	1.2	4.7
P/de Vico	2.3	1.9	0.6	3.8
P/Sanguin	0.7	1.1	9.4	21.4
P/Russel 2	0.7	4.0	13.6	24.8
P/Barnard	0.6	4.3	13.7	25.2

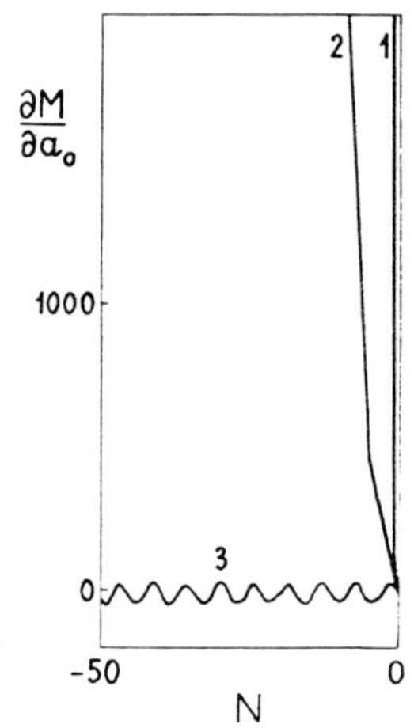

Fig. 1. The derivative $\partial M/\partial a_0$ (in degree/AU) versus the orbit number N in the past for P/Smirnova-Chernykh (1), P/Tuttle (2) and P/Dubiago (3).

applied. The initial orbital elements have been taken from the catalogue (Marsden, 1986).

Uncertainties arise when the equations of motion are solved numerically over long time spans, mainly because of rounding error at each step of the integration, imprecise coordinates for the major planets, errors in the initial orbital elements and nongravitational effects. Therefore, along with this direct integration of the equations of motion, a simplified model was considered: the restricted elliptic problem of three bodies (Sun, Jupiter, comet), averaged over the resonance period. The numerical solution uses procedure (2).

The accuracy of the results were checked by comparing direct numerical integrations of the equations of motion with the solution of the averaged problem. However, close to the epoch when their orbit cross, a planet and the comet may experience an encounter. In order to identify such possible situations we estimated the minimum distance between the comet orbit and the orbits of the outer planets.

The standard interval of integration contained 50 revolutions in the future and 50 revolutions in the past. If a close encounter was registered the integration of the equations of motion was stopped. The long-term evolution of cometary orbits cannot be traced for the case of close encounters (Kazimirchak-Polonskaya, 1978; Milani and Nobili, 1985; Rickman and Froeschlé, 1988). If a comet librated around the resonance with Jupiter, the integration was continued over the boundary of the standard interval till the intersection of the cometary orbit with the orbit of a major planet.

All the comets display chaotic motions, the semi-major axes of their orbits eventually being instable. But there are limited intervals when the cometary orbit is situated far from the planetary orbits and suffers only moderate perturbations. Under these circumstances the motion may have resonant properties of the kind previously discussed (Section 2). Together with typical chaotic motions, there are temporary librations around resonances with Jupiter. The accuracy of the calculation of the long-term evolution depends mainly on the separation rate of neighbouring trajectories. For the libration this rate is substantially less than that for typical chaos.

In Figure 1 we show the variations of the derivative $\partial M/\partial a_0$ (where a_0 is the initial value of a) for P/Dubiago, P/Smirnova-Chernykh, P/Tuttle. P/Smirnova-Chernykh has close encounters with Jupiter. The long-term evolution of the orbit cannot be traced for this case. The orbits of P/Dubiago and P/Tuttle are situated at equal distances from the orbit of Jupiter. But P/Dubiago librates in the 1:5 commensurability with Jupiter and P/Tuttle has a complicated motion near the 7:8, 13:15 and 6:7 resonances. The long-term evolution of cometary orbits may be investigated for cases of libration.

The presence of cometary librations near commensurabilities with Jupiter, though temporary, is therefore very important. The comets which librate during the whole standard interval are listed in Table II (here d_J, d_S, d_U, d_N are the minimum distances between the comet orbit and the orbits of Jupiter, Saturn, Uranus and Neptune, respectively, in the present epoch). Details about these comets and other temporary librations can be found elsewhere (Marsden, 1970; Franklin *et al.*, 1975; Carusi and Valsecchi, 1987; Emel'yanenko, 1987).

Table III contains the comets which are not librating, but the orbits of which are situated outside "the spheres of planetary action" during the whole standard interval. The angular elements of their orbits change regularly.

6. The Motion of Long-Period Comets

The dynamical evolution of long-period comets under the influence of planetary perturbations is an important stage in any theory of cometary origin. A numerical study requires integrations of thousands of cometary orbits over thousands of revolutions. Such calculations are beyond the capabilities of modern computers. So, the diffusion rate of comets was examined only for limited situations (Everhart, 1968; Yabushita, 1979; Fernandez, 1981; Duncan *et al.*, 1987; Stagg and Bailey, 1989). The application of mapping and analytic expansions of the disturbing function simplifies the solution of the problem considerably.

We use the maps (18), (19) for the description of the motion of comets in nearly-parabolic orbits. Dynamics of comets entering the planetary region are strongly chaotic. The assumption of a uniform distribution in the interval $[0, 2\pi]$ for phases $\lambda_P^{(m)}$ gives the following expression for the diffusion rate D_w of the quantity w, when $q < r_P$ for all P.

TABLE IV
The diffusion rate D for the perturbations of Jupiter

I	ω	$q = 2$ AU	$q = 4$ AU	$q = 6$ AU	$q = 8$ AU	$q = 10$ AU
0°		0.49×10^{-6}	0.28×10^{-6}	0.23×10^{-5}	0.91×10^{-7}	0.52×10^{-8}
70°	0^0	0.35×10^{-6}	0.12×10^{-6}	0.23×10^{-6}	0.19×10^{-7}	0.18×10^{-8}
	30	0.30×10^{-6}	0.10×10^{-6}	0.14×10^{-6}	0.13×10^{-7}	0.13×10^{-8}
	60	0.20×10^{-6}	0.66×10^{-7}	0.45×10^{-7}	0.54×10^{-8}	0.81×10^{-9}
	90	0.15×10^{-6}	0.45×10^{-7}	0.24×10^{-7}	0.36×10^{-8}	0.74×10^{-9}
140°	0^0	0.13×10^{-6}	0.14×10^{-7}	0.28×10^{-7}	0.13×10^{-8}	0.15×10^{-9}
	30	0.12×10^{-6}	0.13×10^{-7}	0.16×10^{-7}	0.95×10^{-9}	0.12×10^{-9}
	60	0.87×10^{-7}	0.15×10^{-7}	0.56×10^{-8}	0.51×10^{-9}	0.65×10^{-9}
	90	0.69×10^{-7}	0.16×10^{-7}	0.32×10^{-8}	0.36×10^{-9}	0.40×10^{-10}

$$D_w = \frac{<\Delta w^2>}{P_c} = \frac{4\pi w^{3/2}}{\mu_1^3} \sum_{P=1}^{N_P} \sum_{j'=1}^{\infty} \left[\left(\sum_{\alpha=j'}^{\infty} \sum_{k=j'}^{\alpha} \sum_{s} G_{\alpha,k,s}^{(j',P)} \cos 2s\omega \right)^2 + \left(\sum_{\alpha=j'}^{\infty} \sum_{k=j'}^{\alpha} \sum_{s} G_{\alpha,k,s}^{(j',P)} \sin 2s\omega \right)^2 \right],$$

where P_c is the cometary period. Using the mapping (19) an analogous expression can be written for any $r_P < q$.

The results of calculation in the case of Jupiter ($N_p = 1$) are shown in Table IV for various orbital elements. Here the time unit is equal to the revolution period. Therefore the tabular value of D is connected with D_w by the relation

$$D = \frac{2\pi}{\mu_1 w^{3/2}} D_w$$

and does not depend upon w.

D_w is a periodic function of ω with period π. Since moreover D_w is a symmetrical function with respect to $\omega = \pi/2$ we consider the values $\omega = [0, \pi/2]$ only. In calculations we take into account 8 coefficients of the Fourier series ($j' = 1, 2, \cdots, 8$) and a maximum of α equals 20. The tabulated data show the dependence of the diffusion rate upon the orbital elements in the restricted circular three-body problem. The results of calculations, taking into account the perturbations from 4 planets (Jupiter, Saturn, Uranus, Neptune), are given in Table V for $I = 0^0, 70^0, 140^0$. In this Table $\bar{D}$ is the value D averaged with respect to ω ($N_P = 4$). The dependence of the diffusion rate upon ω is weak.

TABLE V
The diffusion rate ($\bar{D}$) averaged over ω for the perturbations of Jupiter, Saturn, Uranus and Neptune

q (AU)	$I = 0°$	$I = 70°$	$I = 140°$
0.5	0.10×10^{-5}	0.47×10^{-6}	0.33×10^{-6}
2	0.51×10^{-6}	0.26×10^{-6}	0.11×10^{-6}
4	0.29×10^{-6}	0.87×10^{-7}	0.17×10^{-7}
6	0.23×10^{-5}	0.11×10^{-6}	0.13×10^{-7}
8	0.10×10^{-6}	0.12×10^{-7}	0.10×10^{-8}
11	0.64×10^{-7}	0.33×10^{-8}	0.39×10^{-9}
14	0.42×10^{-8}	0.48×10^{-9}	0.43×10^{-10}
17	0.47×10^{-9}	0.92×10^{-10}	0.10×10^{-10}
20	0.13×10^{-8}	0.61×10^{-10}	0.11×10^{-10}
24	0.18×10^{-9}	0.18×10^{-10}	0.20×10^{-11}
28	0.60×10^{-10}	0.85×10^{-11}	0.70×10^{-12}
32	0.54×10^{-9}	0.19×10^{-10}	0.31×10^{-11}
36	0.14×10^{-9}	0.72×10^{-11}	0.75×10^{-12}
40	0.42×10^{-10}	0.31×10^{-11}	0.26×10^{-12}
45	0.11×10^{-10}	0.11×10^{-11}	0.86×10^{-13}
50	0.31×10^{-11}	0.42×10^{-12}	0.32×10^{-13}

More detailed consideration of the complicated dynamics of comets is possible through numerical studies of the motion in addition to the direct use of maps (18), (19). We have investigated the evolution of the distribution function of comets over w for a range of initial values of q, I, ω, w. Initial values of λ_p were distributed uniformly between 0 and 2π. We took into account the perturbations from the four major planets: Jupiter, Saturn, Uranus and Neptune. Table VI gives the calculated distributions based on 1000 comets with initial conditions $a = 3 \times 10^3$ AU, $I = 0$, $\omega = 0$. This table provides the number of comets within each range of w for the different dates T. The data for $T = 0$ correspond to the distribution after the first passage of perihelion.

Inspection of Table VI reveals on interesting fact. At some stage of their evolution, comets achieve a distribution like the Oort cloud which is remarkable for the obvious predominance of the number of comets with $w < 10^{-4}$ AU-1. This distribution arises for all the initial values of $w \leq 10^{-2}$ AU^{-1}. The time of achievement of such state is hardly dependent upon the initial value of w for $q \leq 10$ AU. This fact was discovered by Yabushita (1979) for a simple model of the perturbations having a normal distribution with constant dispersion.

From our investigation it follows that this time does not vary significantly with D_w for $q \leq 10$ AU. This time increases gradually with decreasing of D_w for $q \geq 10$ AU. Moreover, it is necessary to mention that the mean time of capture to a short-period orbits (Stagg and Bailey, 1989) is close to the mean time of formation

TABLE VI
The distribution of comets over w by the planetary perturbations

w in units of 10^{-4} AU^{-1}		0–1	1–2	2–3	3–4	4–5	5–10	10–15	15–20
q (AU)	T (days)								
2	0	30	21	23	19	18	211	256	21
	2×10^8	104	64	16	17	9	21	15	10
	4×10^8	118	18	14	1	4	10	3	5
	6×10^8	98	10	3‘	2	1	7	1	9
	8×10^8	85	3	3	3	0	5	1	1
6	0	59	71	80	60	74	243	21	52
	2×10^8	105	96	19	10	14	20	8	4
	4×10^8	114	14	2	8	1	7	4	1
	6×10^8	97	10	3	1	2	2	0	0
	8×10^8	83	3	2	1	1	3	1	0
15	0	0	0	168	717	115	0	0	0
	8×10^9	345	123	96	53	51	120	60	14
	16×10^9	368	91	36	29	23	63	28	27
	24×10^9	320	39	27	9	4	56	25	19
35	0	0	0	39	961	0	0	0	0
	2×10^{10}	132	234	174	122	118	196	14	0
	4×10^{10}	225	227	141	79	70	147	40	11
	6×10^{10}	302	164	79	67	26	152	56	19
	8×10^{10}	349	120	74	62	46	89	47	3
	10×10^{10}	349	62	59	45	34	72	42	12

of a distribution like the Oort cloud (1–2 Myr for $q \leq 10$ AU). This study leads to the conclusion that the observed distribution (the Oort cloud) is a typical stage of evolution for comets in nearly-parabolic orbits under perturbations by the planets.

7. The Motion of Meteor Streams

The resonant properties of motion in orbits with large eccentricities give rise to special structures of meteor streams. To estimate the effect of gravitational perturbations on the stability of meteor streams, we have considered the evolution of an infinitesimal portion of a stream in the field of the Sun and planets (Emel'yanenko, 1984). The change in the volume of that portion may be written as

$$V = V_0 |\Delta| , \tag{20}$$

where V_0 is the volume at initial time t_0, V is the volume at a subsequent time t and Δ denotes the Jacobian of a transformation from initial coordinates to coordinates

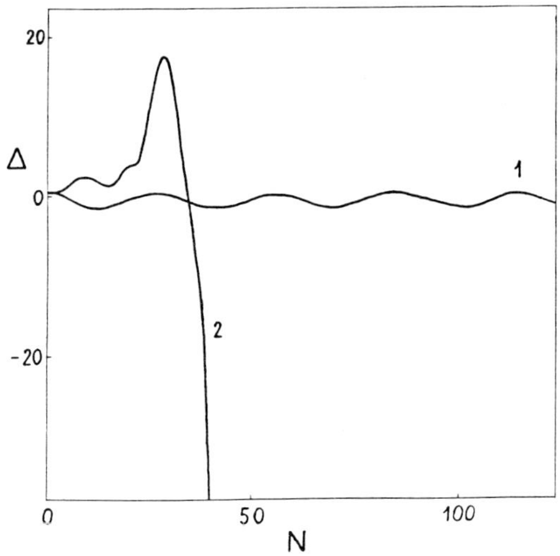

Fig. 2. Variations of the Jacobian Δ versus the orbit number N for two points on the orbit of P/Tuttle (Ursids).

at time t.

In order to evaluate the elements of Δ we must solve a system of 12 second-order differential equations (Emel'yanenko, 1984). According to (20) the value of Δ is a measure of the stream relative contraction or expansion in the neighboorhood of a given point on the orbit. The initial conditions were stipulated such that the value of Δ for the undisturbed stream would remain constant at each perihelion passage. Then by evaluating Δ at each perihelion, when planetary perturbations are taken into account, one can estimate their effect on the stream density.

Figure 2 displays the changes of Δ at two points on the orbit of P/Tuttle. The portion 1 corresponding to the Ursid meteor shower in 1945 moves deep in the libration zone of the 7:8 commensurability. Curve 2 corresponds to the position of the comet itself.

The calculations show again the importance of the librating motion. For the case of libration the variations of the Jacobian are insignificant, and are nearly periodic. After each half period Δ passes through zero, indicating a concentration of particles at that place in the stream.

For the case of typical chaos $|\Delta|$ grows rapidly, i.e., a dispersion of the meteor stream takes place.

The orbits of meteor streams are not sufficiently precise to determine reliably the commensurabilities. Therefore, Emel'yanenko (1988a) determined the parameters of the main resonant zones near the orbits of the basic meteor streams. Outbursts in the activity of old meteor streams are found to be in accordance with the resonant properties of motion. For example, the outbursts in the activity of Lyrids in 1803, 1922, 1982 and Ursids in 1945 took place near centres of resonance.

In order to study in detail the structure of meteor streams as a whole, we consider the behavior of the distribution function of meteor particles. It is necessary to compute the trajectories of many particles for that purpose. The application of mappings is useful in this case as well.

We studied the simplest of models in which the particles are ejected from the comet at the moments of the perihelion passage (Emel'yanenko, 1990a). The orbital periods of many meteor streams are longer than that of Jupiter (for example, Lyrids, Perseids, η-Aquarids, Leonids, Ursids). Dynamics of the comet and the meteor particles were described in these cases by the maps like (18), (19).

The main conclusions are as follows:

When libration occurs, the dispersion speed of meteor streams under the action of planetary perturbations is substantially less than under conditions of typical chaos. The librating portions of the old meteor streams are the most dense and their intersections with the Earth are rare events. The most remarkable example of such dynamical condensation is apparently the Lyrid meteor shower (Emel'yanenko, 1990 c). The last outburst in the activity of Lyrids took place in 1982 (Porubcan and Gevoloni, 1985).

If the motion of a great part of a meteor stream keeps a circulating character, the maximum density corresponds to the saddle point of the system (1). As this property takes place for all values of semi-major axes near the given resonance it must result in a rising activity of the meteor stream in corresponding years. An example of this phenomenon occurs apparently with the Perseid meteor shower (Emel'yanenko, 1990a). The activity of this shower rises to a maximum approximately every 12 years (Badazhanov, 1987).

8. Conclusion

The methods described here, based on an analytic expansion of the disturbing function and our application of mappings for the description of the motion have allowed us to calculate the long-term evolution of orbits with large eccentricities. The efficiency of these methods makes it possible to consider many additional aspects of cometary and meteor dynamics.

References

Badazhanov, P.B.: 1987, *Meteors and their Observations (in Russian)*, Nauka, Moskva

Boda, K.: 1931, 'Entwicklung der Storungfunction and Ableitungen in Reihen, welche für beliebige Exzentrizitaten und Neigungen konvergieren', *Astron. Nachrichten* **243**, 17–42

Carusi, A. and Valsecchi, G.B.: 1987, 'Dynamical Evolution of Short-Period Comets', *Publ. Astron. Inst. Czechosl. Acad. Sci.* **67**, 21–28

Charlier, C.L.: 1927, *Die Mechanik des Himmels*, Walter de Gruyter, Berlin-Leipzig

Chirikov, B.V.: 1979, 'A Universal instability of Many-Dimensional Oscillator Systems', *Phys. Rep.* **52**, 263–379

Chirikov, B.V. and Vecheslavov, V.V.: 1986, *Chaotic Dynamics of Comet Halley*, Inst. of Nuclear Physics Preprint 86-184, Novosibirsk

Duncan, M., Quinn, T. and Tremaine, S.D.: 1987, 'The Formation and Extent of the Solar System Comet Cloud', *Astron. J.* **94**, 1330–1338

Elenevskaya, N.B.: 1962, 'Expansion of the Disturbing Function for an Eccentricity near 1', *Bull. Inst. Teor. Astron. (in Russian)* **8**, 444–456

Emel'yanenko, V.V.: 1984, 'On the Evolution of the Density of Meteor Streams due to Planetary Perturbations', *Pis'ma Astron. Zh. (in Russian)* **10**, 315–318

Emel'yanenko, V.V.: 1987, 'On the Dynamics of Short-Period Comets', *Kinematika i Fizika Nebesnykh Tel (in Russian)* **3**, 52–56

Emel'yanenko, V.V.: 1988a, 'The Motion of Meteor Streams near the Commensurabilities with Jupiter', *Pis'ma Astron. Zh. (in Russian)* **10**, 651–658

Emel'yanenko, V.V.: 1988b, 'Peculiarities of Cometary Resonance Motion', *Kinematika i Fizika Nebesnykh Tel (in Russian)* **4**, 55–59

Emel'yanenko, V.V.: 1990a, 'The Resonance Model of a Meteor Stream', *Kinematika i Fizika Nebesnykh Tel (in Russian)* **6**, 58–65

Emel'yanenko, V.V.: 1990b, 'The Motion of Nearly-Parabolic Comets under the Weak Perturbations of Planets', *Pis'ma Astron. Zh. (in Russian)* **16**, 737–744

Emel'yanenko, V.V.: 1990c, 'Dynamics of the Lyrid Meteor Stream', *Astron. Vestnik (in Russian)* **24**, 308–313

Emel'yanenko, V.V.: 1991, 'An Expansion for the Secular and Resonant Parts of the Disturbing Function in the Theory of Long-Period Comets Motion', *Pis'ma Astron. Zh. (in Russian)* **17**, 857–864

Everhart, E.: 1968, 'Change in Total Energy of Comets Passing through the Solar System', *Astron. J.* **73**, 1039–1052

Everhart, E.: 1974, 'Implicit Single-Sequence Methods for Integrating Orbits', *Celest. Mech.* **10**, 35–55

Fernandez, J.A.: 1981, 'New and Evolved Comets in the Solar System', *Astron. Astrophys.* **96**, 26–35

Franklin, F.A., Marsden, B.G. and Williams, J.G.: 1975, 'Minor Planets and Comets in Libration about the 2:1 Resonance with Jupiter', *Astron. J.* **80**, 729–746

Giacalgia, G.E.O.: 1972, *Perturbations Methods in Non-linear Systems*, Springer-Verlag, New York-Heidelberg-Berlin

Kazimirchak-Polonskaya, E.I.: 1978, 'Capture of Comets by Jupiter and Some Regularities in Secular Evolution of Cometary Orbits', *Problemi Issledovania Vselennoi (in Russian)* **7**, 340–383

Krasinsky, G.A.: 1973, in N.S. Samoilova-Yachontova ed., ed(s)., *Basic Equations of Planetary Theory in Minor Planets*, Nauka, Moskva, 81–107

Lichtenberg, A.J. and Lieberman, M.A.: 1983, *Regular and Stochastic Motion*, Springer-Verlag, New York-Heidelberg-Berlin

Marsden, B.G.: 1970, 'On the Relationship between Comets and Minor Planets', *Astron. J.* **75**, 206–217

Marsden, B.G.: 1986, *Catalogue of Cometary Orbits*, Smithsonian Astrophys. Observatory, Cambridge, Massachusetts

Milani, A. and Nobili, A.M.: 1985, in A. Carusi and G.B. Valsecchi, Eds., ed(s)., *Errors in Numerical Integrations and Chaotic Motions*, Dynamics of Comets: Their Origin and Evolution, 215–226

Petrosky, T.Y.: 1986, 'Chaos and Cometary Clouds in the Solar System', *Physics Letters* **A 117**, 328–382

Petrovskaya, M.S.: 1970, 'Expansion of the Negative Powers of Mutual Distances between Bodies', *Celest. Mech.* **3**, 121–128

Porubcan, V. and Gevolani, G.: 1985, 'Unusual Display of the Lyrid Meteor Shower in 1982', *Prace Astronomiskeho Observatoria Na Skalnatom Plese* **13**, 247–253

Rickman, H. and Froeschlé, C.: 1988, 'Cometary Dynamics', *Astron. Observatory Uppsala Preprint* **29**,

Schubart, J.: 1968, 'Long-Period Effects in the Motion of Hylda-Type Planets', *Astron. J.* **73**, 99–103

Stagg, C.R. and Bailey, M.E.: 1989, 'Stochastic Capture of Short-Period Comets', *Monthly Notices Roy. Astron. Soc.* **241**, 507–541

Yabushita, S.: 1979, 'A Statistical Study of the Evolution of the Orbits of Long-Period Comets', *Monthly Notices Roy. Astron. Soc.* **187**, 445–462

EVOLUTION OF SHORT-PERIOD METEOROID STREAMS

P.B. BABADZHANOV and YU.V. OBRUBOV
Institute of Astrophysics, Dushanbe, 734670, U.S.S.R.

Abstract. The dynamical evolution of meteoroid streams associated with comets Encke, Halley, Machholz 1986 VIII and asteroid Phaethon is discussed. It is shown that the planetary perturbations can greatly increase the streams thickness and each stream may produce several couples of meteor showers active in different seasons of the year. The theoretical and observed data are in a satisfactory accordance.

Key words: Meteoroid streams, comets, meteor showers.

1. Introduction

Meteoroid streams are assumed to be formed by the process of cometary decay. Whenever the Earth passes through meteoroid streams, meteor showers are observed. During the last decade a possible relationship of some meteor showers and Apollo, Amor and Aten asteroids was established (Babadzhanov and Obrubov, 1983; Whipple, 1983; Kresak and Stohl, 1990; Olsson-Steel, 1990). It is very likely that corresponding asteroids are the nuclei of "defunct" comets.

Investigators have distinguished about 20 annual major meteor showers. The list of these showers, dates of maximum activity, the geocentric radiant (α – the right ascension and δ – the declination) and velocities v_g (in km s^{-1}), maximum visual zenithal rate per hour N (or maximum radar echo rate per hour for daytime showers) are given in Table I.

The data of the periodical Andromedid, Draconid and Leonid meteor storms of maximum zenithal rate of more than tens of thousands of meteors per hour, are also given in the Table. Until the second half of the 20th century the associations of 9 major showers with comets were established according to the similarity of their orbital elements. Moreover, it was considered as a rule that each meteoroid stream (or comet) could produce only one or two meteor showers (Kramer, 1953; Hasegawa, 1958; Drummond, 1981, 1982; Olsson-Steel, 1987, 1988). The parent bodies of 8 meteor showers were revealed lately owing to the discovery of new comets and asteroids (Whipple, 1983; McIntosh, 1990) or on the basis of dynamical evolution of these bodies and meteoroid streams (Fox *et al.*, 1984; Jones and Hawkes, 1986; Babadzhanov and Obrubov, 1987, 1992; McIntosh, 1990). These investigations took into account the effects of the differential ejection velocities from cometary nucleus. Moreover, it was established that a meteoroid stream could produce from two up to eight discrete annual meteor showers (Babadzhanov and Obrubov, 1987).

Celestial Mechanics and Dynamical Astronomy **54:** 111–127, 1992.

TABLE I
The annual major meteor showers

Shower	Max.	α^0	δ^0	v_g	N	Parent body
Quadrantids	Jan.3	231	+50	41	140	*P*/Machholz 1986 VIII
Lyrids	Apr.22	271	+34	48	12	Thatcher 1861 I
η Aquarids	May 3	336	0	66	30	*P*/Halley 1910 II
D.Arietids	June 7	43	+23	39	60	*P*/Machholz 1986 VIII
ζ Perseids	June 7	62	+23	27	40	*P*/Encke 1971 II
β Taurids	June 29	86	+19	30	30	*P*/Encke 1971 II
$S.\delta$ Aquarids	July 29	333	–16	41	30	*P*/Machholz 1986 VIII
$N.\delta$ Aquarids	Aug.12	339	–5	42	20	*P*/Machholz 1986 VIII
α Capricornids	July 30	307	–10	23	30	Adonis (2101)
$S.\iota$ Aquarids	Aug.5	333	–15	34	15	???
$N.\iota$ Aquarids	Aug.20	327	–6	31	15	???
Perseids	Aug.12	46	+57	59	70	*P*/Swilft-Tuttle 1862 III
κ Cygnids	Aug.18	286	+59	25	5	???
Day Sextantids	Sept.29	152	0	32	30	Phaethon (3200)
Draconids	Oct.9	262	+54	20	storm	*P*/Giac.-Zinner 1933 III
Orionids	Oct.21	94	+16	66	30	*P*/Halley 1910 II
S. Taurids	Nov.3	50	+14	27	7	*P*/Encke 1971 II
N. Taurids	Nov.13	58	+22	29	<7	*P*/Encke 1971 II
Leonids	Nov.17	152	+22	71	storm	*P*/Tempel-Tuttle 1866 I
Andromedids	Nov.27	25	+44	16	storm	*P*/Biela 1846 II
Dec Phoenicids	Dec.5	15	–55	12	20	*P*/Blanpain 1819 IV
Geminids	Dec.14	112	+32	34	70	Phaethon (3200)
Ursids	Dec.22	217	+76	33	20	*P*/Machholz 1986 VIII

The ejection velocity V of a meteoroid particle from a cometary nucleus at a distance r_{AU} from the Sun may be written down as (Whipple, 1951):

$$V = 3.28\sqrt{\left(\frac{4}{\rho\delta r^{3/4}} - 0.52R_c\right) R_c} \quad \mathrm{m\,s^{-1}}\,, \tag{1}$$

where R_c is the radius of a cometary nucleus in km, ρ and δ are particle's radius and density both in CGS units. This velocity V is considerably less that the comet's orbital velocity and so the differences between the orbits of a comet and its meteoroid ejecta are very small.

The maximum differences between orbital semi-major axis $\Delta_1 a$ and the eccentricity $\Delta_1 e$ of the ejected meteoroid from those of the comet will be (Plavec, 1955):

$$\begin{aligned} \Delta_1 a &= \pm 6.72 \times 10^{-5} a_c^2 V \sqrt{2/r - 1/a_c} \quad \mathrm{AU}\,, \\ \Delta_1 e &= \pm 6.72 \times 10^{-5} r e_c V \sqrt{2/r - 1/a_c}\,, \end{aligned} \tag{2}$$

where a_c and e_c are the semimajor axis and the eccentricity of the comet's orbit respectively. The other orbital elements of released particules differ only slightly from the cometary ones and, thus, these differences may be neglected.

Just after the release from the comet nucleus, the meteoroids are exposed to the solar radiation pressure. This pressure leads to an increase in the orbital period and to variations in the eccentricity (Radzievsky, 1951):

$$\begin{aligned} \Delta_2 a &= 5.76 \times 10^{-5} (\rho\delta)^{-1} a_c \, (2a_c/r - 1) \quad \text{AU} \, , \\ \Delta_2 e &= 5.76 \times 10^{-5} (\rho\delta)^{-1} p_c e_c^{-1} \, (1/r - 1/a_0) \, , \end{aligned} \tag{3}$$

where $\rho_c = a_c(1 - e_c^2)$ is the semi-latus rectum of the comet's orbit in AU.

The initial dispersion of the meteoroid stream orbits arises therefore through the dispersion due to ejection velocities and light pressure. A young meteoroid stream is very flat, narrow at the perihelion and broad at the aphelion because the ejection of particles is more intensive near the perihelion.

Due to the differences in the planetary perturbing action on the stream meteoroids of different semi-major axes and eccentricities the rate of long-period variations in the orbital elements will be different. Thus, the orbital dispersion, which may be defined by the dispersion in the arguments of perihelia increases with time. Multiple ejections of particles from the parent body during many revolutions cause meteoroid orbits of equal sizes to be at different evolutionary stages distinguished by their arguments of perihelia. This process can essentially increase the meteoroid stream in size and, first of all in its thickness. The stream may thicken and take such a shape as to cause the start of several discrete active showers at different solar longitudes. The discreteness of these related showers is a result of selectivity in the groundbased observations.

2. Evolution of meteoroid streams

Here, the evolution of four meteoroid steams producing the Geminid, Quadratid, Taurid, Eta Aquarid and Orionid annual meteor showers is investigated in some detail.

2.1. GEMINIDS

The relationship between the asteroid Phaethon (3200) and the Geminids is already well-recognized (Whipple, 1983; Fox *et al.*, 1984; Babadzhanov and Obrubov, 1986, 1987; Jones and Hawkes, 1986). As far as only comets were recognized as producer of meteoroid streams, and a model for the formation of the Geminids based on a collision between two rock like bodies could not explain the observed distribution of aphelion distances of Geminid meteors (Hunt *et al.*, 1986), it is very likely that Phaethon is a remnant of a cometary nucleus. If the Geminid meteoroid stream was formed as a result of a cometary decay then under planetary perturbations it takes the shape presented in Figure 1.

The characteristic feature of the Geminid shape is a very large thickness. This shape differs greatly from the traditional notion of a meteoroid stream which was usually supposed to be like an elliptic ring of relatively small thickness. Thus, it turns out that after 20 millennia of the stream's existence (the maximum stream age)

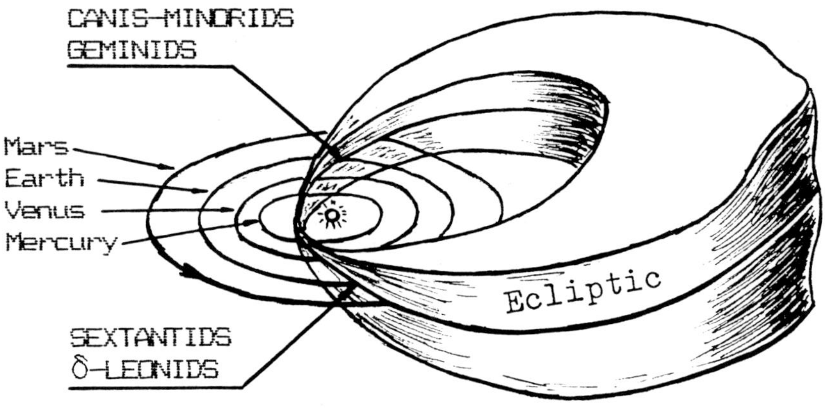

Fig. 1. The spatial shape of the Geminid stream model.

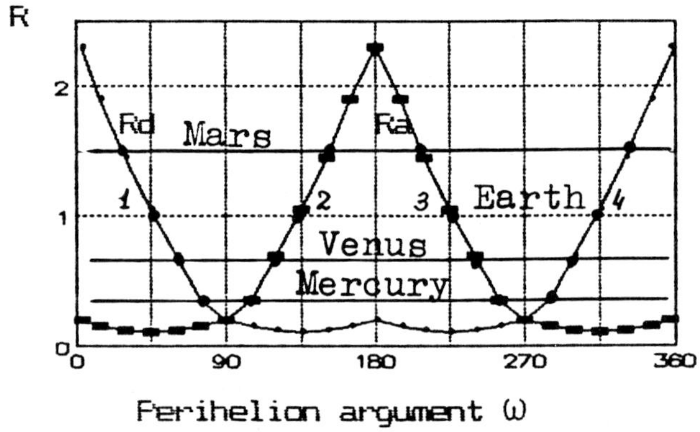

Fig. 2. The radial difference (AU) of the ascending and descending nodes (Ra, Rd) of Phaethon's orbit versus the argument of perihelion ω. The intersection 1 of Phaethon's and Earth's orbits correspond to the δ leonids, 2 – to the Canis-Minorids, 3 – to the Daytime Sextantids and 4 – to the Geminids.

the dispersion of arguments of perihelia reaches 316°. This dispersion is sufficient to fill out practically the total volume of the Geminid meteoroid stream presented in Figure 1 (Babadzhanov and Obrubov, 1986, 1987).

Figure 2 shows secular variations in the radius-vectors of the ascending (Ra) and descending (Rd) orbital nodes of Phaethon's orbit versus the argument of perihelion ω. As is seen from this Figure Phaethon's orbit may intersect the Earth's orbit at four values of the argument of perihelion. The dispersion of arguments of perihelia, 20 millennia after the stream formation, embraces all possibles values at which the intersections with the Earth's orbit take place. Thus the stream may be expected to produce four annual meteor showers. Table II presents the theoretical geocentric radiants, velocities and dates of activity of these showers calculated for

TABLE II
Theoretical and observed geocentric radiants of Geminid stream's meteor showers (V_g is in km s^{-1})

Shower	Theoretical				Observed				D
	Date	α	δ	V_g	Date	α	δ	V_g	
Geminids	Dec.14	113°	33°	34	Dec.14	112°	32°	34	0.05
Canis-Minorids	Dec.14	110	11	34	Dec.10	109	12	39	0.04
δ Leonids	Oct.5	170	16	34	not observed				
D.Sextantids	Oct.5	161	-4	34	Oct.8	157	-8	30	0.15

Phaethon's orbits intersecting the Earth's orbit. According to the radiant positions these showers can be called the Geminids, the Canis-Minorids, the Sextantids and δ-Leonids. The first two showers are nighttime whilst the second two are daytime. Among these four showers three have been observed. The observed data of these showers are given in Table II and agree satisfactorily with the theoretical calculations. The D-criterion of Southworth and Hawkins (1963) expressing the total difference between the theoretical and observed orbits in terms of difference of their orbital elements a, e, i, ω and Ω is also given. The observed orbit for the Geminids is taken from the catalogue of Cook (1973) and for the D.Sextantids from the catalogue of Sekanina (1976). As seen from this table, given the usual criterion for membership ($D < 0.2$), the theoretical and the observed orbits of meteor showers produced by the Phaethon meteoroid stream are in excellent accord.

2.2. Quadrantids

The Quadrantids are one of the most interesting meteor showers. In May 1986 D. Machholz discovered a new comet which got the designation 1986 VIII. It might be a parent comet of the Quadrandid meteoroid stream (McIntosh, 1990; McBride and Hughes, 1990). Due to the short orbital period (5.25 yr) and small perihelion distance (0.127 AU) this comet is a unique object for investigations of physical evolution of cometary nuclei. The fact that the comet Machholz was discovered only in 1986 is of interest as well. Thus Sekanina (1990) and Green *et al.* (1990) assumed the comet Machholz was dormant for a long time before 1986.

It seems very likely that *P*/Machholz was an active comet in the past. Hence, the meteoroids were ejected from the comet's rotating nucleus in different directions and at different velocities. To model the meteoroid stream evolution one should choose the initial moment of the stream formation and give the distribution of meteoroid's orbits. The initial moment is taken to be 4500 B.C. for two reasons:

(1) It is at about this epoch that we find the greatest similarity between *P*/Machholz and the mean Quadrantid osculating orbit.

(2) The radius-vector of the ascending node of the comet's orbit was at this epoch close to the orbit of Jupiter.

The first condition implies that the ejection velocities responsible for the Quadrantid shower formation were minimized. The second condition is based on the possible capture of the comet by Jupiter into a short-period orbit.

It is known that the meteoroids' orbits in the same stream differ greatly from the comet's orbit by their semimajor axes and eccentricities. Thus, the initial distribution of the meteoroid orbits was taken to be:

$$\begin{aligned}
a &= 2.70\text{AU} \quad e = 0.691 \quad E = 30°,\ 270°,\\
a &= 2.80\text{AU} \quad e = 0.702 \quad E = 10,\ 70, 130, 190, 250, 310,\\
a &= 3.03\text{AU} \quad e = 0.725 \quad E = 0,\ 60, 120, 180, 240, 300,\\
a &= 3.20\text{AU} \quad e = 0.740 \quad E = 20,\ 80, 140, 200, 260, 320.
\end{aligned}$$

For all 20 orbits of these meteoroids we took $q = 0.834$ AU, $i = 69°$, $\Omega = 283.3°$ and $\omega = 170.7°$. The comet Machholz had such elements in 4500 B.C. This distribution, except for the eccentric anomaly E, corresponds to meteoroid ejection velocities of up to 300 m s^{-1} along and opposite to the comet velocity at perihelion, and represents a young flat meteoroid stream.

For all particles we then calculated the perturbations due to Jupiter and Saturn beginning from 4500 B.C. till 3000 A.D. It turned out that all intersections of particles' orbits with the Earth's orbit remained fixed. It was found that the Earth-crossing orbits could be divided into two groups: the first group includes the orbits with high inclinations (from 67° to 80°) and perihelion distances from 0.9 to 1.0 AU; the second group – the orbits with moderate inclinations (from 19° to 40°) and small perihelion distances – from 0.03 to 0.12 AU.

Each group may be divided into four subgroups according to their perihelion arguments. In the first group the perihelion arguments cluster around $\omega = \pm 5°$ and $180 \pm 6°$, and in the second group – around $\omega = \pm 26°$ and $180 \pm 30°$. So there are eight subgroups of Earth-crossing orbits, i.e. eight meteor showers, which can be produced by *P*/Machholz. This fact is illustrated by Figure 3, which presents the dependence of the radius-vectors of the nodes of *P*/Machholz orbits versus the argument of perihelion.

The names and ranges of orbital elements and geocentric radiants of these eight showers are given in Table III. The observed orbital elements and geocentric radiants for six of the eight meteor showers based on the data of different authors (Kascheev *et al.*, 1967; Sekanina, 1970; Cook, 1973) are also given there. The satisfactory agreement between the theoretical and observed radiants and orbits confirms a possible generic relationship between these showers and comet Machholz. The κ–Velid and Carinid meteor showers which may only be observed from the Southern Hemisphere have not yet been detected.

Figure 4 shows the increase in the dispersion of the perihelion arguments of the simulated *P*/Machholz's meteoroid stream under planetary perturbations. As seen 7.5 millennia after meteoroid ejection the dispersion becomes more than 360° (Figure 4). By this time all eight meteor showers must manifest their activity. The arguments of perihelia of six observed meteor showers cover the range from

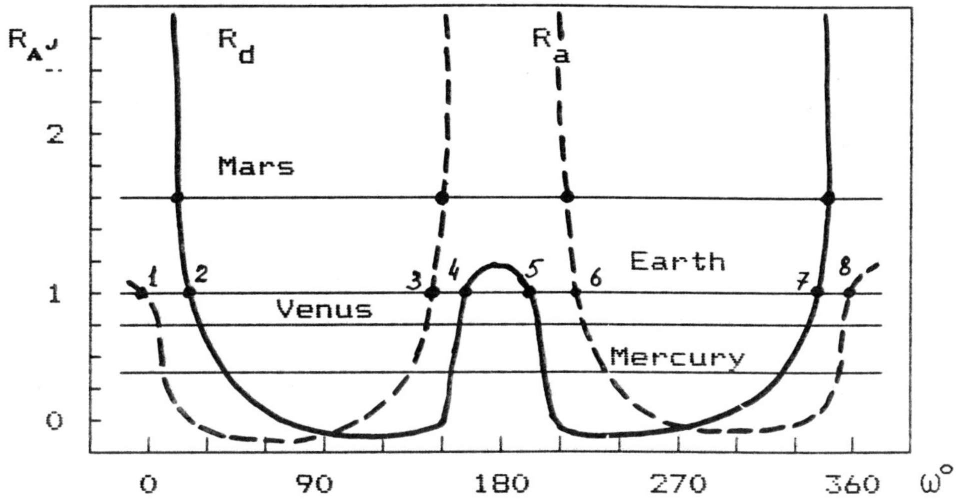

Fig. 3. The dependence of the radius-vectors of the nodes of *P*/Machholz's orbits. The intersection (1) corresponds to the κ-Velids, (2) Daytime Arietids, (3) Southern δ-Aquarids, (4) Quadrantids, (5) Ursids, (6) Daytime α-Cetids, (7) Northern δ-Aquarids, (8) Carinids.

TABLE III

Theoretical and observed geocentric radiants and velocities of the *P*/Machholz's meteor showers. See the note at the end of the paper concerning the two unobserved meteor showers

	Date	α	δ	V_g
Theoretical				
Quadrantids	Jan.2	223–243	[+42, +54]	38–43
Ursids	Dec.31	215–227	[+48, +60]	38–46
Carinids	Jan.2	149–164	[−64, −46]	41–49
κ-Velids	Jan.3	142 – 160	[−61, −50]	41–49
$N.\delta$-Aquarids	Jul.30	316–340	[−10, +2]	40–44
$S.\delta$-Aquarids	Jul.30	334–351	[−18, −13]	39–44
$D.$Arietids	Jun.12	42–50	[+22, +25]	40–44
α-Cetids	Jun.6	39–53	[+8, +13]	39–44
Observed				
Quadrantids	Jan.3	219–232	[+48, +55]	40-43
Ursids	Dec.22	190–226	[+58, +76]	33-40
Carinids	not observed			
κ-Velids	not observed			
$N.\delta$-Aquarids	Aug.12	337–346	[−5, +3]	40–42
$S.\delta$-Aquarids	Jul.29	339–351	[−19, −14]	40–44
$D.$Arietids	Jun.8	43–50	[+22, +26]	39–44
α-Cetids	Jun.9	44–53	[+6, +12]	37–39

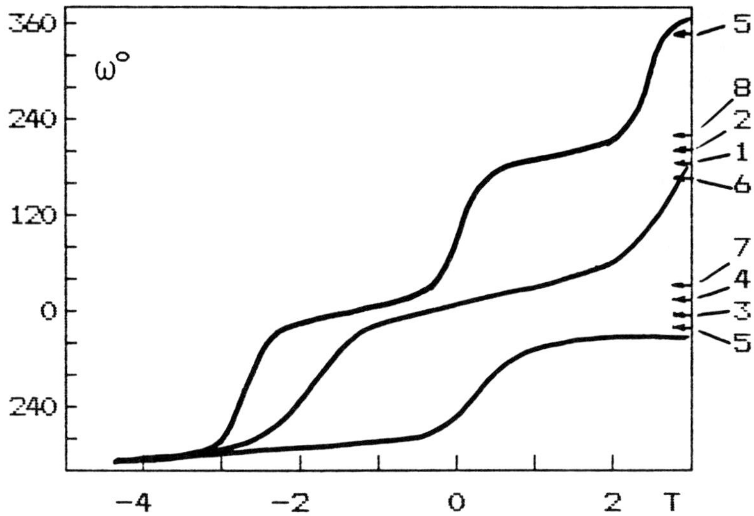

Fig. 4. The increase of the dispersion of ω in the simulated *P*/Machholz's meteoroid stream. The shower perihelion arguments are indicated by arrows: (1) Quadrantids, (2) Ursids, (3) Carinids, (4) κ–Velids, (5) Northern δ-Aquarids, (6) Southern δ-Aquarids, (7) Daytime Arietids, (8) α-Cetids. (*T*) time in millennia.

20° to 340°. Apparently, the real dispersion in ω of the stream is not less than that observed. It is very likely that the dispersion of ω embraces all possible values from 0 to 360°. Since according to our model the dispersion of 360° in the arguments of perihelia has been formed in the course of about 7.5 millennia, this interval may be looked upon as an estimate of the *P*/Machholz's meteoroid stream age.

According to Babadzhanov *et al.* (1991a), to ensure the observed activity of six meteor showers, the comet Machholz must have a maximum initial mass of $1.4 \times 10^{18} g$ corresponding to a radius of about 7 km if the average density of the cometary nucleus is equal to 0.8 g cm^{-3}. According to Sekanina (1990), the current radius of the nucleus of *P*/Machholz does not exceed 2.7 km. If one assumes that the comet was inside Jupiter's orbit during these 7–8 millennia, then taking into account its small perihelion distance, it is not difficult to explain the decrease of its radius from 7 to 2.7 km or even less.

2.3. Meteoroid Complex of *P*/Encke.Taurids

The *P*.Encke meteoroid stream is intersected twice by the Earth and produces at least 4 meteor showers: the nighttime Northern and Southern Taurids and daytime β Taurids and ζ Perseids.

The orbit of *P*/Encke may intersect the Earth's orbit at four different values of the argument of perihelion. If we take into account the initial dispersion in orbital elements of meteoroids ejected at $V \leq 300$ m s^{-1} then about 5000 years are required to produce the above mentioned four meteor showers (Babadzhanov *et al.*,

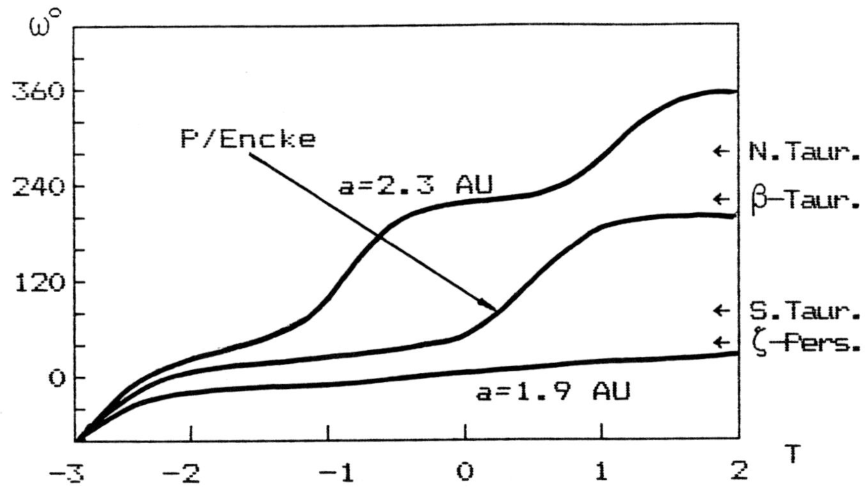

Fig. 5. The increase in the dispersion of the arguments of perihelia in *P*-Encke meteoroid *T* in millennia. The observed values of the arguments of perihelia are marked by arrows for all four meteor showers.

TABLE IV

Theoretical and observed geocentric radiants of *P*/Encke's meteor showers. The data for the observed radiants are taken from the catalogues of Cook (1973) (denoted C), Sekanina (1976) (denoted S) and Lebedinets *et al.* (1972) (denoted L)

Shower	Theoretical radiants				Observed radiants				D	Cat.
	Date	α	δ	V_g	Date	α	δ	V_g		
N.Taurids	Nov.2	52	23	31	Nov.13	58	22	29	0.10	C
S.Taurids	Nov.2	54	15	31	Nov.3	50	14	27	0.16	C
ζ Perseids	Jul.3	84	27	31	Jul.4	84	24	29	0.13	S
β Taurids	Jul.3	85	20	31	Jun.28	86	19	30	0.07	C
N.Piscids (Ass. 475)	Sept.19	10	8	31	Sept.17	7	9	30	0.04	L
S.Piscids (Ass. 477)	Sept.19	13	1	31	Sept.16	12	4	31	0.15	L
Assoc. 189	May 19	38	19	31	May 20	39	20	28	0.06	L
Assoc. 190	May 19	40	12	31	May 20	42	10	29	0.06	L
N.χ-Orionids	Nov.22	73	27	31	Nov.24	78	31	32	0.18	L
S.χ-Orionids	Nov.22	74	19	31	Nov.23	76	12	28	0.16	L
Assoc. 365	Jul.24	107	26	31	Jul.29	116	25	29	0.14	L
Assoc. 307	Jul.24	106	19	31	Jul.20	103	15	36	0.18	L

1990; Steel *et al.*, 1991). Figure 5 shows the secular increase in the dispersion of the arguments of perihelia in *P*/Encke meteoroid stream over 5 millennia. Theoretical and observed radiants of these showers are in satisfactory accordance (Table IV).

The longitudes of Taurid's perihelia change with time almost linearly and at the first approximation they may be given as (Babadzhanov *et al.*, 1990):

$$\Pi = \Pi_0 + (8.5a - 12.1) \cdot t \, , \tag{4}$$

where Π_0 is the longitude of perihelion of the comet's orbit at the moment of particle ejection, t – time in millennia beginning from the moment of ejection, a – the semimajor axis in AU.

Some authors regard the nighttime Piscids and χ Orionids, which have Northern and Southern branches, as members of the *P*/Encke meteoroid complex as well. In order to explain the membership of the Piscids and χ Orionid to *P*/Encke's meteoroid complex one needs to assume a dispersion in the longitudes of perihelia over the range 110° to 185°. According to Equation (4) depending on the dispersion in the semi-major axes the necessary dispersion in Π may be formed over 8–18 millennia.

As the *P*-Encke's meteoroid stream age is probably sufficiently large and the Piscid and χ Orionid meteor showers have Northern and Southern branches, so the corresponding daytime two showers should exist. A search indicates that all 12 showers of *P*-Encke's meteoroid stream exist. The theoretical and observed radiants of these showers are presented in Table IV.

2.4. Eta-Aquarids and Orionids

The fact that *P*-Halley is a progenitor of the Eta-Aquarids and the Orionid meteor showers is almost beyond doubt because the lifetime of *P*-Halley in the inner regions of the Solar system is long enough for the formation of large orbital dispersion in the stream.

To study the orbital evolution of the particles ejected from the comet Halley three distinct perihelion passages of the comet (in 9025 B.C., 5028 B.C. and 1404 B.C.) have been chosen in the time span of 11000 years (Babadzhanov *et al.*, 1991b). Investigations into the orbital evolution of meteoroids ejected from the comet nucleus at velocities equal to or less than 100 m s^{-1} show that around the orbit of the parent comet a quasi-diffusion mixing process of meteoroids takes place. The meteoroids released at different time can have the same ascending nodes. This process can change the particles flux density along the orbit of the *P*-Halley meteoroid stream as observed during the activity periods of the Eta-Aquarid and Orionid showers.

There is another interesting result of our calculations, namely, the stability of comet's orbit and that of the particles ejected from it. Over the 11000 years period under consideration the majority of orbital elements change within relatively narrow ranges (Babadzhanov *et al.*, 1991b).

Within the considered period the meteoroid orbits intersected the Earth's orbit many times. Figure 6 shows the solar longitudes versus time of the intersections of such meteoroid orbits with the Earth's orbit. The period of visibility of Orionids and Eta-Aquarids of different ejections are also shown in this figure. Moreover, the observed solar longitudes in the period of activity of showers at present and accord-

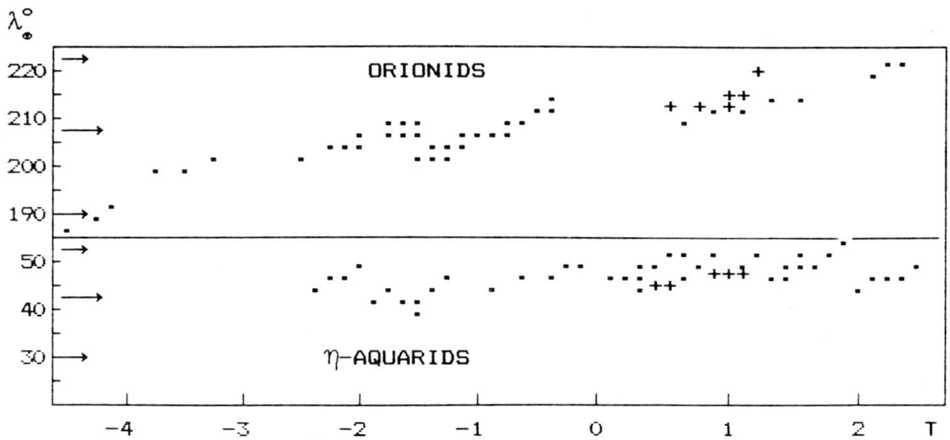

Fig. 6. Solar longitudes λ_0 at the shower activity periods of the η-Aquarids and Orionids versus time T (in millennia). The arrows correspond to the beginning, maximum and the end of current shower activity periods. Date of Imoto and Hasegawa (1958) are marked by " + ".

ing to the data of ancient observations (Imoto and Hasegawa, 1958) are indicated. Figure 6 shows the satisfactory agreement between theoretical and observed longitudes. As is seen beginning from 2250 B.C. the dispersion in the stream meteoroid orbits becomes so large that the activity of both the Orionids and Eta-Aquarids now begins. At about 1000 A.D. the Eta-Aquarids shower was caused by the particles ejected in 1403 B.C., and the Orionids by the particles ejected in 5028 B.C. Based upon these data we conclude that the stream age (as well as that of P-Halley's) is not less than six millennia.

It is interesting to note that our estimation of the secular displacements of the maximum activity dates, i.e. 0.4°/100 years for Orionids and 0.2°/100 years for Eta-Aquarids (Babadzhanov and Obrubov 1982), are in satisfactory agreement with the observations and results of McIntosh and Jones (1988).

The theoretical and observed geocentric radiants of the Eta-Aquarids and Orionids and their daily motions are given in Figure 7. A systematic increase in the right ascension of Orionid's radiant with time is connected with the corresponding changes of the date of shower activity. For the Eta-Aquarids this effect is manifested rather faintly (Figure 7). According to our model the theoretical geocentric velocities of meteoroids are equal to 66–67 km s^{-1} and no dependence on time or date of observations has been revealed. The theoretical and observed dates of shower activity (solar longitudes at the periods of shower activity) as well as the theoretical radiants are in satisfactory agreement with the observed Orionids and Eta-Aquarid radiants.

The agreement between the P-Halley meteoroid stream simulation and the observational data testifies as to the correctness of our model and eliminates any doubt as to the reality of a generic relationship between the Orionids, the Eta-

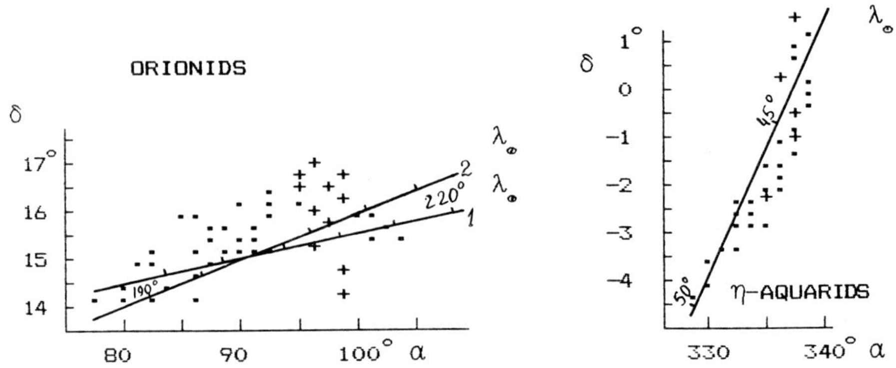

Fig. 7. Theoretical (points) and observed (crosses) geocentric radiants of P-Halley meteor showers. The ephemerides of radiants for the Orionids are given according to McIntosh and Jones (1988): (1) theoretical ephemeris, (2) photographic observations of faint meteors. The ephemerides of radiants for the η-Aquarids are given according to Lindblad (1990).

Aquarids and P-Halley. Furthermore, the large visibility periods of these showers and the duration of their activities are explained.

2.5. Other Short-Period Meteoroid Streams

Simulation of the orbital evolution of the Geminid, Quadrantid and Taurid meteoroid streams show that their qualitative features such as the number of meteor showers and stream sizes may be determined by the following integrals of motions

$$(1 - e^2)\cos^2 i = C_1 = \text{const.}, \tag{5}$$

$$e^2(0.4 - \sin^2 i \sin^2 \omega) = C_2 = \text{const.}, \tag{6}$$

where C_1 and C_2 are constants to be determined from the observed orbital elements. Integral (5) follows from Tisserand's criterion if a = const. (Moiseev 1945). Integral (6) is derived by Lidov (1961). Using these integrals it is possible to estimate the ranges of secular variations of the eccentricity and inclination. If we add the condition of intersection of the given orbit with the Earth orbit:

$$a(1 - e^2) = 1 \pm e\cos\omega \tag{7}$$

to (5,6) and using the known semimajor axis a, and the constants C_1 and C_2, we can determine the orbital elements of a particle crossing the Earth's orbit (Babadzhanov and Obrubov, 1987).

From Cook's (1973) list we have chosen the short-period meteoroid streams with orbital period less than 6 years and estimated their thickness formed under planetery perturbations: at perihelion (H_q), at aphelion (H_Q), in some middle part (H_m) and at the distance of 1 AU from the Sun (H_1). The results are listed in Table V which contains also the theoretical and observed number N_s of showers

TABLE V
Sizes of short-period meteoroid streams, theoretical and observed number of meteor showers

Stream	H_Q AU	H_m AU	H_q AU	H_1 AU	L AU	N_s Earth Th.	N_s Earth Obs.	N_s Theoretical Me.	N_s Theoretical Ve.	N_s Theoretical Ma.
Quadrantids	4.0	4.4	0.33	0.90	0.04	8	6	8	8	4
δ Cancrids	–	–	–	–	0.10	2	2	–	2	2
Virginids	0.52	0.41	0.03	0.15	1.03	4	3	4	4	4
δ Leonids	0.96	0.73	0.14	0.24	0.42	4	4	–	–	4
Camelopard.	0.51	0.43	0.24	0.24	0.06	4	2	–	–	4
σ Leonids	0.14	0.15	0.03	0.04	0.39	4	3	–	–	4
δ Draconids	3.40	2.64	0.31	0.68	0.21	6	4	–	8	4
μ Virginids	1.98	1.53	0.18	0.57	0.47	4	2	–	4	4
α Scorpiids	0.43	0.25	0.02	0.11	0.45	4	3	4	4	4
α Bootids	2.64	2.04	0.44	0.64	0.25	4	1	–	–	4
Φ Bootids	0.98	0.82	0.58	0.63	0.15	4	2	–	–	4
τ Herculids	1.99	1.54	0.41	0.46	0.16	4	2	–	–	4
Θ Ophiuchids	0.72	0.56	0.06	0.17	0.10	4	3	–	4	4
α Capricor.	1.09	0.86	0.14	0.30	0.16	4	4	–	4	4
ι Aquarids	1.10	0.65	0.05	0.27	0.92	4	4	4	4	4
κ Cygnids	3.81	2.95	0.45	1.12	0.62	8	3	–	8	4
Taurids	0.85	0.64	0.07	0.27	1.80	4(12)	4(12)	4	4	4
κ Aquarids	0.10	0.14	0.03	0.04	0.11	4	2	–	–	4
Ann.Androm.	0.78	0.61	0.08	0.16	0.52	4	2	–	4	4
Geminids	2.00	1.50	0.05	1.00	0.19	4	3	4	4	4
Dec.Phenicids	1.54	1.19	0.27	0.32	0.002	2	2	–	–	4
δ Arietids	0.21	0.19	0.05	0.07	0.04	4	3	–	–	4
					All:	102	72	28	62	86

produced by each stream and the stream width L along the Earth's orbit, derived from the duration of shower activity. We see that the overwhelming majority of meteoroid streams may produce 4 meteor showers each. The Quadrantids, the χ-Cygnids and probably the δ-Draconids belong to meteoroid streams which may produce 8 meteor showers (Babadzhanov 1990; Babadzhanov and Obrubov, 1987, 1989). The 22 meteoroid streams under consideration may theoretically produce 102 meteor showers. The existence of 72 is confirmed by photographic and radar observations.

According to our model of meteoroid stream evolution each stream contains particles whose orbits intersect not only the Earth's orbit but those of other planets, producing meteor showers in their atmospheres. Thus the Quadrantid meteoroid stream produce 4 meteor showers in the atmosphere of Mars, 8 meteor showers in the atmosphere of Venus and 8 meteoroid showers bombarding the surface of Mercury. The theoretical numbers of meteor showers produced by short-period meteoroid streams on Mars, Venus and Mercury are presented in Table V.

3. Relationships of meteoroid streams with comets and asteroids

The orbital elements of related showers and meteor storms and their parent bodies are given in Table VI.

The relationship of the Draconids, Andromedid and Leonid meteor storms and that of Lyrid, Perseid, Geminid, Eta-Aquarid and Orionid meteor showers with their parent bodies has been established from the similarity of their current orbits. The possible relationship between the Geminids and asteroid Phaethon, as well as the Eta-Aquarids, Orionids and comet Halley were confirmed through studies of their orbital evolution over long intervals of time. The parent bodies of the remaining meteor showers were found from investigations of the orbital evolution of their parent bodies and stream dynamics. Thus, only two parent bodies, those of the κ Cygnids, and Northern and Southern ι Aquarid meteor showers remain to be found.

4. Conclusions

The observed structure of meteor showers and the results of theoretical studies of meteoroid streams allow us to distinguish the following stages of meteoroid stream evolution:

(1) A compact cloud of meteoroids is present only in the vicinity of a parent cometary nucleus. Whenever the Earth passes through this cloud a meteor storm is observed (the Andromedids, the Draconids).

(2) Around the cometary orbit there is an irregular and less dense stream component but near a cometary nucleus a compact cloud of particles exists yet. Whenever the Earth passes through such stream a weak meteor shower is observed annually, whereas meteor storms are observed periodically (the Leonids).

(3) Meteoroids are distributed almost evenly in the stream volume. The expansion of stream volume is determined by the increase of the dispersion in the arguments of perihelia. Practically there are no changes in the observed meteor rate from year to year. The shower should have northern and southern branches and the stream should be intersected by the Earth before and after the perihelion (the Geminids, the Quadrantids).

(4) Meteoroids are distributed in the stream evenly but the spatial density of the meteoroid stream is low. The activity of twins as well as of the northern and southern shower branches is observed when the Earth passes through the stream. Further, the expansion of the stream volume is determined by the increase of dispersion in the longitudes of perihelia (Taurids).

(5) At the last stage of evolution the flux density of a meteoroid stream decreases strongly owing to the broadening of the stream and impacts of stream particles with the sporadic micrometeoroids. Meteor showers produced by the stream become indistinguishable from the sporadic background.

The duration of the first two stages is defined by the dispersion in the meteoroid

TABLE VI
Orbital elements of major meteor showers and meteor storms and their parent bodies

Related showers and their parent bodies	a AU	e	q AU	ω	Ω degrees	i	π
Quadrantids	3.1	0.683	0.977	170	283	72	93
Ursids	4.0	0.76	0.968	195	281	63	116
Northern δ Aquarids	2.6	0.97	0.07	332	139	20	111
Southern δ Aquarids	2.9	0.98	0.07	153	305	27	98
Daytime Arietids	1.6	0.94	0.09	29	77	21	106
α Cetids	1.3	0.95	0.06	202	258	20	100
P-Machholz 1986 VIII	3.0	0.958	0.127	14	94	60	108
Lyrids	28	0.968	0.919	214	32	79	246
Comet Thatcher 1861 I	56	0.983	0.921	213	31	80	245
η Aquarids	13	0.958	0.560	95	42	164	138
Orionids	15	0.962	0.571	82	28	164	110
P/Halley	18	0.967	0.587	111	57	162	168
Northern Taurids	2.6	0.861	0.359	292	230	2.4	162
Southern Taurids	1.9	0.806	0.375	113	40	5.2	153
β Taurids	2.2	0.85	0.34	246	276	6	162
ζ Perseids	1.6	0.79	0.34	59	78	0	137
P/Encke	2.2	0.847	0.339	186	334	12	160
Perseids	28	0.965	0.953	152	139	114	290
P/Swift-Tuttle 1862 III	24	0.960	0.963	153	139	114	292
Geminids	1.4	0.896	0.142	324	261	24	225
Canis Minorids	1.6	0.945	0.089	151	78	33	229
Daytime Sextantids	1.2	0.87	0.16	213	4	22	217
Phaethon (3200)	1.3	0.890	0.143	322	265	22	227
α Capricornids	2.5	0.77	0.59	269	127	7	36
χ Sagittarids	2.0	0.783	0.434	108	280	4	28
ε Aquarids	2.0	0.741	0.518	85	315	9	40
χ Capricornids	1.7	0.789	0.359	242	144	7	26
Adonis (2101)	1.9	0.764	0.448	42	351	1	33
December Phaonicids	3.0	0.68	0.98	0	74	14	73
P/Blanpein	3.0	0.699	0.892	350	79	9	69
Andromedids 1885	3.5	0.76	0.86	222	247	13	109
P/Biela 1852 III	3.5	0.756	0.861	223	247	13	110
Draconids 1933	3.5	0.717	0.996	172	196	31	8
P/Giacobini-Zinner 1946 V	3.5	0.717	0.996	172	196	31	8
Leonids 1833	12	0.915	0.985	172	234	163	47
P/Tempel-Tuttle 1965 IV	10	0.904	0.982	173	232	163	46

orbital periods due to ejections from the parent body and the action of radiation pressure. The planetary perturbations and the Poynting-Robertson effect are of a great importance at the third and fourth stages of stream evolution. The joint action of planetary perturbations, solar radiation and collisions with the sporadic micrometeroids prevail at the fifth stage of meteoroid stream evolution.

Note

After the presentation of this paper at the Kyoto Conference Mr. K. Ohtsuka, member of the Nippon Meteor Society, informed us that the Catalogue of meteor orbits from Equatorial Survey contains a few κ-Velid meteors. We thank Mr. K. Ohtsuka for his useful information. Actually, there are more than ten meteors belonging to the κ-Velids and Carinids in this Catalogue. So, all eight theoretical predicted showers of *P*/Machholz's meteoroid stream exist and manifest their activity.

Acknowledgments

We are grateful to Drs. S.V.M. Clude and V. Zappala for helpful comments. P.B.B. is thankful to the Local Organizing Committee for support without which attendance to the conference would not have been possible.

References

Babadzhanov, P.B.: 1990, in C.I. Lagerkvist, H. Rickman, B.A. Lindblad and M. Lindgren, eds., ed(s)., *Asteroids, Comets, Meteor III*, Reprocentralen HSC, Uppsala, 497–503

Babadzhanov, P.B. and Obrubov, Yu.V.: 1982, in W. Fricke and G. Teleki, eds., ed(s)., *Sun and Planetary Systems*, Dord. Reidel, 401–402

Babadzhanov, P.B. and Obrubov: 1983, in C.I. Lagerkvist and H. Rickman, eds., ed(s)., *Asteroids, Comets, Meteor III*, Uppsala Universitet, Uppsala, 411–417

Babadzhanov, P.B. and Obrubov, Yu.V.: 1986, *Doklady AN SSSR* **290**, **1**, 53–57

Babadzhanov, P.B. and Obrubov, Yu.V.: 1987, in Z. Ceplecha and P. Pecina, eds., ed(s)., *Interplanetary Matter*, Astron. Inst. of Czechoslovakia, Praha, 141–150

Babadzhanov, P.B. and Obrubov, Yu.V.: 1989, *Highlights of Astronomy* **8**, 287–293

Babadzhanov, P.B. and Obrubov, Yu.V.: 1992, in A. Harris and E. Bowell, eds., ed(s)., *Asteroids, Comets, Meteors 1991*, in press,

Babadzhanov, P.B., Obrubov, Yu.V. and Machmudov, N.: 1990, *Astron. Vestnik* **24**, **1**, 18–28

Babadzhanov, P.B., Obrubov, Yu.V. and Pushkarev, A.N.: 1991a, *Astron. Vestnik* **25**, **1**, 82–92

Babadzhanov, P.B., Obrubov, Yu.V. and Pushkarev, A.N.: 1991a, *Astron. Vestnik* **25**, **1**, 82–92

Babadzhanov, P.B., Hajduk, A., Obrubov, Yu.V. and Pushkarev, A.N.: 1991b, *Astron. Vestnik* **25**, **2**, 208–216

Cook, A.F.: 1973, in C.L. Hemenway, P.M. Millman and A.F. Cook, eds., ed(s)., *Evolutionary and Physical Properties of Meteoroids*, US GPO, Washington D.C., 183–191

Drummond, J.D.: 1981, *Icarus* **47**, 500–517

Drummond, J.D.: 1982, *Icarus* **49**, 143–153

Fox, K., Williams, I.P. and Hughes, D.W.: 1984, *Mon. Not. R. Ast. Soc.* **208**, **3**, 11P–15P

Green, D.W.E., Rickman, H., Porter, A.C. and Meech, K.J.: 1990, *Science* **247**, 1063–1067

Hasegawa, I.: 1958, *Documentation des Observateurs* **3**, 163–177

Hunt, J., Fox, K. and Williams, I.P.: 1986, in C.I. Lagerkvist, B.A. Lindblad, H. Lundstedt and H. Rickman, eds., ed(s)., *Asteroids, Comets, Meteor III*, Uppsala Universitet, Uppsala, 549–553
Jones, J. and Hawkes, R.L.: 1986, *Month. Not. R. Astron. Soc.* **223, 3**, 479–486
Imoto, S. and Hasegawa, I.: 1958, *Smithson. Contr. Astrophys.* **5**, 131–144
Kascheev, B.L., Lededinets, V.N. and Lagutin, M.F.: 1967, *Meteor Phenomena in the Earth's atmosphere*, Nauka, Moskva, 260 pp
Kramer, E.N.: 1953, *Izvestiya Astron. Obs. Odessa* **3**, 163–247
Kresak, L. and Stohl, J.: 1990, in C.I. Lagerkvist, H. Rickman, B.A. Lindblad and M. Lindgren, eds., ed(s)., *Asteroids, Comets, Meteor III*, Uppsala Universitet, Uppsala, 379–388
Lebedinets, V.N., Korpusov, V.N. and Sosnova, A.K.: 1972, *Trudy IEM* **1 (34)**, 88–171
Lidov, M.: 1961, *Iskusstvennie Sputniki Zemli* **8**, Nauka, Moskva 5–45
Lindblad, B.A.: 1990, in C.I. Lagerkvist, H. Rickman, B.A. Lindblad and M. Lindgren, eds., ed(s)., *Asteroids, Comets, Meteor III*, Uppsala Universitet, Uppsala, 551–553
McBride, N. and Hughes, D.W.: 1990, in C.I. Lagerkvist, H. Rickman, B.A. Lindblad and M. Lindgren, eds., ed(s)., *Asteroids, Comets, Meteor III*, Uppsala Universitet, Uppsala, 555–558
McIntosh, B.A.: 1990, *Icarus* **86**, 299–304
McIntosh, B.A. and Jones, J.: 1988, *Month. Not. Roy. Astron. Soc.* **235**, 673–693
Moiseev, N.D.: 1945, *Trudy Gos. Astorn. Inst. Mosk. Univer.* **15**, 75–99
Olsson-Steel, D.: 1987, *Australian J. Astron.* **2**, 21
Olsson-Steel, D.: 1988, *Australian J. Astron.* **2**, 93
Olsson-Steel, D.: 1990, in C.I. Lagerkvist, H. Rickman, B.A. Lindblad and M. Lindgren, eds., ed(s)., *Asteroids, Comets, Meteor III*, Uppsala Universitet, Uppsala, 159–162
Plavec, M.: 1955, *Bull. Astron. Inst. Czech.* **6, 1**, 20–23
Radzievsky, V.V.: 1951, *Soviet Astron. J.* **28, 5**, 363–372
Southworth, R.B. and Hawkins, G.S.: 1963, *Smith. Contr. to Astrophys.* **7**, 261–286
Sekanina, Z.: 1970, *Icarus* **13**, 475–493
Sekanina, Z.: 1976, *Icarus* **27**, 265–321
Sekanina, Z.: 1990, *Astron. J.* **99**, 1268–1278
Steel, D.I., Asher, D.J. and Clube, S.V.M.: 1991, *Mon. Not. Roy. Astr. Soc.* **251**, 632–648
Whipple, F.L.: 1951, *Astrophys. J.* **113, 3**, 464–474
Whipple, F.L.: 1983, *IAU Circ.* **388**,

HISTORICAL VARIATION IN THE METEOR FLUX AS FOUND IN CHINESE AND JAPANESE CHRONICLES

ICHIRO HASEGAWA
Otemae Junior College, Inano 2-chome, Itami, Hyogo 664, Japan

Abstract. Numbers of meteors recorded in Chinese and Japanese histories are counted. Two distinct maxima in Chinese records are found in the 11th and 15th centuries, and the latter is also recorded in Japan. Of those records, numbers of bright meteors with sound and great fireballs that appeared in the daytime are also investigated.

Correlations between the meteor numbers and the apparitions of naked-eye comet likely to be found, and seasonal variations in the meteor flux recorded during nineteen centuries show two maxima in July-August and October-November, the latter may be related to the Taurid complex.

Key words: Meteor flux, comets, taurid complex.

1. Introduction

The variations in the meteor flux would have probably been related to the evolution of the solar system, and catastrophic collisions of meteoritic bodies with the earth are considered to have important effects on the geological and biological histories on the earth (Clube and Napier, 1982, 1989).

Chinese, Korean and Japanese historical records contain the description of many celestial phenomena such as comets, meteor showers and meteorite falls. Imoto and Hasegawa (1958) and Zhuang (1966) compiled historical records of meteor showers, and Ho (1962), and Ho and Ang (1970) published catalogues of comets and novae in Chinese sources, and Hasegawa (1980) collected world-wide records of naked-eye comets. Dall'olmo (1978) compiled meteor records from the European medieval sources. Rasmussen (1990, 1991) discussed on the periodicity of meteoric events and suggested the accretion of cosmic bodies disintegrated around the earth.

Recently in China, a complete astronomical records, the *General Compilation of Chinese Ancient Astronomical Records* (Beijing Observatory 1988, for short CAR in this paper) has been published. In the present paper, the number of records of individual meteor apparition collected in CAR is reported, and the number of Japanese records of meteor and those of the world-wide naked-eye comets are also investigated for comparison. Japanese meteor records before AD 1600 are taken from Kanda's catalogue (1935), and the ones during AD 1601 and AD 1867 are quoted from Ohsaki's work (1990). Korean meteor records are not studied here.

Celestial Mechanics and Dynamical Astronomy **54:** 129–142, 1992.

TABLE I

Number of meteors recorded in Chinese histories

Month	BC	AD 1	101	201	301	401	501	601	701
		100	200	300	400	500	600	700	800
January	0	2	0	0	2	5	7	1	1
February	2	3	0	0	1	6	5	1	1
March	2	4	2	0	2	5	4	2	3
April	2	4	1	0	1	3	8	0	4
May	0	3	2	0	1	8	4	1	1
June	0	1	0	0	1	7	2	5	1
July	0	1	1	1	2	12	7	2	4
August	0	0	2	1	1	13	8	0	0
September	2	1	0	1	3	3	2	1	2
October	2	3	0	1	1	5	5	3	5
November	0	4	2	0	4	16	3	3	1
December	1	1	2	0	4	7	9	3	2
Unknown	5	1	0	1	4	1	6	2	0
Total	16	28	12	5	27	91	70	24	25

Month	801	901	1001	1021	1041	1061	1081	1101	1121
	900	1000	1020	1040	1060	1080	1100	1120	1140
January	3	2	8	3	7	16	20	5	1
February	6	5	1	10	10	20	15	3	1
March	2	4	6	3	9	23	19	3	4
April	4	7	6	6	10	23	13	2	4
May	5	10	6	9	21	24	25	5	2
June	4	6	10	11	8	27	15	2	2
July	7	11	9	7	35	74	43	6	5
August	9	5	9	5	25	43	27	4	3
September	10	9	9	4	35	55	20	3	2
October	5	9	8	26	41	79	58	3	1
November	5	5	6	8	14	60	44	2	3
December	5	4	6	5	10	16	16	3	6
Unknown	5	3	0	0	0	0	0	1	0
Total	70	80	84	97	225	460	315	42	34

Month	1141	1161	1181	1201	1221	1241	1261	1281	1301
	1160	1180	1200	1220	1240	1260	1280	1300	1320
January	0	2	0	4	1	0	1	1	0
February	1	1	0	2	3	0	1	0	0
March	0	2	1	1	2	0	2	0	0
April	0	2	0	3	3	0	1	0	0
May	1	0	1	0	3	4	2	0	0
June	0	3	0	3	1	0	3	1	0
July	2	9	1	10	8	5	6	0	1
August	0	10	2	3	4	10	2	0	0
September	1	13	0	4	3	1	1	0	0
October	1	6	4	3	3	5	3	1	0
November	0	9	0	2	5	3	0	0	1
December	0	5	0	5	2	0	1	0	0
Unknown	0	0	0	0	0	0	0	1	0
Total	6	62	9	40	38	28	23	4	2

TABLE II
Number of meteors recorded in Chinese histories

Month	1321 1340	1341 1360	1361 1380	1381 1400	1401 1420	1421 1440	1441 1460	1461 1480	1481 1500	1501 1520
January	0	1	4	9	21	47	19	11	10	8
February	0	4	6	5	20	41	25	16	12	13
March	0	0	11	15	21	66	24	20	4	18
April	0	0	11	8	26	58	22	22	4	11
May	0	0	6	8	22	52	18	14	10	10
June	1	1	5	6	19	38	19	17	8	8
July	1	1	12	12	27	83	27	16	9	18
August	1	2	12	10	45	72	14	10	8	9
September	0	0	7	9	34	69	31	13	16	14
October	1	1	10	16	61	99	51	36	18	15
November	1	2	11	9	51	91	39	13	11	9
December	0	1	7	6	26	43	27	11	8	16
Unknown	1	2	0	1	0	0	1	0	0	0
Total	6	15	102	114	373	759	317	199	118	149

Month	1521 1540	1541 1560	1561 1580	1581 1600	1601 1620	1621 1640	1641 1660	1661 1680	1681 1700	1701 1720
January	2	4	6	1	3	1	8	9	2	5
February	1	2	3	4	3	2	7	13	5	1
March	0	1	3	7	5	4	6	7	5	5
April	2	1	0	7	7	4	5	6	4	1
May	4	8	1	5	4	4	10	8	0	1
June	10	4	2	3	8	4	10	8	2	2
July	5	8	5	11	3	8	11	9	5	3
August	9	3	1	11	6	11	14	18	7	4
September	9	6	1	6	5	2	20	21	4	4
October	17	6	3	8	9	3	9	13	5	1
November	8	5	1	4	4	3	16	10	4	6
December	7	3	4	2	4	6	8	11	8	1
Unknown	0	0	0	1	0	0	0	0	0	0
Total	74	51	29	69	61	52	124	133	51	34

Month	1721 1740	1741 1760	1761 1780	1781 1800	1801 1820	1821 1840	1841 1860	1861 1880	1881 1900
January	2	7	11	9	3	3	9	7	5
February	3	3	12	4	2	5	4	7	6
March	0	4	12	1	4	8	5	7	3
April	5	1	10	7	4	5	11	7	4
May	5	7	9	16	6	1	5	13	5
June	0	9	11	9	0	6	12	9	18
July	6	10	7	22	2	8	16	17	7
August	5	8	29	37	13	14	20	29	12
September	1	6	21	16	5	11	11	10	6
October	5	3	10	15	2	1	7	12	4
November	2	3	32	21	6	5	16	21	8
December	4	3	16	9	0	5	6	9	13
Unknown	0	0	0	0	0	0	0	0	0
Total	38	64	180	166	47	72	122	148	91

TABLE III

Number of bright meteors with sound and fireballs appearing in daytime and recorded in Chinese histories

Month	BC	AD 1 100	101 200	201 300	301 400	401 500	501 600	601 700	701 800
January	0	1	0	0	0	0	2	0	0
February	0	0	0	0	0	1	2	1	0
March	0	0	1	0	1	0	1	2	2
April	1	1	0	0	1	1	0	0	1
May	0	1	0	0	1	1	1	0	0
June	0	0	0	0	0	2	1	2	1
July	0	0	0	1	0	0	1	0	0
August	0	0	0	0	0	0	0	0	0
September	1	0	0	0	3	0	0	0	0
October	0	0	0	0	1	0	1	1	1
November	0	2	0	0	3	5	0	1	0
December	0	0	1	0	3	1	2	1	0
Unknown	3	1	0	0	2	0	4	0	0
Total	5	6	2	1	15	11	15	8	5

Month	801 900	901 1000	1001 1020	1021 1040	1041 1060	1061 1080	1081 1100	1101 1120	1121 1140
January	0	0	2	0	0	0	1	1	0
February	1	1	0	0	3	1	0	0	0
March	1	2	0	1	0	0	0	0	0
April	0	2	2	0	0	0	2	0	0
May	4	2	0	2	0	0	1	0	1
June	2	3	4	0	0	0	1	1	0
July	4	1	1	1	2	3	1	0	1
August	0	1	3	0	0	0	0	0	0
September	3	2	2	0	0	0	0	0	0
October	4	2	1	0	1	0	0	0	0
November	1	1	2	1	1	0	0	1	0
December	1	2	2	0	0	1	0	1	0
Unknown	1	1	0	0	0	0	0	0	0
Total	22	20	19	5	7	5	6	4	2

Month	1141 1160	1161 1180	1181 1200	1201 1220	1221 1240	1241 1260	1261 1280	1281 1300	1301 1320
January	0	0	0	3	0	0	0	0	0
February	0	0	0	0	0	0	1	0	0
March	0	0	0	0	2	0	0	0	0
April	0	0	0	1	2	0	1	0	0
May	0	0	1	0	0	1	0	0	0
June	0	0	0	2	0	0	0	1	0
July	1	0	0	0	4	2	1	0	0
August	0	2	1	1	1	2	0	0	0
September	0	0	0	1	1	0	0	0	0
October	0	1	1	2	1	1	0	1	0
November	0	0	0	0	0	1	0	0	0
December	0	1	0	1	2	0	0	0	0
Unknown	0	0	0	0	0	0	0	1	0
Total	1	4	3	11	13	7	3	3	0

TABLE IV

Number of bright meteors with sound and fireballs appearing in daytime and recorded in Chinese histories

Month	1321	1341	1361	1381	1401	1421	1441	1461	1481	1501
	1340	1360	1380	1400	1420	1440	1460	1480	1500	1520
January	0	1	0	0	1	3	0	0	4	4
February	0	3	0	0	0	2	0	0	6	4
March	0	0	0	1	0	0	0	0	2	8
April	0	0	1	0	1	1	0	0	3	4
May	0	0	0	0	0	0	0	1	6	5
June	0	1	1	0	0	0	0	1	4	2
July	1	0	0	0	0	1	1	0	3	4
August	0	2	1	1	0	0	1	0	3	5
September	0	0	0	0	0	1	2	0	2	7
October	0	0	0	3	0	1	1	1	8	3
November	1	1	1	0	0	1	1	0	4	7
December	0	1	0	0	0	0	2	1	4	6
Unknown	0	1	0	1	0	0	1	0	0	0
Total	2	10	4	6	2	10	9	4	49	59

Month	1521	1541	1561	1581	1601	1621	1641	1661	1681	1701
	1540	1560	1580	1600	1620	1640	1660	1680	1700	1720
January	1	2	2	0	3	0	5	6	1	4
February	0	0	3	3	2	2	2	5	0	1
March	0	1	2	3	4	2	2	3	3	3
April	2	1	0	4	2	1	4	1	0	0
May	2	2	1	2	1	2	7	5	0	1
June	3	2	2	2	2	1	7	3	1	2
July	2	3	1	7	1	4	8	5	5	0
August	1	0	0	3	2	2	6	7	4	3
September	2	1	1	3	1	0	11	10	1	2
October	5	1	2	1	5	2	4	6	3	0
November	4	4	1	2	1	1	13	2	1	4
December	2	1	3	1	4	6	8	7	5	0
Unknown	0	0	0	0	0	0	0	0	0	0
Total	24	18	18	31	28	23	77	60	24	20

Month	1721	1741	1761	1781	1801	1821	1841	1861	1881
	1740	1760	1780	1800	1820	1840	1860	1880	1900
January	2	3	1	4	3	2	8	7	1
February	2	2	0	1	1	3	4	4	3
March	0	0	1	1	2	5	4	5	2
April	2	1	0	2	3	4	5	5	3
May	2	4	1	2	2	1	3	6	4
June	0	4	2	5	0	3	5	6	10
July	4	3	0	3	1	3	9	12	3
August	2	1	1	2	8	6	9	14	7
September	1	0	1	2	4	8	7	8	5
October	2	3	1	0	1	1	5	8	1
November	2	0	2	3	6	2	12	15	5
December	4	1	2	0	0	3	3	5	7
Unknown	0	0	0	0	0	0	0	0	0
Total	23	22	12	25	31	41	74	95	51

TABLE V
Number of meteors recorded in Japanese histories

Month	AD	637 800	801 900	901 1000	1001 1100	1101 1200	1201 1300	1301 1400	1401 1500
January		3	5	0	1	2	3	3	3
February		1	5	1	0	3	2	3	9
March		1	6	0	1	3	4	2	4
April		1	1	0	2	1	5	3	8
May		0	4	0	0	0	2	3	2
June		1	5	1	0	6	3	3	4
July		0	6	2	1	6	6	7	13
August		0	10	1	1	5	4	8	9
September		1	9	0	0	5	3	3	6
October		1	8	1	1	8	4	5	19
November		1	2	2	1	4	3	4	6
December		0	7	0	0	4	5	3	8
Unknown		0	0	1	0	2	0	0	2
Total		10	68	8	8	49	44	47	93

Month	AD	1501 1600	1601 1650	1651 1700	1701 1750	1751 1800	1801 1850	1851 1867
January		6	1	1	5	1	2	1
February		5	1	4	2	2	2	1
March		2	1	4	5	4	1	2
April		6	2	9	4	5	1	0
May		3	4	2	2	2	0	3
June		2	6	8	3	3	0	1
July		4	3	5	5	2	2	1
August		3	3	7	6	3	1	1
September		1	2	0	6	3	2	0
October		3	4	2	1	1	3	0
November		1	2	2	3	3	3	2
December		5	0	3	1	3	2	2
Unknown		0	0	0	0	0	0	0
Total		41	29	47	43	32	19	14

2. Records of Individual Meteors

Among several sections in CAR, there are three sections of the meteorite falls, the meteoric showers and the individual meteors. Some records of meteorite falls have got mixed with the records of meteor. So, after the strict collations, the records of meteorite fall are removed, and the records of meteor shower are also excluded from our statistics. Finally, we have 5661 records of meteors in Chinese histories

TABLE VI
Number of bright meteors with sound and fireballs appearing in daytime and recorded in Japanese histories

Month	AD	637 800	801 900	901 1000	1001 1100	1101 1200	1201 1300	1301 1400	1401 1500
January		1	0	0	1	0	2	0	0
February		0	0	0	0	1	0	0	3
March		1	1	0	0	1	0	0	1
April		0	0	0	1	1	2	0	2
May		0	0	0	0	0	0	0	0
June		0	1	1	0	0	0	1	0
July		0	3	0	1	1	2	2	2
August		0	1	0	0	0	0	1	2
September		0	1	0	0	1	0	0	2
October		0	0	0	0	0	0	1	4
November		0	2	1	0	0	1	1	1
December		0	0	0	0	1	2	0	2
Unknown		0	0	0	0	1	0	0	0
Total		2	9	2	3	7	9	6	19

Month	AD	1501 1600	1601 1650	1651 1700	1701 1750	1751 1800	1801 1850	1851 1867
January		2	1	1	0	0	0	0
February		1	0	0	1	1	0	0
March		0	0	1	0	1	0	0
April		0	0	4	2	0	0	0
May		1	1	1	0	0	0	0
June		0	0	0	1	0	0	0
July		0	0	0	1	0	1	0
August		1	0	1	0	0	0	0
September		0	0	0	1	0	1	0
October		1	0	0	1	0	0	0
November		0	0	0	0	1	0	0
December		2	0	0	0	0	0	0
Unknown		0	0	0	0	0	0	0
Total		8	2	8	7	3	2	0

during BC 651 and AD 1900.

In Tables I and II, the numbers of individual meteor record in Chinese histories are given in the monthly bin. During AD 1 and AD 1000, the period of time is taken as one hundred-year, while during AD 1001 and AD 1900, it is twenty-year.

A lot of celestial phenomena are recorded in the Sung and the Ming dynasties.

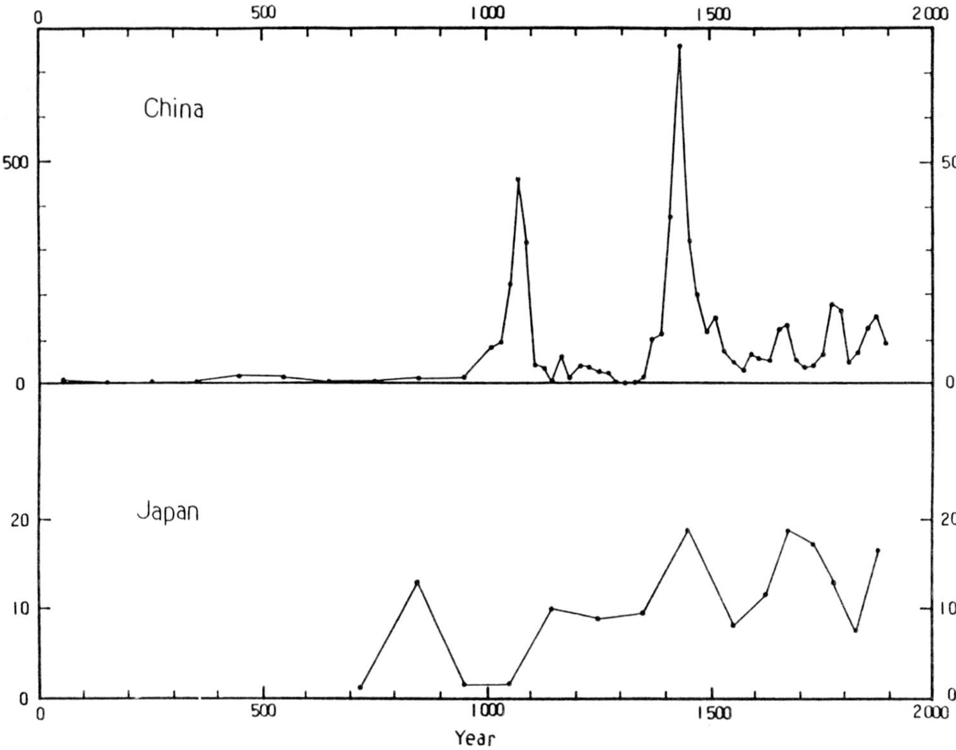

Fig. 1. The number of meteor records every twenty years, compiled from Tables I and II for Chinese, and from Table V for Japanese observations respectively.

The Sung dynasty continued from AD 960 to AD 1279, and meteor records increased in number remarkably since AD 1001, just at the beginning of the dynasty. This increase may depend on the changes in the principles or the methods of the meteor observation in the Sung dynasty. Another great increase is in the Ming dynasty, which was founded in AD 1368.

These variations in the Chinese meteor records are illustrated in the upper part of Figure 1, and show two distinct maxima around AD 1070 and AD 1430, and some other maxima are also visible around AD 1670 and AD 1770.

The Japanese records of meteor are given in Table V and in the lower part of Figure 1. It is likely obvious that some maxima of meteor apparitions are observed both in China and Japan at the same time, especially around AD 1450 and AD 1650, and a coincidence of the maximum around AD 1450 seems to be the most real. But, it is likely impossible to determine the exact scales of the meteor flux from those of historical records.

We have some other data of the bright meteors accompany with sound and the fireballs appeared in the daytime. These numbers recorded in China are included in Tables I and II, and the extracted ones are given in Tables III and IV and in the

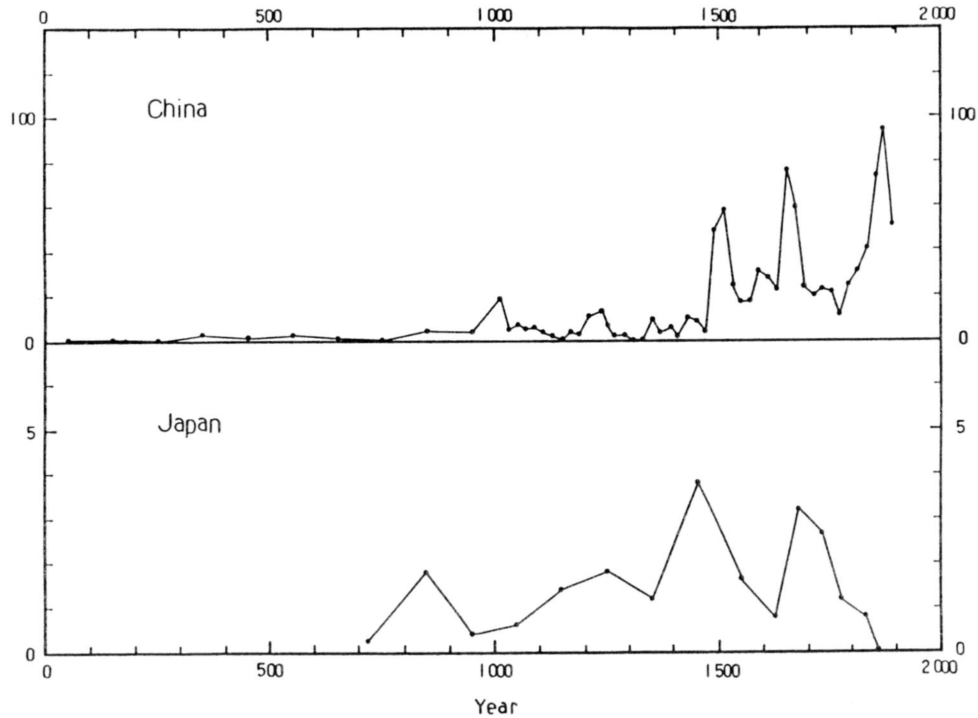

Fig. 2. The number of bright meteors with sound and fireballs appeared in the daytime every twenty years, compiled from Tables III and IV for Chinese, and from Table VI for Japanese observations respectively.

upper part of Figure 2. Japanese data of the bright meteors and the daytime fireballs are also given in Table VI and in the lower part of Figure 2.

Contrary to an ordinary meteor, a bright meteor with sound and a large fireball can be observed in the wide regions of the country, so CAR includes them which are not observed at the national observatory in the palace. Therefore, the numbers of bright meteor records given in Tables III and IV, and those of meteorite falls in Table VII, are not treated as uniform data through long years. Because the adjustment factors should be taken into account, we must consider that the discussions on these records given here anticipate further thorough studies.

Monthly variations of the records of individual meteors in Chinese histories are compiled from the data given in Tables I and II, and are illustrated in Figure 3. The conspicuous maxima in July-August and November -December are likely as same as ones observed in the recent years. It is considered that this gives another evidence as to the reality of the Chinese record of the meteors.

It is noted that the November-December maximum may be related to the Taurid complex. According to the Chinese records of more than 1500 fireballs translated by

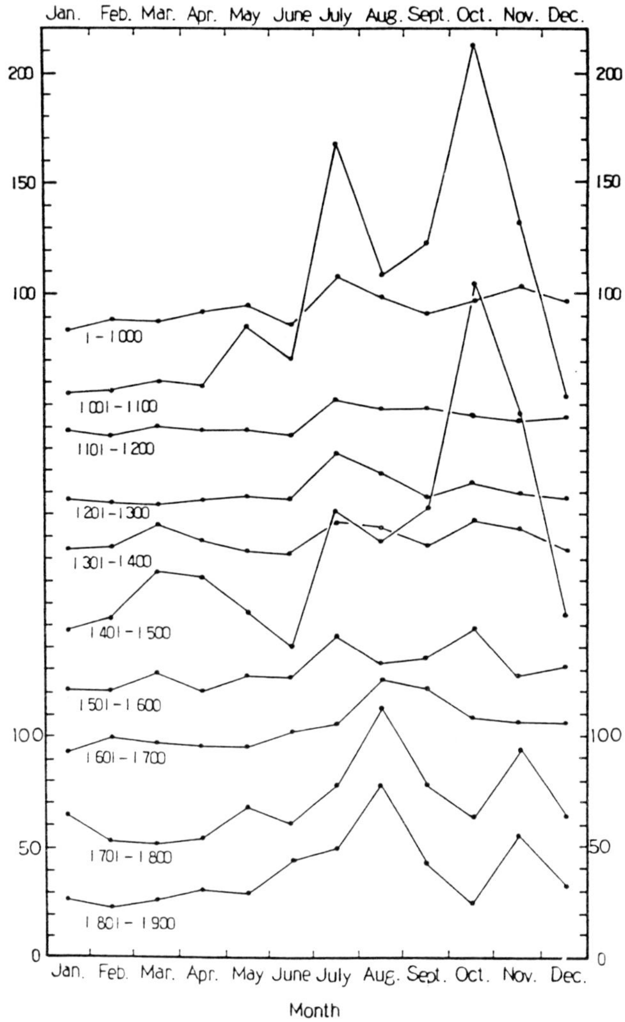

Fig. 3. The monthly variations of meteors recorded in China during AD 1-1000 and all centuries after AD 1001, compiled from Tables I and II. The zero points on the ordinate for each curve are taken arbitrary to avoid overlap.

Biot (1848), Astapovic and Terenteva (1968) derived a conclusion that the Taurids were the most active shower in the 11th century. The Chinese records of great flux of meteors during the 11th century were also discussed by Bailey *et al.* (1990), and were considered to have arisen from a fragmentation in the Taurid meteor stream. According to Tables I and II and Figure 1, there may be enhancements of meteor apparitions probably relate to the Taurids in the following periods of year;

$$401 - 500,\ 1041 - 1100,\ 1401 - 1480,\ 1641 - 1680 \text{ and } 1761 - 1800.$$

TABLE VII
Number of meteorite falls in China

Month	AD 1 100	101 200	201 300	301 400	401 500	501 600	601 700	701 800	801 900	901 1000
January	0	0	0	0	0	1	1	0	0	0
February	0	0	1	2	0	0	0	0	0	0
March	0	0	1	1	0	0	0	0	0	0
April	0	1	0	0	0	1	0	0	0	0
May	0	0	0	0	1	0	1	0	0	0
June	0	0	1	1	0	2	0	0	0	0
July	2	1	0	0	0	0	0	0	0	0
August	0	0	0	0	0	0	2	0	0	0
September	0	0	0	0	0	0	1	0	0	0
October	0	1	0	0	0	0	0	0	0	0
November	0	0	0	0	0	0	1	0	0	0
December	0	0	0	0	0	0	1	0	0	1
Unknown	0	0	1	2	1	2	2	0	1	1
Total	2	3	4	6	2	6	9	0	1	2

Month	1001 1100	1101 1200	1201 1300	1301 1400	1401 1500	1501 1600	1601 1700	1701 1800	1801 1900
January	0	0	0	0	0	1	2	2	3
February	0	0	0	1	1	2	3	1	3
March	1	1	0	0	2	2	6	2	4
April	0	0	0	0	1	4	4	2	6
May	1	0	0	1	1	6	7	4	8
June	0	0	0	0	1	5	8	3	6
July	0	0	1	2	3	3	1	2	9
August	0	2	0	1	2	3	0	2	7
September	0	0	0	0	0	3	6	3	3
October	0	0	0	1	1	10	4	0	3
November	0	0	0	1	1	3	4	2	6
December	0	0	0	2	0	1	1	1	2
Unknown	1	0	1	6	6	12	10	1	14
Total	3	3	2	15	19	55	56	25	74

It is interesting to note that the last one is corresponding to the epoch of the discovery of Comet Encke.

3. Relations between Meteorite Falls and Naked-eye Comets

In Table VII, the numbers of meteorite falls recorded in Chinese histories are given, and yearly variations of them are illustrated in Figure 4. In these records

TABLE VIII
Number of observed naked-eye comets, NC from Equation (1) and its residual

	Year	Obs.	NC	$O-C$		Year	Obs.	NC	$O-C$
BC	200–101	39	31.2	+7.8	AD	801–900	52	46.3	+5.7
BC	100–1	32	32.2	-0.2		901–1000	53	48.5	+4.5
AD	1–100	25	33.3	-8.3		1001–1100	60	50.7	+9.3
	101–200	32	34.5	-2.5		1101–1200	41	53.2	-12.2
	201–300	46	35.8	+10.2		1201–1300	52	55.7	-3.7
	301–400	29	37.2	-8.2		1301–1400	65	58.3	+6.7
	401–500	39	38.8	+0.2		1401–1500	70	61.1	+8.9
	501–600	42	40.5	+1.5		1501–1600	80	64.0	+16.0
	601–700	41	42.3	-1.3		1601–1700	54	67.0	-13.0
	701–800	30	44.2	-14.2		1701–1800	63	70.2	-7.2

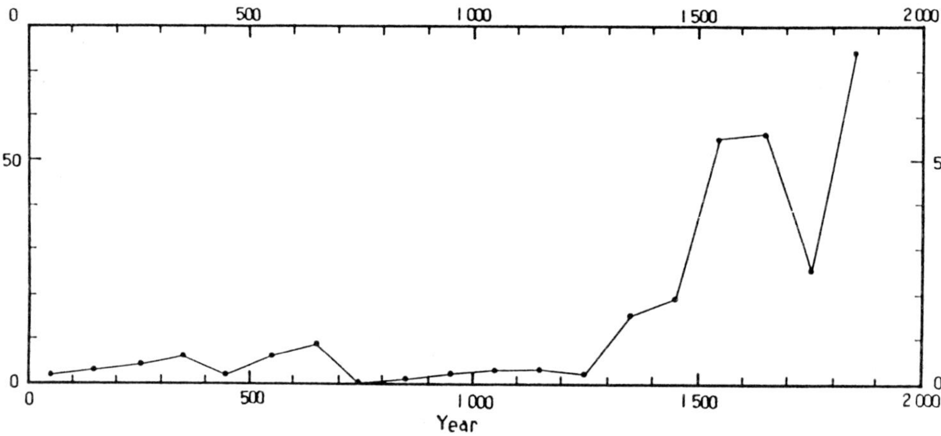

Fig. 4. The number of meteorite falls every century recorded in China, compiled from Table VII.

of meteorite fall in China, a maximum is shown during AD 1500 and AD 1700, however, we are not sure that it is real or not, and the monthly variations of meteorite falls derived from Table VII show no conspicuous features.

Hasegawa (1980) collated extensively comet records in the world, and compiled a general list of naked-eye comets. In Table VIII, the numbers of naked-eye comets every 100 years are given, and shown in Figure 5.

It is clear that there is an increasing trend in the numbers of naked-eye comets, and a least square solution for the trend gives the following expression. In the equation, C denotes an integer of AD/100 (except BC 200-101 and BC 100-1, for which case, $C = -2$ and -1 respectively), NC, the mean number of naked-eye

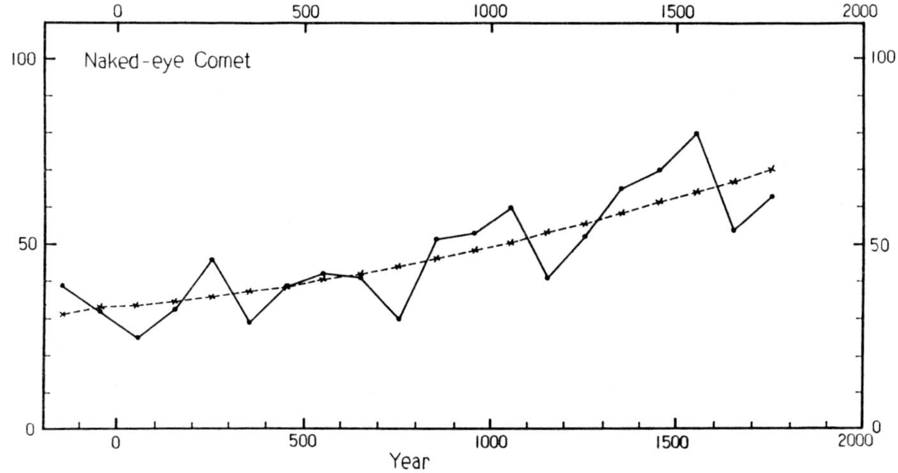

Fig. 5. The number of naked-eye comets every century. A dotted line shows a least-square solution for the secular trend computed from the Equation (1).

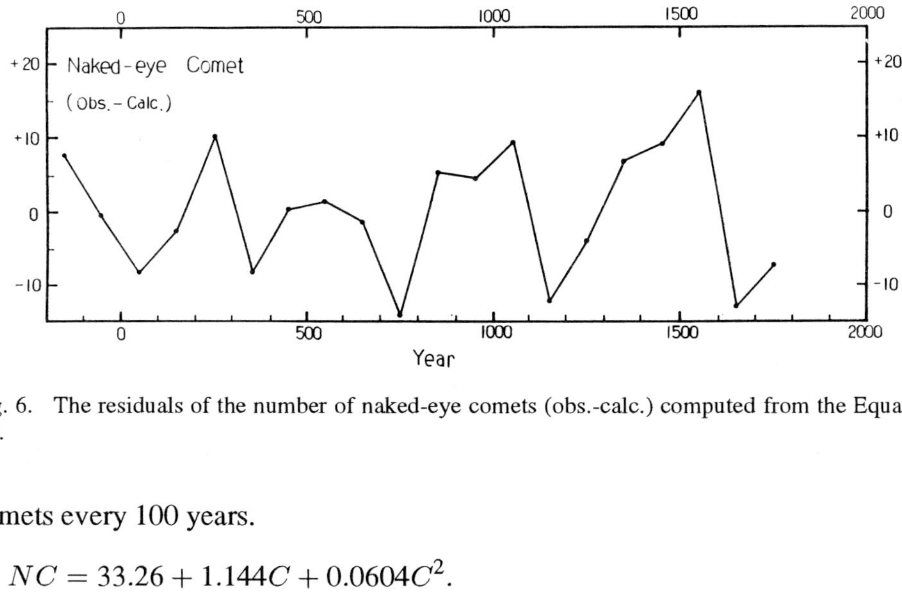

Fig. 6. The residuals of the number of naked-eye comets (obs.-calc.) computed from the Equation (1).

comets every 100 years.

$$NC = 33.26 + 1.144C + 0.0604C^2. \tag{1}$$

The residuals from the Equation (1) are also given in Table VIII and illustrated in Figure 6. It is likely that, in the numbers of naked-eye comets, the periodic variations are recognized, and the maxima exist around AD 1000, AD 1500 and so on. These maxima seem to correspond to those of the meteor records in Chinese histories given in the previous section, and both periods of variations, nearly 500 years, likely to be in good agreement with each other. Because the solar system moves only about 0.03 light years in the galactic space in 500 years, it is doubtful whether this periodicity has some meaning.

4. Conclusion

Both variations in the meteor flux found in Chinese and Japanese historical sources show some corresponding maxima during AD 1000 and AD 1900, and the Taurids are considered to be the most active in these meteor records.

The meteor flux also seems to correspond to the apparitions of bright comets, and the period of variations may be nearly 500 years.

Acknowledgements

I would like to thank Dr S.V.M. Clube for his encouragement and useful suggestions in carrying out this work.

References

Astapovic, I. S., and Terenteva, A. K.: 1968, in *Physics and Dynamics of Meteors*, I.A.U. Symp. No.33, ed. L. Kresak and P. M. Millman (Reidel, Dordrecht), p.308.
Bailey, M. E., Clube, S. V. M. and Napier, W. M.: 1990, *The Origin of Comets*, Pergamon, London.
Beijing Observatory: 1988, *General Compilation of Chinese Ancient Astronomical Records*.
Biot, E.: 1848, Mem. Acad. des Sci. de l'Inst. Nat. de France, Paris, **10**, 129.
Clube, S. V. M. and Napier, W. M.: 1982, *The Cosmic Serpent*, Farber & Farber, London.
Clube, S. V. M. and Napier, W. M.: 1989, *Cosmic Winter*, Basil Blackwell, Oxford.
Dall'olmo, U.: 1978, *J. History Astron.*, **9**, 123.
Hasagawa, I.: 1980, *Vistas Astron.*, **24**, 59.
Ho Peng Yoke: 1962, *Vistas Astron.*, **5**, 127.
Ho Peng Yoke and Ang Tian-Se: 1970, *Oriens Extremes*, **1/2**, 63.
Imoto, S. and Hasagawa, I.: 1958, *Smithson. Contrib. Astrophys.*, **2**, 131.
Kanda, S.: 1935, *Japanese Historical Records of Celestial Phenomena*, Tokyo.
Ohsaki, S.: 1990, in preparation.
Rasmussen, K. L.: 1990, *Quart. J. Roy. Astron. Soc.*, **31**, 95.
Rasmussen, K. L., 1991, *Quart. J. Roy. Astron. Soc.*, **32**, 25.
Zhuang Tian-Shan, 1966, *Acta Astron. Sinica*, **14**, 37. (English translation: 1977, *Chinese Astron.*, **1**, 197).

EPISODES OF TERRESTRIAL GEOLOGIC ACTIVITY DURING THE PAST 260 MILLION YEARS: A QUANTITATIVE APPROACH

MICHAEL R. RAMPINO
Earth Systems Group, New York University, NY 10003, U.S.A. & NASA, Goddard Institute for Space Studies, New York, NY 10025, U.S.A.

and

KEN CALDEIRA
Department of Geosciences and Earth System Science Center, The Pennsylvania State University, University Park, PA 16802, U.S.A.

Abstract. Records of major geologic events of the past ~260 Myr including: biologic extinction events, ocean-anoxic and black-shale events, major changes in sea level, major evaporite (salt) deposits, continental flood-basalt eruptions, first-order discontinuities in sea-floor spreading, and major mountain building events, have been aggregated and analyzed with moving-window and spectral techniques that facilitate recognition of clustering and possible cyclicity. Significant clustering of events suggests a model in which changes in rates and directions of sea-floor spreading ("ridge jumps") are associated with episodic rifting, volcanism, mountain building, global sea level and changes in the composition of the earth's atmosphere via the carbon cycle. Variation in atmospheric CO_2 affects global climate, ocean circulation and marine productivity. The geologic data formally show a statistically significant underlying periodicity of 26.6 Myr for the Mesozoic and Cenozoic (the exact period differs with minor changes in geologic dating). Phase information suggests that the most recent maximum of the cycle occurred within the last 9 Myr, and may be close to the present time. The quasi-regular pulses of activity might be related to internal earth processes. However, a similar periodicity in impact craters and in galactic dynamics, and a one-to-one correlation among mass extinctions, large impact events, and flood-basalt volcanism, indicate an extraterrestrial pacemaker.

Key words: Episodes, geologic activity, correlation, biologic extinctions, periodicity, extraterrestrial impacts.

1. Introduction and Data Compilation

We have searched the literature for published coherent data sets of various major geologic events of the last 260 million years. The results of our survey produced the seven independent sets of data shown in Figure 1 and listed in Table I (where the dates have been placed in 10-Myr bins). An important question in a study of this kind concerns possible subjectivity and bias in choosing the data sets for analysis. We have intentionally refrained from making major changes or "corrections" in any of the data sets in order to preserve their independence. The data are largely from compilations already in the literature prior to the recent renewed interest in periodicity in the geologic record, and were not initially collected for the purpose of time-series analysis. Note also that the data sets are a mixture of stratigraphic and radiometric age determinations. Where stratigraphic dates were involved, events were redated using the Palmer-DNAG time scale [1]. A study using continuous

Celestial Mechanics and Dynamical Astronomy **54:** 143–159, 1992.

TABLE I
Geo-Event data for the last 260 Myr (see text for sources)

	Mass Extinct.	Anoxic Events	Evapor. Depos.	Flood Basalts	Sea-Fl. Spread.	Sequen. Bound.	Orogenic Events
0–9	1.6		5		2		0.6, 2.5, 4, 5
10–19	11.2			17	10, 17	16	12.5
20–29							25
30–39	36.6		36.6	35		30, 36.6	
40–49					40		40
50–59					58	52	
60–69	66			62, 66		60.6	65
70–79					77		
80–89		84				86, 88	80, 87
90–99	91	91		92	94	97	
100–109							100
110–119	113	113		110	112	113	
120–129							
130–139				130, 135		138	
140–149	144		144		148	144	145
150–159		156				154	155
160–169		163					
170–179	176			170		173	
180–189							
190–199	193	193		190		196	
200–209		208	208	200		208	
210–219	216						
220–229						220	220
230–239						230	
240–249	245	245					
250–259			258	250			250

time series, and an intercomparison of various data sets are in preparation. We study the last 260 Myr here as the dating in this time period is best.

Dates of biologic extinctions listed here are for all 11 peaks in the per-genera extinction rate based on all 17,500 genera in the latest (1989) Sepkoski data set ([2], Figure 2) for 49 sampling intervals, with the dating here after the Palmer-DNAG time scale [1]. We chose this extinction data set because it is the most conservative and complete available, and does not contain any ranking of the extinction events according to severity.

Times of widespread stratigraphic sequence boundaries, commonly interpreted as indicating major low stands of sea level [3], were taken from the most recent detailed study by Hubbard [4] of rifted margins in the North Atlantic, South Atlantic and Arctic Oceans, and re-dated using the Palmer chronology. It is worth noting

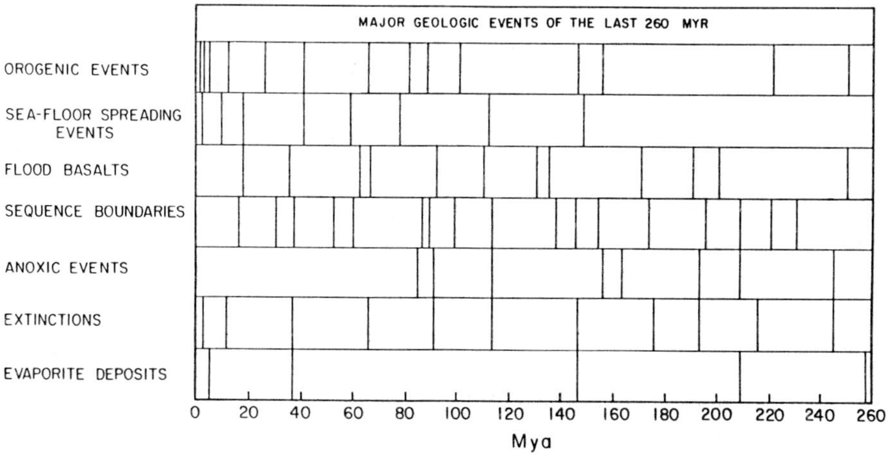

Fig. 1. Time chart of the major geological events of the last 260 million years as compiled here. See text for sources.

that despite correlations over a wide area, Hubbard disagrees with the conclusions of the Exxon group [3] and others that major sequence boundaries are globally synchronous and represent short-term lows in global sea level. For comparison, we have also analyzed the 21 "major sequence boundaries" of the global sea-level curve of Haq *et al.*, using their chronology [3].

Dates of initiation times of episodes of major continental flood-basalt volcanism were taken from Rampino and Stothers [5]. We have specifically omitted data from oceanic plateaus that may represent ocean flood basalts because of uncertainties in dating of initiation times. The data in Rampino and Stothers [5] are absolute ages based on a compilation of more than 900 radiometric and isotopic age determinations. No stratigraphic age data were used. We have added the date of an additional continental flood basalt, the Madagascar Basalts, now suspected of being of significant original volume. These basalts are intercalated with Turonian sediments (~91 Myr) [6]. The estimated initiation time of the Brito-Arctic Basalts is determined as 62 Myr from radiometric dates [5]. A later episode at ~57 Myr may be an independent pulse of North Atlantic volcanism, but has not been analyzed separately here.

The estimated times of initiation and the relatively brief duration (<1 to a few Myr) of flood-basalt volcanism episodes have been confirmed by a number of new precision $^{40}Ar/^{39}Ar$ and U/Pb age determinations, such that the initiation times of most flood basalts now have estimated errors of less than ±3 million years. Recently, the largest known flood basalt eruptions of the past 260 Myr, the Siberian Traps, have been accurately dated by the $^{40}Ar/^{39}Ar$ method at 249±3 Myr [64].

Dates of major orogenic events (mountain building events) are those listed and dated in the summary contained in the latest edition of the Geologic Time Table

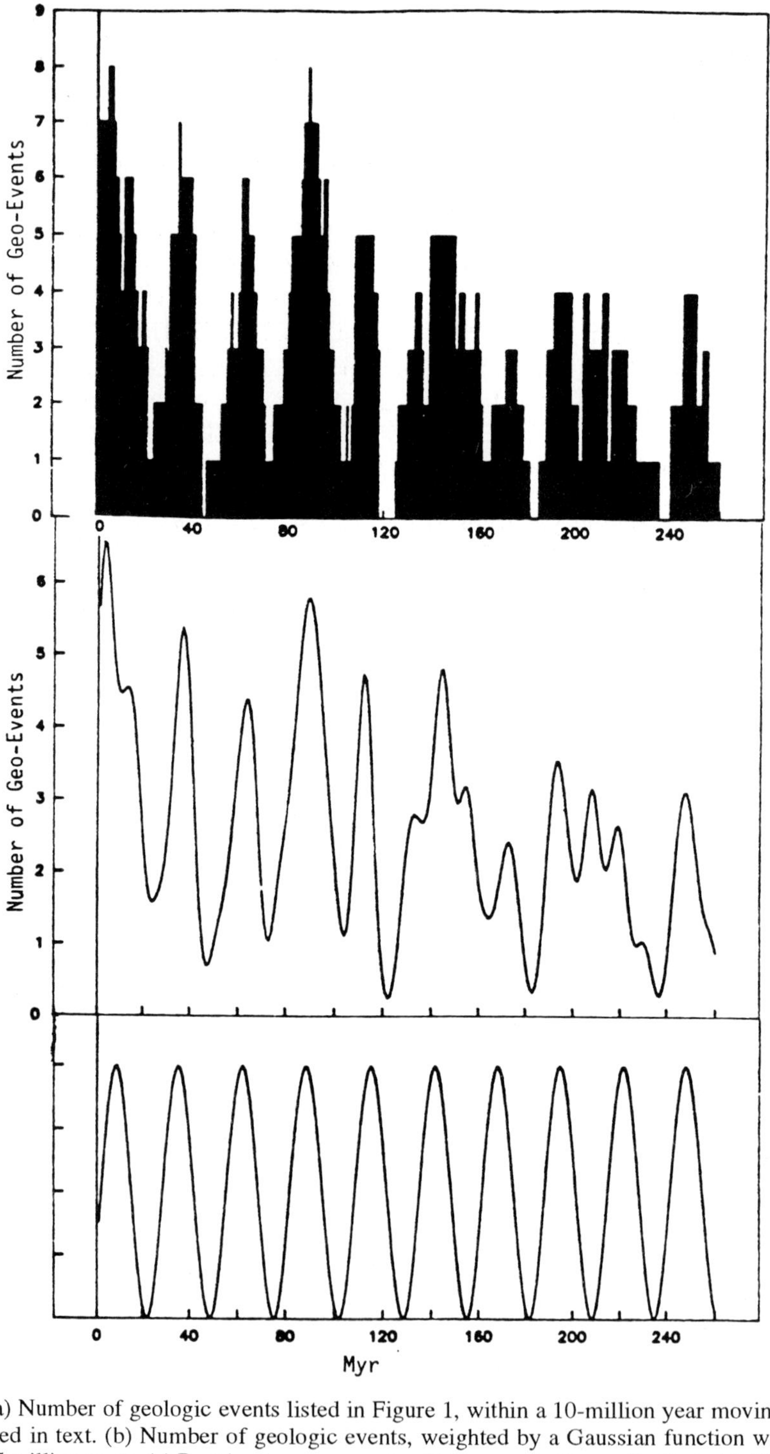

Fig. 2. (a) Number of geologic events listed in Figure 1, within a 10-million year moving window as described in text. (b) Number of geologic events, weighted by a Gaussian function with a scale length of 5 million years. (c) Best least-squares cosine fit to curve in Figure 2b. Period is 26.6 million years.

by Haq and Van Eysinga [7], and these were compared with many other sources that give similar episodes. Orogenies have been interpreted from radiometric dates and from unconformities in the geologic record. No error estimates are given in the Table itself, but most orogenic events in the past 260 Myr are dated to within a stage ($<$6 Myr). No assumption is made here that these orogenies are globally synchronous events, but the persistence of similar ages for deformation events in many parts of the world suggests some large-scale effect. Recent detailed studies support the idea that orogenies may be characterized by brief pulses of peak activity (e.g., the Nevadan Orogeny deformation is now dated narrowly at 155 ± 3 Myr) [8]. The best data set might be a global compilation of the best radiometric ages, but this does not yet exist. A recent summary of more than 6,000 age determinations for the last 100 Myr in the western U.S., however, shows peaks at about 30 to 40 Myr and 65 to 75 Myr [ref. 68], and a compilation of Mesozoic radiometric ages from Japan shows distinct peaks at $\sim$60, 90 and 120 Myr [9].

The dates of major discontinuities in sea-floor spreading, and thus plate movement, that may be global in scale, were taken from the summary paper by Schwan [10] (using data from a number of earlier papers on sea-floor spreading discontinuities from marine magnetic anomalies), with an additional recent discontinuity at 2 Myr [11]. Independent studies give essentially the same dates for widespread spreading discontinuity events, often associated with "jumps" in spreading ridge axes (along with some additional regional variations) [12,13,14]. Small random differences in dating are largely due to the rounding off of dates and/or the specific time scale used. We therefore chose not to make alterations in the original Schwan data set, except to use means of his short ranges where applicable (Table I).

Times of open ocean "anoxic" and/or widespread black-shale events in shallow seas on the platforms are from a compilation by Leary and Rampino [15] from a search of the literature, dated with the Palmer time scale. An additional anoxic event was added, possibly only regionally significant, in the Coniacian-Santonian [16], dated at 84 ± 4.5 Myr (end Santonian; Palmer time scale). For consistency, and to avoid interpolation between dated stage boundaries, the few anoxic events that occur within a stage are assigned to the next stage boundary (e.g., the Aptian/Albian); this produces only small changes in the ages. Estimated errors are those associated with the stage boundaries. Finally, epochs of major global evaporite deposition come from the compilation of Ronov [17], and are here dated after Palmer.

The raw data are shown plotted as line spectra in Figure 1. This diagram shows the manner in which the geologic data were originally assembled for the present study, as we hoped that possible correlations or cyclicity might be apparent from such a chart. Examination of the figure should dispel any concerns that the data were chosen consciously or unconsciously because of obvious correlations and clustering, or a visible display of periodicity. No obvious cyclicity is detectable. The detection of clustering of the geologic events and of the possible underlying periodicity described in the present study were accomplished through rigorous

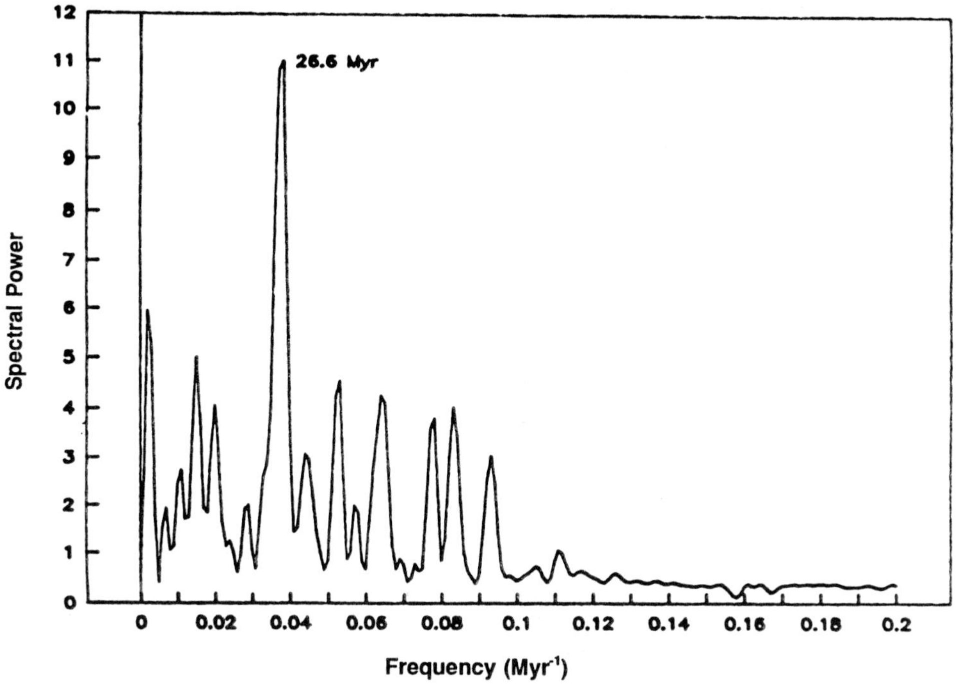

Fig. 3. Spectral power computed as described in text from the data in Table I. The peak is at a period of 26.6 million years.

standard statistical analyses as reported below.

Some authors have argued that tests for periodicity in the geologic record should wait until some unspecified time in the future when "high-quality" data bases become available [18]. Error analysis suggests, however, that rather narrow Gaussian error distributions are associated with each of the dates in Figure 1. Furthermore, analysis using the available data would seem to be the only way to determine the quality of present data in terms of providing answers to geological questions, and to establish the nature and accuracy of additional data required to characterize the geologic record. As early as 1927, Arthur Holmes assembled evidence pertaining to long-term periodicity in the geological record [19]. More than 60 years have elapsed since that study, and the results of a number of recent investigations of possible geological periodicity (see below) indicate that geologic data have improved to the point where quantitative methods can now be brought to bear in testing this hypothesis with meaningful and statistically significant results.

As a case in point, we analyzed three different sets of data related to sea-level fluctuations in the last 250 million years—sequence boundaries in the Atlantic/Arctic Ocean [4], major global sequence boundaries [3], and extreme low sea levels, from three different sources, with essentially similar results. A major improvement in dating will come in the Paleozoic, and this will eventually allow

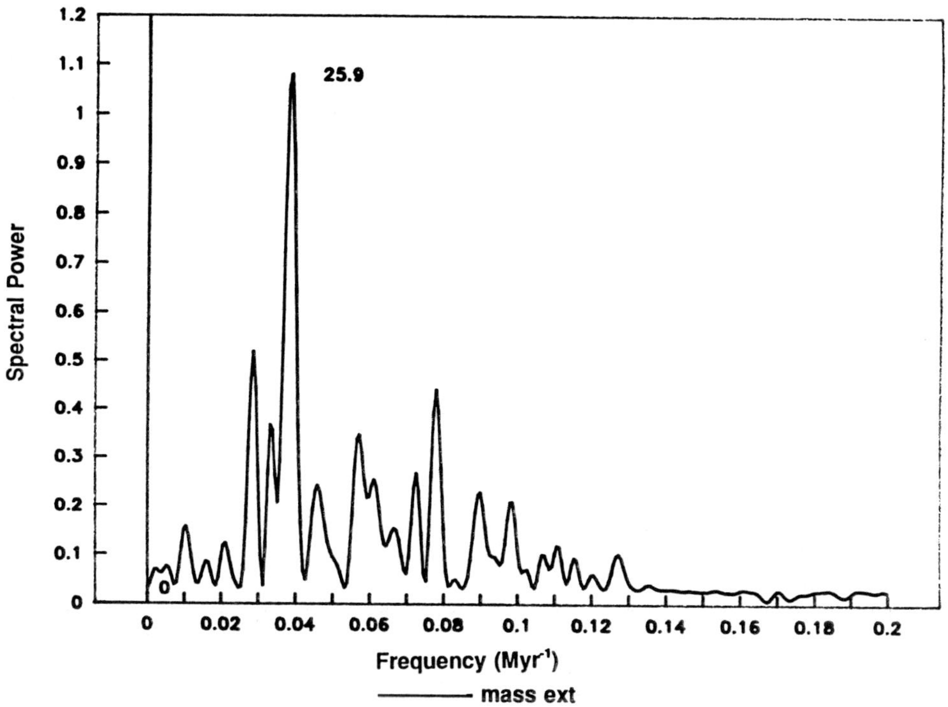

Fig. 4. Spectrum of mass extinction events.

tests of possible periodicity and clustering over a period more than twice as long.

2. Results of Statistical Analyses

Because the geologic-event data (Table I) may contain dating errors, phase lags, anti-correlations, unrelated events, and non-periodic components, simple inspection of the data is an inadequate method for identifying possible clustering or periodicity (e.g., Figure 1). We have adopted, therefore, standard quantitative methods for identifying clustering of the events listed in Table I. Initially the data were aggregated, and the number of events in the aggregated record that fall within a 10-Myr moving window, centered on the 0.5 Myr values, were computed at intervals of 1 Myr (Figure 2a). Peaks in the clustering of dates about every 25 to 30 Myr become apparent. In between the peaks, the number of geological events drops off sharply and often reaches zero. In order to test the effects of window length on the location of peaks, we used varying window lengths from 2 to 15 Myr. The peaks remained stable in position despite variations in the length of the window. An alternative method we used was to treat each event in the record as a delta function, and then to apply a Gaussian smoothing function with a scale-length of 5 Myr to the resultant function. The Gaussian filter acts to remove high frequency noise, and permits improved visual recognition of any dominant period (Figure

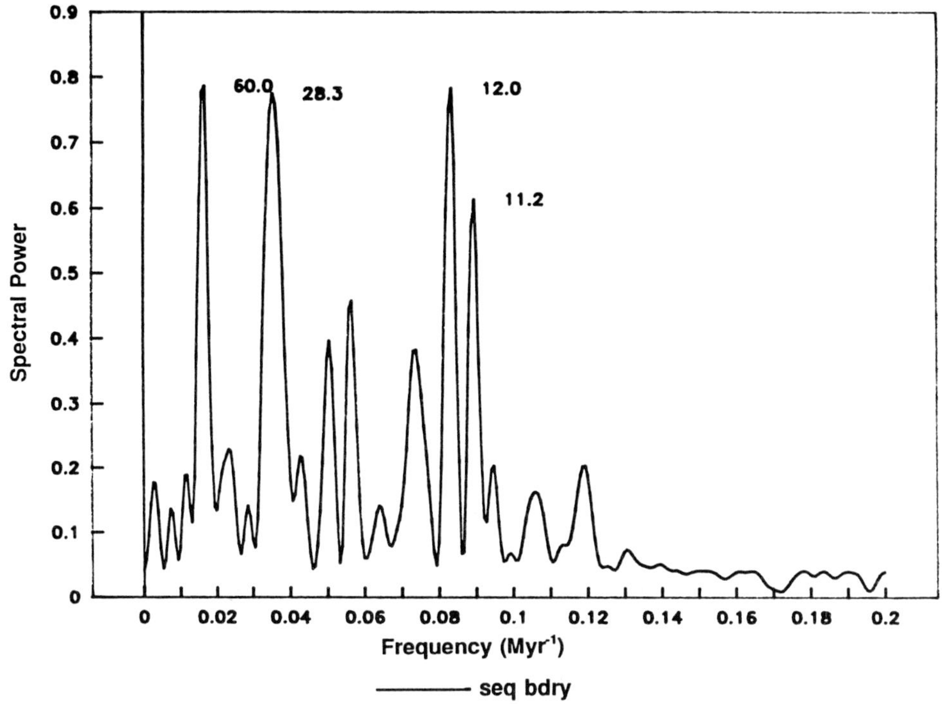

Fig. 5. Spectrum of stratigraphic sequence boundaries (sea-level indicators).

2b). A simple least-squares curve fit of a sine wave to the filtered record suggests a dominant underlying periodicity or quasi-periodicity of 26.6 Myr, with the most recent maximum of the 26.6 Myr cycle occurring 8.9 Myr (but see below) (Figure 2c).

While the moving-window and the Gaussian filter techniques facilitate visual recognition of the clustering and possible periodicity in the geologic-event data, other techniques are needed to measure the magnitude and relative significance of any underlying periodicity. In a standard time-series analysis, we computed the Fourier transform of the autocovariance function of the time-series data, rounded to the nearest million years, using a Tukey window with a bandwidth of 4.5 Myr [20]. The highest peak in the Fourier power spectrum occurs at 26.6 Myr (Figure 3). The Fourier transform of the time-series data suggests a phase with the most recent maximum occurring 8.7 Myr. Estimates of significance of this result cannot be purely statistical because the selection of the universe against which the significance is to be determined implicitly or explicitly contains assumptions about possible alternative data sets. Nonetheless, some indication as to the significance of the result can be obtained by comparing the computed Fourier transform with the Fourier transform of pseudo-data sets with similar statistical properties. To do this we computed the Fourier transform of 1,000 pseudo-data sets each containing the

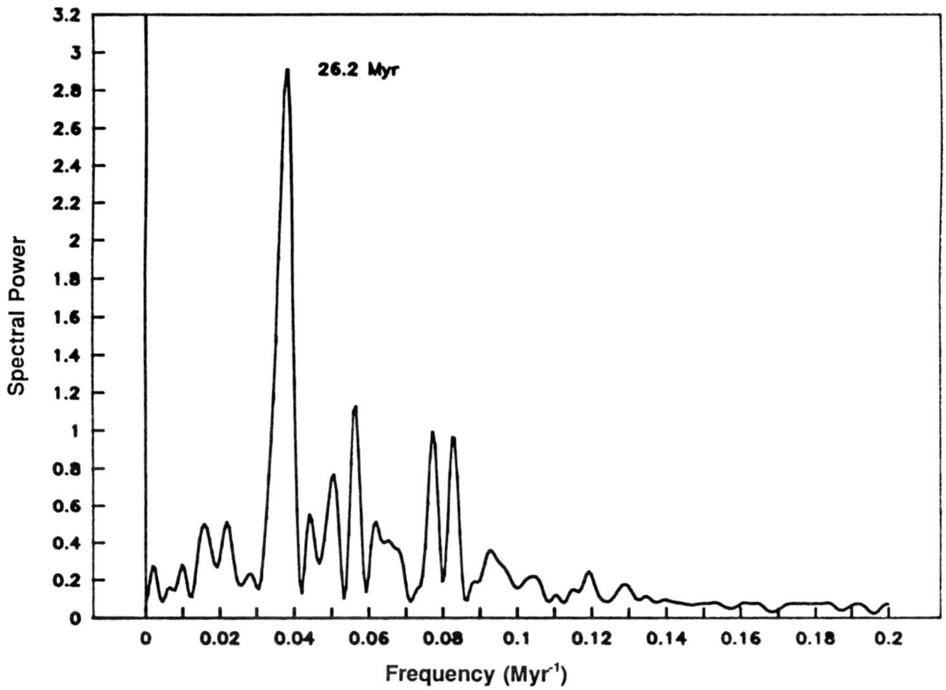

Fig. 6. Spectrum of combined extinction events and stratigraphic sequence boundaries.

same number of pseudo-randomly dated pseudo-events over the same time-interval (0 to 260 Myr).

Two questions that may be asked are: What is the probability of having a Fourier transform of a pseudo-data series with greater spectral power at 26.6 Myr than the actual data set? and: What is the probability of having a Fourier transform of a pseudo-data series with greater spectral power at some period in the range of 10 to 65 Myr, than the actual data set had at 26.6 Myr? Based on analysis of the 1,000 pseudo-data sets, we conclude that the probability of randomly generating higher spectral power at 26.6 Myr is about 0.01%, and the probability of generating higher spectral power in any period between 10 and 65 Myr is about 4.5%.

From visual inspection (Figure 2), it can be seen that the periodicity is more regular during the Cenozoic and Cretaceous Periods (the last 140 Myr) than it is during the Jurassic and Triassic Periods. This may be due in part to the significantly larger dating errors in the pre-Cretaceous record [1]. The dominant period in the Cenozoic and Cretaceous portion of the record is 26.5 Myr, with the most recent maximum at 7.9 Myr. While the relative power at the dominant period is greater for the 140 Myr record than it is for the 260 Myr record, the statistical significance is actually less due to the reduced record length. Analysis of 1,000 pseudo-data sets with 49 events distributed over 140 Myr shows that, within the period range

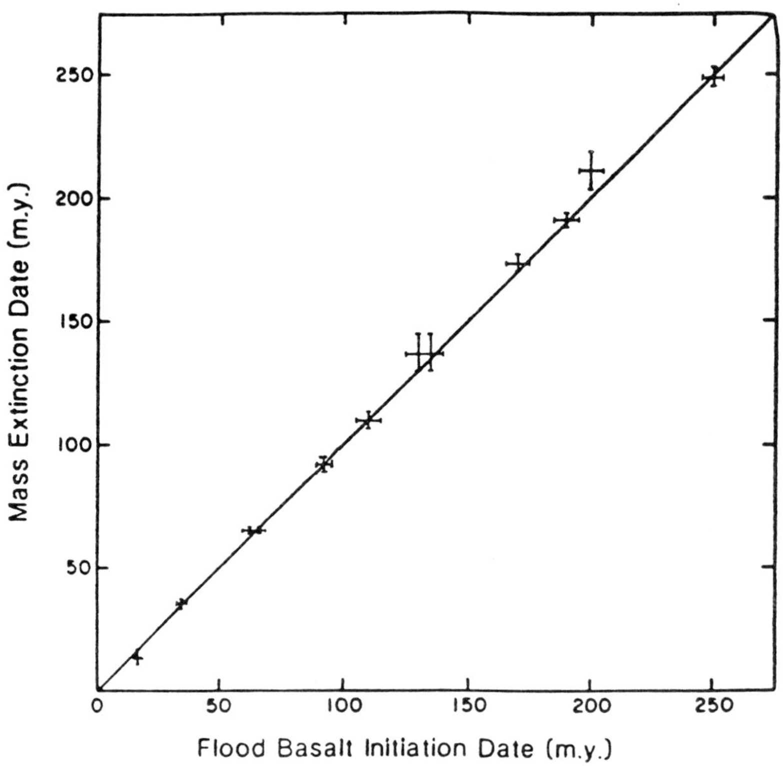

Fig. 7. Correlation between extinction events and flood-basalt eruptions during the last 250 Myr (from 53, with the addition of the Madagascar Basalts and the end-Cenomanian extinction event).

from 10 to 65 Myr, the probability of generating any peak larger than that of the actual record is about 6.3%.

We tested for robustness of the period in several ways. Initially, we re-analyzed the data while eliminating each type of geologic event from the aggregated record one at a time (Table II). When this was done, the dominant frequency remained at 26.6 ± 0.4 Myr, which emphasizes the fact that this frequency is not dependent on any one particular type of geo-event time series. It is noteworthy that the dominant period of all geologic events with the extinction events removed (26.7 Myr) is very close to the periodicity seen in the extinction events themselves (25.9 Myr) (Figure 4). Sequence boundaries (sea level) show a spectrum with strong peaks at 28.3 and 60 Myr (twice the 28.3 Myr period) (Figure 5). In the higher frequency portion of the spectrum, peaks occur at 11.2 and 12 Myr. When the extinction and sequence-boundary data sets are combined and analyzed, the result is a very strong periodicity of 26.2 Myr, whereas the other peaks become much weaker or disappear (Figure 6). When we then analyzed the remaining aggregated data sets, we found the signal at around 26 million years to be rather weak (although a strong

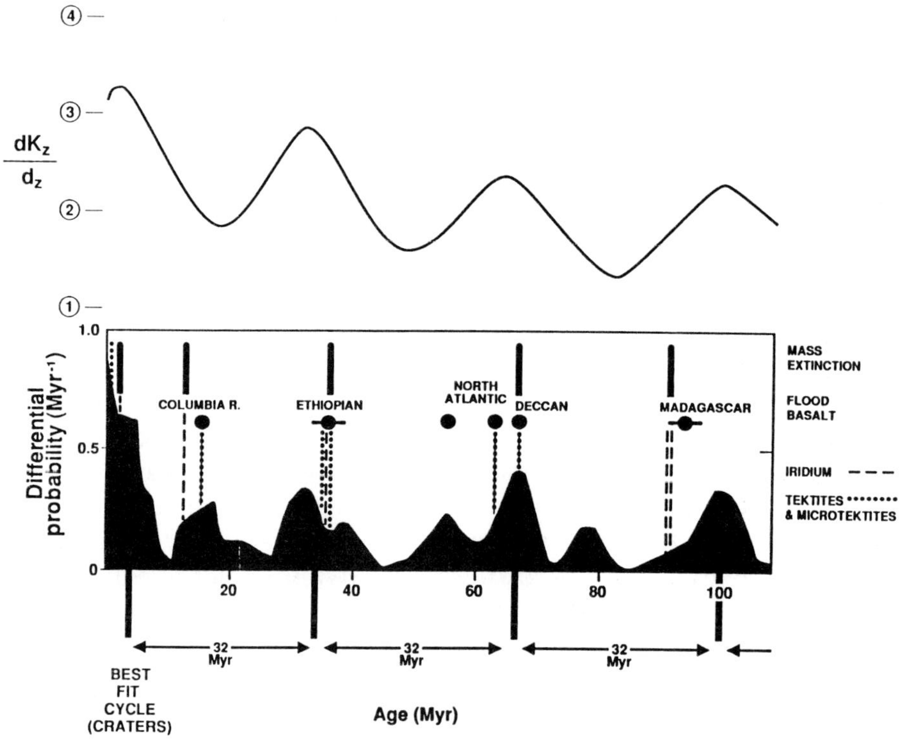

Fig. 8. Record of dated terrestrial impact craters (plotted as a probability function per Myr) of the last 110 Myr (59, with addition of Yucatán crater). Also plotted are the times of extinction events, flood basalts, iridium anomalies, and microtektite/tektite layers (see text). The variation of galactic tide (dKz/dz) at the Sun (from a proportional model, with zpresent=10 pc); shows approximate times of plane crossing, where dKz/dz is a maximum (M. Valtonen, pers. comm.).

TABLE II

Geologic events, with dominant periods (listed in the order in which they add power to the 26.6 million year spectral peak of the aggregate), phase, and spectral power at 26.6 million years. Dominant spectral peaks for each type of geologic event are listed in order of spectral power

Geologic Events	Dominant Period		Phase at 26.6 Myr	
	of Events (10^6 y)	of Aggregate minus Events	Value	Power
Extinctions	26.0	26.9	9.0	0.94
Sea Levels	28.2	26.5	9.3	0.73
Flood Basalts	23.1, 15.4, 26.2	26.7	9.5	0.53
Orogenic Events	30.6	26.2	8.4	0.24
Sea-Floor Spread.	18.4, 26.7	26.6	9.8	0.07
Ocean Anoxic Ev.	39.7, 26.1	26.9	3.6	0.37
Evaporite Deposits	28.2	26.5	11.7	0.04

peak at 52 Myr might reflect the incompleteness of the other data). This shows that most of the power at 26 Myr comes from the extinction and sequence-boundary data (Table II). These alone constitute 28 out of the 77 data points (36%) in the analysis, and are probably the most complete data sets used.

The weaker periodicity in the other data sets probably reflects, at least partially, the lesser number of events in those compilations. It is important to note, therefore, that the results of the moving window analysis shows that the other five types of geologic events tend to cluster with the extinctions/sequence boundaries over the last 260 million years, and are following a similar time history. Thus, although these five data sets together show a weaker periodicity, the fact that they track the more strongly periodic data sets suggests that they too are paced by a similar underlying cycle.

We also computed the phase for each event type at the period 26.6 Myr by computing the complex Fourier transform of the Gaussian filtered time-series data (Table II). The most recent maxima in the 26.6 Myr harmonic for extinctions, sequence boundaries, flood basalts, orogenic events, and sea-floor spreading events in the last 260 Myr fall in the range of 9.1 ± 0.7 Myr. It should be pointed out that the specific value obtained for the phase is probably significantly affected by the fact that the most recent part of the record is actually composed of two overlapping peaks—one very close to the present time, and the second at about 17 Myr (Figure 2a). These two peaks (or sub-peaks) apparently contribute to the determination of a mean phase of about 8 to 9 Myr for a 26.6 Myr period.

It has been argued that periodicity could be an artifact of problems in the identification and dating of geologic events, or of the geologic time scale itself [21, 22]. However, as Raup and Sepkoski [23] have pointed out, inaccurate dates and misplaced events should act to degrade the sample in a direction toward randomness, and away from any regular signal. The inclusion of less accurate data makes the statistical testing more conservative, and uncertainties in the data cannot explain the discovery of a significant periodicity. Raup and Sepkoski [23] estimated the effect of the structure of the geologic time scale on detection of periodicity in extinctions, and found that the Harland-1982 time scale [24] introduced only a small bias towards a best-fitting period of 26 Myr. Stothers [25] discovered a weak 28 Myr period in the Palmer time scale used here.

In order to further test possible dependence of periodicity on the particular time scale used, we repeated the time-series analysis described above after redating all possible events using the recent Harland-1989 timescale [26]. We found a statistically significant periodicity of 26.7 Myr, with the most recent maximum of this harmonic at 8.2 Myr, which represents only a very slight shift of the dominant period from the analysis using the Palmer time scale. These results argue that the strong 26 Myr underlying periodicity in the clusters of geologic events described here is not an artifact of the quality of the data or of the specific time scale used in the analysis.

The so-called "time-scale effect" discussed by Stigler and Wagner [27, 28] depends on the existence of some inherent or artifactual periodicity in the geologic time scale. For the Harland-1982 timescale, they suggest that non-linear interpolation of stage lengths in the Mesozoic (6, 6, 6, 7 Myr, repeated several times) contributes strongly to the preference for a 26-million year periodicity. They find, however, that if linear interpolation of stage lengths is used, and a few other minor changes in the time scale are made, the preference for a 26 Myr period disappears [28]. The new Harland-1989 time scale has realistic Mesozoic stage lengths varying from 0.8 to 13.9 Myr. Fourier analysis of the stage boundaries over the last 245 million years from the Harland-1989 time scale shows a high spectral peak at 30 Myr, but no strong preference for a periodicity of $\sim$26 Myr. Since we detect the same 26 Myr periodicity in geologic events when using the Harland-1989 time scale, and no peak at 30 Myr, we conclude that the period is most likely not an artifact of the time scale. We also stress, however, that since the geologic time scale is not arbitrary, but was constructed as a reflection of the timing of important geologic events (e.g., extinctions, stratigraphic sequence boundaries), detection of a periodicity in the ages of stage boundaries in the time scale could reflect a real periodicity or quasi-periodicity in the geologic record.

3. Discussion

Major peaks in the aggregated geologic-event time series (Figure 2) correspond approximately to the *Pliocene* to *Mid-Miocene* (1.6–16.6 Myr), the *Late Eocene* (36.6 Myr), the *Cretaceous/Tertiary boundary* (65 Myr), the *Cenomanian/Turonian boundary* (91 Myr), the Aptian/Albian boundary ($\sim$113 Myr), the Valanginian/Hauterivian boundary to Late Tithonian ($\sim$131 to $\sim$144 Myr), the Bajocian/Bathonian boundary ($\sim$176 Myr), the Pliensbachian/Toarcian boundary ($\sim$193 Myr), the *Late Norian* ($\sim$208 Myr), the late Carnian ($\sim$225 Myr) and the *Permian/Triassic boundary* ($\sim$245 Myr; see [65]) (with dating after Palmer time scale).

It is not surprising that these times have been recognized in previous studies as the critical boundaries in the Cenozoic/Mesozoic—they are all times of reported extinctions of marine and/or non-marine organisms [2, 29]. The italicized boundaries are associated with iridium anomalies, suggesting the possibility of concurrent impacts, and/or shocked minerals, providing unequivocal evidence of hyper-velocity impacts [30, 31].

The possibility of a $\sim$30 Myr underlying periodicity in the geologic record has a long history [32]. Only recently, however, have geological phenomena of various kinds been subjected to formal times-series analysis. Raup and Sepkoski [29] originally identified 12 extinction events at the family level over the last 250 Myr, and reported a 26.4 Myr periodicity in the extinctions, with the most recent maxima of the cycle at 10 Myr. Rampino and Stothers [33] reported a period of 30$\pm$1 Myr for the 9 most severe extinctions.

Periods of 26 to 31 Myr have been derived using different subsets of extinction events and various methods of time-series analysis [1, 29, 34-36], although the statistical significance of the derived periodicity has been a subject of debate [27, 37].

Using a linear time-series analysis technique, Rampino and Stothers [38] found a period of 33±3 Myr in the timing of 18 orogenic events of the last 600 million years after Stille [39]. Compilations of radiometric ages of kimberlite and carbonatite intrusions over similar intervals showed weaker periodicities of 35±1 and 34±5 Myr, respectively.

Major sea-floor spreading events over the last 180 Myr, and sea-level fluctuations, in the form of the extreme low sea-level stands from the "Vail Curve" over the past 200 Myr, revealed an underlying periodicity of about 33 Myr [38].

A Fourier analysis of the more recent Haq *et al.* sea-level curve [3] for the last 200 Myr also shows a strong 33 Myr cycle [40]. Our own analysis of the 21 major sequence boundaries of the Haq *et al.* curve gives a strong spectral peak at 34.5 Myr. An approximately 30 million year period was determined for flood-basalt eruptions [5]. Recently, Shackleton and Imbrie examined oxygen-isotope records of climate for the past 130 Myr and found a significant periodicity of about 30 Myr [41], supporting earlier qualitative estimates of a ~30-Myr climatic periodicity during the Phanerozoic [42, 43]. Several independent analyses of geomagnetic reversals over the last 165 Myr give a high spectral peak at about 30 Myr [44, 45, 46].

4. Conclusions and Implications

The analyses reported here provide evidence for clustering of major geological events, and additional evidence for an underlying periodicity or quasi-periodicity of about 30 Myr in the geologic record [38]. Our results suggest a model in which pulses of geological activity, which are manifested as changes in rates and directions of sea-floor spreading and major plate reorganizations, lead to fluctuations in continental rifting, volcanism, mountain building, global sea level and changes in the composition of the earth's atmosphere, especially atmospheric carbon dioxide, through perturbations of the carbon cycle [66]. In this scenario, pulses of activity affect global climate, ocean circulation and marine productivity. Veizer [47] has shown a good correlation between Phanerozoic sea-level fluctuations and $\delta^{34}S$, $\delta^{13}C$ and ^{87}Sr/^{86}Sr of seawater on a 10^7 year time scale. This suggests that ocean chemistry is also closely tied to pulses of plate tectonic activity.

The tectonic oscillation has been labelled "pulsation tectonics" [48, 49], and could be a result of internal core/mantle dynamical processes, that may affect the geomagnetic field as well. Loper and McCartney [50], and more recently, Larson [67] suggested that hot mantle plumes or super-plumes rising from the D" Layer near the core/mantle boundary could lead to pulses of geologic activity. Loper and others [51] calculated that the D" Layer could become unstable about every 22 to 30 Myr, leading to quasi-periodic plume rise. However, because of the uncertainties in lower mantle viscosity and plume-rise times, this estimate may be uncertain by more than an order of magnitude [52, 53].

Independent time-series analyses of terrestrial impact craters [33, 38, 54, 55, 56] have provided evidence of a significant 28 to 32 Myr periodicity, even under quite stringent significance criteria [56], with the most recent maximum of the cycle varying from about 2 to 10 Myr depending on the subsets of impact craters used [38, 54; but see 57]. The similarity of the impact cratering period to the dominant period detected in geological phenomena led to the renewed suggestion [38] that terrestrial tectonic/climatic processes might be modulated in some way by large-body impacts [69]. Periodic or quasi-periodic impacts could come from showers of comets [33]. Extinction events may be caused by the immediate effects of large-body impacts [58], perhaps by the cumulative effects of multiple impacts during comet showers [59].

By contrast, extinctions have traditionally been associated with "terrestrial" geological events including enhanced volcanism, orogenesis, climate change, sea-level fluctuations, ocean anoxia, and ocean-salinity variations [60]. The results of the present study, especially the moving window analysis, suggest an inter-relationship and clustering in time among major geologic processes not unexpected considering the plate tectonic paradigm. The results also argue for a common underlying cause for the pulses of activity, such as internal earth dynamics or large-body impacts, or perhaps from some interaction of the two [61]. The excellent correlation between extinctions and the times of flood-basalt eruptions (Figure 7), and the increasing number of coincidences of extinction events with evidence of extraterrestrial impacts, as noted above, argue that energetic impacts are providing a driving force, or at least a strong modulation, for terrestrial geologic processes.

The similarity between the dominant underlying periodicity found in the geologic events and the $\sim$30 Myr half-period of the Solar System's oscillation about the plane of the Milky Way Galaxy [62] is suggestive of a cosmic pacemaker related to the solar sytem's crossing of the galactic plane (Figure 8). A record of the most severe mass-extinction events of the last 800 Myr ([70], with additional data) suggests a longer cycle of $\sim$185 Myr, which may also have underlying astronomical causes. We note that Trendall [63] recently proposed a new unit of geological time (the "Holmes"), equivalent to 10^{15} seconds or about 31.7×10^6 years, which approximates the cyclic pulse detected here and elsewhere.

Acknowledgements

We thank M.A. Arthur, M. Bailey, S.V.M. Clube, A.G. Fischer, P. Leary, M. Leinen, S. Self, R.B. Stothers, S.A. Tighe, T. Volk, M. Valtonen, P.L. Ward, and S. Yabushita for discussions and preprints, and D.M. Raup and S. M. Stanley for critical reviews. Support was provided by NASA Grants NGT-5070 and NAGW-1697, and the Center for Global Habitability at Columbia University.

References

1. Sepkoski, J.J., Jr. (1989) *Jour. Geol. Soc. London* 146, 7–19.
2. Palmer, A.R. (1983) *Geology* 11, 503.
3. Haq. B.U., Hardenbol, J. & Vail, P.R. (1987) *Science* 235, 1156–1167.
4. Hubbard, R.J. (1988) *American Assoc. Petrol. Geol. Bull.* 72, 49–72.
5. Rampino, M.R. & Stothers, R.B. (1988) *Science* 241, 663–668.
6. Mahoney, J., Nicollet, C., Dupuy, C. (1991) *Earth Planet. Sci. Lett.* 104, 350–363.
7. Haq, B.U. & Van Eysinga, F.W.B. (1987) *Geologic Time Table* (Elsevier, Amsterdam).
8. Schweikert, R.A., Bogen, N.L., Girty, G.H., Hanson, R.E. & Merguerian, C. (1984) *Geol. Soc. Amer. Bull.* 95, 967–979.
9. Matsumoto, T. (1977) *Can. Jour. Earth Sci.* 14, 551–561.
10. Schwan, W. (1980) *American Assoc. Petrol. Geol. Bull.* 64, 359–373.
11. Pollitz, F.F. (1986) *Nature* 320, 738–741.
12. Rona, P. A. & Richardson, E.S. (1978) *Earth and Planetary Sci. Lett.* 40, 1–11.
13. Rich, J.E., Johnson, G.L., Jones, J.E. & Campsie, J. (1986) *Paleoceanography* 1, 85–95.
14. Engebretson, D., Cox, A. & Gordon, R. (1985) *Geol. Soc. America Spec. Paper* 206, 1–59.
15. Leary, P.N. & Rampino, M.R. (1990) in *Extinction Events in Earth History* (eds. Kauffman, E.G. & Walliser, O.H.) 45–55 (Springer-Verlag, Berlin).
16. Arthur, M.A., Dean W.E. & Stow, D.A.V. (1984) in *Geol. Soc. Lond. Spec. Pub. 15* (eds. Stow, D.A.V. & Piper, D.J.W.) 527–560.
17. Ronov, A.B. (1983) *The Earth's Sedimentary Shell* (American Geological Institute, Falls Church VA).
18. Baksi, A.K. (1990) *Geology* 18, 983–986.
19. Holmes, A. (1927) *The Age of the Earth : An Introduction to Geological Ideas* (Benn, London).
20. Jenkins, G.M. & Watts, D.G. (1968) *Spectral Analysis and its Applications* (Holden-Day, San Francisco).
21. Hoffman, A. (1985) *Nature* 315, 659–662.
22. Hoffman, A. & Ghiold, J. (1984) *Geol. Mag.* 12, 1–4.
23. Raup, D.M. & Sepkoski, J.J., Jr. (1986) *Science* 231, 833–836.
24. Harland, W.B., Cox, A.V., Llewellyn, P.G., Pickton, C.A.G., Smith, A.G. & Walters, R. (1982) *A Geologic Time Scale* (Cambridge Univ. Press, Cambridge).
25. Stothers, R.B. (1989) *Geophys. Res. Lett.* 16, 119–122.
26. Harland, W.B., Armstrong, R.L., Cox, A.V., Craig, L.E., Smith, A.G. & Smith, D.G. (1989) *A Geologic Time Scale* (Cambridge Univ. Press, Cambridge).
27. Stigler, S.M. & Wagner, M.J. (1987) *Science* 238, 940–945.
28. Stigler, S.M. & Wagner, M.J. (1988) *Science* 241, 96–99.
29. Raup, D.M. & Sepkoski, J.J., Jr. (1984) *Proc. Nat. Acad. Sci. USA* 81, 801–805 (1984).
30. Orth, C.J. (1990) in *Mass Extinctions, Processes and Evidence* (ed. Donovan, S.K.) (Columbia, New York).
31. Kerr, R.A. (1991) *Science* 251, 161.
32. Rampino, M.R. & Stothers, R.B. (1986) in *The Galaxy and the Solar System* (eds. Smoluchowski, R., Bahcall, J.N. & Matthews, M.S.) 241–259 (Univ. of Arizona, Tucson).
33. Rampino, M.R. & Stothers, R.B. (1984) *Nature* 308, 709–712.
34. Thomson, K.S. (1976) *Nature* 261, 578–580.
35. Connor, E.F. (1986) in *Patterns and Processes in the History of Life* (eds., Raup, D.M. & Jablonski, D.) 119–147 (Springer-Verlag, Berlin).
36. Raup, D.M. & Sepkoski, J.J., Jr. (1988) *Science* 241, 94–96.
37. Hoffman, A. (1989) *Jour. Geol. Soc. Lond.* 146, 21–35.
38. Rampino, M.R. & Stothers, R.B. (1984) *Science* 226, 1427–1431.
39. Stille, H. (1944) *Abh. Preuss.-Akad. Wiss., Math.-Nat. Kl.* 3, 1–80.
40. Negi, J.G., Tiwari, R.K. & Rao, K.N.N. (1990) *Terra Nova* 2, 138–141.
41. Shackleton, N.J. & Imbrie, J. (1990) *Clim. Change* 16, 217–230.
42. Dorman, F.H. (1968) *Jour. Geol.* 76, 297–313.
43. Fischer, A.G. & Arthur, M.A. (1977) *Soc. Econ. Paleont. and Mineral. Special Pub.* 25, 19–50 (1977).

44. Negi, J.G. & Tiwari, R.K. (1983) *Geophys. Res. Lett.* 10, 713–716.
45. Raup, D.M. (1985) *Nature* 314, 341–343.
46. Stothers, R.B. (1986) *Nature* 322, 444–446.
47. Veizer, J. (1985) *Geophys. Monograph* 32, 595–601 (Amer. Geophys. Union, Washington).
48. Sheridan, R.E. (1983) *Tectonophysics* 94, 169–185.
49. Sheridan, R.E. (1987) *Paleoceanog.* 2, 97–118.
50. Loper, D.E. & McCartney, K. (1986) *Geophys. Res. Lett.* 13, 1525–1528.
51. Loper, D.E., McCartney, K. & Buzyna, G. (1988) *Jour. Geol.* 96, 1–15.
52. Courtillot, V. & Besse, J. (1987) *Science* 237, 1140–1147.
53. Stothers, R.B. & Rampino, M.R. (1990) *Geol. Soc. America Spec. Paper* 247, 9–18.
54. Alvarez, W. & Muller, R.A. (1984) *Nature* 308, 718–720.
55. Stothers, R.B. (1988) *Observatory* 108, 1–9.
56. Yabushita, S. (1991) *Mon. Not. Royal Astron. Soc.* 250, 481–485.
57. Grieve, R.B., Sharpton, V.L., Rupert, J.D. & Goodacre, A.K. (1987) *Proc. 18th Lunar Planet. Sci. Conf.*, 375–382.
58. Alvarez, L.W., Alvarez, W., Asaro, F. & Michel, H.V. (1980) *Science* 208, 1095–1108.
59. Hut, P., Alvarez, W., Elder, W.P., Hanson, T., Kauffman, E.G., Keller, G., Shoemaker, E.M. & Weissman, P. (1987) *Nature* 329, 118–126.
60. Donovan, S. K. ed. (1989) *Mass Extinctions, Processes and Evidence* (Columbia, New York).
61. Shaw, H.R. (1987) *Eos* 68, 1651–1665.
62. Bahcall, J.N. & Bahcall, S. (1985) *Nature* 316, 706–708.
63. Trendall, A.F. (1991) *Geology* 19, 195.
64. Renne, P.R. & Basu, A.R. (1991)*Science* 253, 176–179.
65. Holser, W.T. & Schonlaub, H.P., eds., (1991) *The Permian-Triassic Boundary in the Carnic Alps of Austria (Gartnerkofel Region)*, Abhand. der Geologisch. Bundesanstalt, Band 45, 1–232.
66. Caldeira, K. & Rampino, M.R. (1991) *Geophys. Res. Lett.* 18, 987–990.
67. Larson, R. (1991) *Geology* 19, 547–550.
68. Ward, P.L. (1991) *Jour. Geophys. Res.* 96, 12,479–12,496.
69. Clube, S.V.M. & Napier, W.M. (1982) *Earth Planet. Sci. Lett.* 57, 251–262.
70. Sepkoski, J.J., Jr. (1986) in Raup, D.M. & Jablonski, D., eds., *Pattern and Process in the History of Life* (Springer-Verlag, Berlin), 277–295.

PERIODICITY IN THE CRATER FORMATION RATE AND IMPLICATIONS FOR ASTRONOMICAL MODELING

S. YABUSHITA
Department of Applied Mathematics & Physics, Kyoto University, Kyoto 606, Japan

Abstract. Claim for periodicity in the crater formation rate is reinvestigated using a criterion proposed by Broadbent, and data sets of Rampino and Stothers and of Grieve are shown to satisfy the periodicity criterion ($P \simeq 30$ Myr).

On the other hand, currently observed impactors are mainly asteroids, while long and short periodic comets whose fluxes may vary by external disturbances occupy only a small fraction. Using a Monte Carlo simulation, constraints are obtained for the dispersion σ(Myr) from an exact periodicity and for the periodic components (F_{lp}) in the signals for their periodicity to be detected. It is found that for $\sigma = 5, 6$ and 7 Myr, F_{lp} would have to be 40% or greater, 60% or greater and 80% or greater, respectively. These constraints are used to discuss whether the giant molecular cloud perturbations can give rise to the periodicity in the impact events. The amplitude of the solar Z-motion need to be some 100pc for $\sigma = 6$ Myr, which requires the periodic component (SP and LP comets, if the former originate from the latter) to be 60%, while for $\sigma = 7$ Myr, the periodic component need to be 80%. The GMC perturbation model consistent with the periodicity appears to be the one where the amplitude is 100pc and the periodic component $\simeq 60\%$ of the impactors. If SP comets mainly originate from a source such as the hypothetical Kuiper belt, the GMC perturbation would not be consistent with the periodicity.

Key words: Crater formation, asteroid, comet, giant molecular cloud.

1. Introduction

Whether there is periodicity in the rate of crater formation on planetary surfaces is an important problem in that it may drastically alter our understanding of the earth evolution and give clue to the immediate environment. Seyfert and Sarkin (1979) noted several periods of strong tectonic activities which appear to be triggered by impacts of large bodies. Rampino and Stothers (1986) argued that the rate of crater formation exhibits a period of 30 Myr and that the period is due to the sun's crossing of the galactic mid-plane, when large-molecular clouds perturb the Oort cloud of comets, thus increasing the flux of potential impactors. Accepting the 30 Myr periodicity, Clube and Napier (1982, 1986) favor the capture of comets as the solar system passes through spiral arms or large molecular clouds which are concentrated near the galactic mid-plane. Recently, the present author (Yabushita 1991) adopted a criterion proposed by Broadbent (1955, 1956) to investigate whether the claimed periodicity is statistically significant. It has been found that the data sets of crater ages given by Rampino and Stothers (1986) and by Grieve (1987) satisfy the Broadbent criterion for periodicity at $P \simeq 30$ Myr. The criterion, which will be explained in Section 2 is more stringent than the ones adopted earlier and the fact that some of the data sets of crater ages support the periodicity hypothesis

Celestial Mechanics and Dynamical Astronomy **54:** 161–178, 1992.

ought to be taken seriously. Although some authors (Heissler and Tremaine, 1986; Wolfendale, 1989) question the periodicity, the periodicity of crater ages appears to be established.

On the other hand, it has been argued that only a small fraction of impactors are long period comets. Weissman (1988) showed that some 12 % of the impactor flux of large craters is long-periodic or short-periodic comets, while Olsson-Steel (1987) derived 6% as the contribution from LP comets. The rest are due to short-periodic comets (12%) and asteroids (82%). Since only the flux of LP comets can be directly modulated by external disturbances such as the molecular cloud or the galactic tide, a problem naturally arises as to whether a periodic variation in the LP flux can give rise to the detected periodicity in the crater formation rate, unless one introduces a hypothetical planet X (Matese and Whitmire, 1986) or an unseen solar companion, Nemesis. It is the purpose of the present paper to present a method aimed at investigating the problem here stated and to discuss whether molecular cloud/spiral arm perturbations of the LP comet flux can account for the periodicity in the crater age distribution.

In the following, an account is given of the adopted criterion to test the periodicity hypothesis, and results will be reviewed as to whether the hypothesis can be accepted. We will then describe how the criterion can be used to give constraints on the component in the impactors which are periodic in time. In so doing, the short-periodic comets play important roles and problems relating to their capture from the flux of long-periodic comets are discussed. Then the present status of the periodicity hypothesis in view of our knowledge with galactic environment is summarized. Finally, it is concluded that if a majority of asteroids are short-periodic comets in disguise, our knowledge of the galactic environment appears consistent with the detected periodicity. If, however, asteroids and SP comets do not originate from the known source of LP comets, it seems difficult to associate the periodicity with the molecular cloud/spiral arm perturbations.

2. Testing the Periodicity

Before explaining the method whereby investigation is made regarding whether periodic variations in the LP flux can give rise to the detected periodicity in the crater formation rate, it seems appropriate to give an account on how one can regard a derived periodicity statistically significant or otherwise, and to present the results so far obtained.

Let $t_1, t_2, \cdots t_N$ be the ages of N craters. If there were an exact periodicity with period P, there should be a relation of the form

$$t_i = \alpha + n_i P \qquad i = 1, 2, \cdots N \tag{2.1}$$

where n_i is an integer $(0, 1, 2, \cdots)$; α corresponds to the age of the latest crater. In general, such an exact periodicity would not be expected, because there would be errors in age determination and disturbance which displace the times of crater formation from exactly periodic dates. The problem then is to find a criterion whereby one can judge whether the displacements from an exact periodicity are sufficiently large to make the hypothesis of periodicity unacceptable or not. The displacement from an exact periodicity can be measured by the quantity s/P ;

$$\frac{s^2}{P^2} = \frac{1}{N}\sum_{i=1}^{N} q_i^2 \ , \ q_i = t_i - (\alpha + n_i P) \tag{2.2}$$

where the integers $n_i(= 0, 1, 2, \cdots)$ are so chosen that $\mid q_i \mid < P/2$ for a given set of values of α and P. The quantity s^2/P^2 varies by changing the adopted values of α and P. By finding a set of values of α and P which give minimum values of s/P, a period P and the epoch α are obtained. This method is equivalent to the one proposed by Stothers (1979). Broadbent (1955, 56) showed that for the detected period to be statistically significant, the following inequality should hold true ;

$$\sqrt{N}\left(\frac{1}{3} - \frac{s^2}{d^2}\right) > 1 \ , \ d = P/2 \, . \tag{2.3}$$

For uniform random distribution of t_1, t_2, $\cdots t_N$, this inequality is satisfied once in more than 1000 trials. This requirement seems more stringent than ordinary criteria adopted in statistics, but since the period P is derived from the data, and not given *a priori*, such a stringent criterion ought to be adopted. The criterion was used, for instance, to investigate whether energy levels of atomic nuclei are quantized (multiples of a basic unit).

Let P and s_m be the derived period and the displacement which is minimum for the given set of data t_1, t_2, $\cdots t_N$. Of the impactors, let F_{lp} be the fraction of LP comets which are assumed to be periodic with period P. Then, $NF_{lp} \ (= N_{lp})$ craters will be due to LP comets while the rest will be due to asteroids. For the sake of simplicity we do not make distinction between short-periodic comets and asteroids (The term asteroid is used to designate the component of the impactors which are uniformly random in time).

Generate a set of random numbers T_1, T_2, $\cdots T_{N-N_{lp}}$ and add another set of numbers

$$T_i = \alpha + n_i P + \xi_i \, , \, i = N - N_{lp} + 1 \, , \ N - N_{lp} + 2, \cdots N, \tag{2.4}$$

where ξ_i is a random number with mean zero ($\overline{\xi} = 0$). This random number allows for the situation that the LP component of the impacts does not take place with exact periodicity. We will then have a set of values T_1, T_2, $\cdots T_N$. This is a set of crater ages which one would expect to obtain if the fraction F_{lp} of the N craters were due to periodic showers of LP comets, and the rest due to the impactor component which is uniformly random in time. Calculate the minimum value of the quantity s^2/P^2 for the set T_1, T_2, $\cdots T_N$. We designate the minimum value by $\overline{s}_m$. $\overline{s}_m$ may be greater than the value specified by equating the both sides of the inequality (2.3). We denote the critical value by s_c. In this case, it will be natural to conclude that the artificially generated set $T_1, T_2, T_3, \cdots T_N$ is less periodic than that required by the hypothesis. In the other case (namely $\overline{s}_m < s_c$), one may conclude that the artificially generated series may be regarded periodic. Since the quantity $\overline{s}_m$ depends on the fraction F_{lp} which is the periodic component in the series $T_1, T_2, T_3, \cdots T_N$, it would be possible to calculate a critical value of F_{lp} such that $\overline{s}_m$ equals s_c. In this way, it should be possible to decide whether a periodic variation of the observed fraction of the long-period comets can account for the detected periodicity in the crater formation rate. Note that s/P can be made as small as one pleases by taking small values of P, so that a minimum value of P ought to be specified. It should not be smaller than, say, the average spacing of the ages.

There is some arbitrariness in how to generate the random numbers $\xi's$ in Equation (2.4) ; especially as to the standard-deviation. We will assume that $\xi's$ obey a Gaussian distribution with zero mean and standard deviation σ (expressed in million years).

3. Periodicities Obtained Earlier

Before carrying out calculations regarding the constraint on the part which varies periodically, it seems worthwhile to review periodicities detected from various data sets of crater ages.

Alvarez and Muller (1984) claimed to have detected a period of 26 million years in the age of 11 craters and their work stimulated other workers to investigate the periodicity or otherwise in the crater ages. The present author (Yabushita, 1991) found that the data set does not satisfy the criterion proposes by Broadbent. There is also a data set presented by Shoemaker and Wolfe (1986), but it does not satisfy the Broadbent criterion. The data set of 65 craters listed by Rampino and Stothers (1986) satisfies the criterion. For the data set, $s/P = 0.222$ (Yabushita, 1991). For the data set of Grieve (1987), the following groups have been tested to see if they satisfy the criterion (Table I).

In Figure 1 is plotted s/P for craters with $D < 10$ km.

The epoch of the latest bombardment, α lies between 1 and 5 Myr before present for all of the significant period. This indicates that the solar system is probably in a comet shower, or has experienced it in a recent past.

To summarize the results so far given, some data sets of crater ages which are claimed to be periodic do not satisfy the Broadbent criterion, but there are data sets which do. Admittedly, the Broadbent criterion is not necessarily the one to be adopted in testing a hypothesis of periodicity. But it is one of not many criteria which are simple and natural. Therefore, it seems appropriate to use it in testing the periodicity hypothesis. It is interesting to note that small craters (diameter< 10 km) better satisfy the adopted criterion. When some young craters are removed, a periodicity at 16.5 Myr has been detected. This may have significance in the work

TABLE I
Data sets of crater ages

Data sets	No.of craters	Period (Myr)	s/P	Broadbent criterion
Rampino and Stothers	65	~~32~~ 31	0.222	satisfied.
Grieve				
age $\leq$ 600 Myr	98	30	0.223	satisfied.
age $\leq$ 600 Myr, $D \geq$ 10 km	41	31	0.26	not satisfied.
age $\leq$ 600 Myr, $D <$ 10 km	57	30	0.206	satisfied.

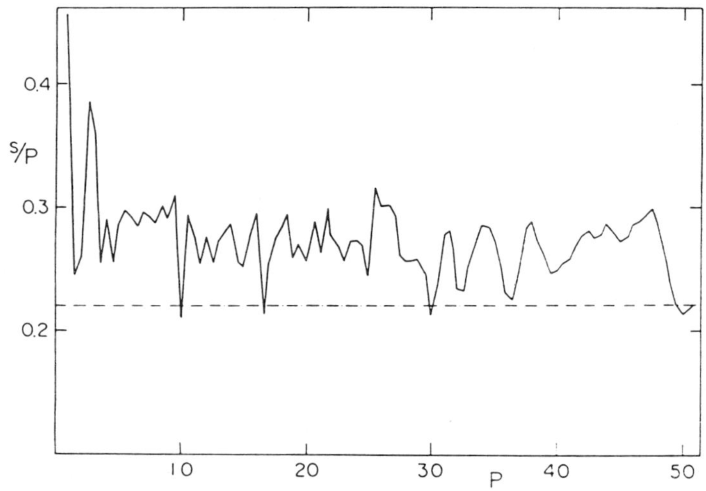

Fig. 1. s/P plotted against P (Myr). Data set taken from Grieve (1987). Craters with $D < 10$ km and ages < 600 Myr are considered. The dotted line gives the level below which the Broadbent criterion is satisfied. Note that there are significant periods at $P = 16.5$, 30 and 50 Myr. The period, $P = 10$ Myr is an artifact due to several crater ages being expressed in multiples of 10 Myr.

of Clube and Napier who regard 15 Myr periodicity as a genuine galactic signal.

Since crater ages in some data sets satisfy the stringent requirement for periodicity, they may be regarded periodic. It will thus be meaningful to discuss whether one can generate similar periodic distribution of ages by superimposing a periodic component on a steady random component or otherwise. This will be done in the next section.

4. Fraction of Periodic Component

Here, we present the result of simulations described in Section 2. A number of runs have been performed to calculate the minimum value of $\overline{s}_m$ for the time sequence

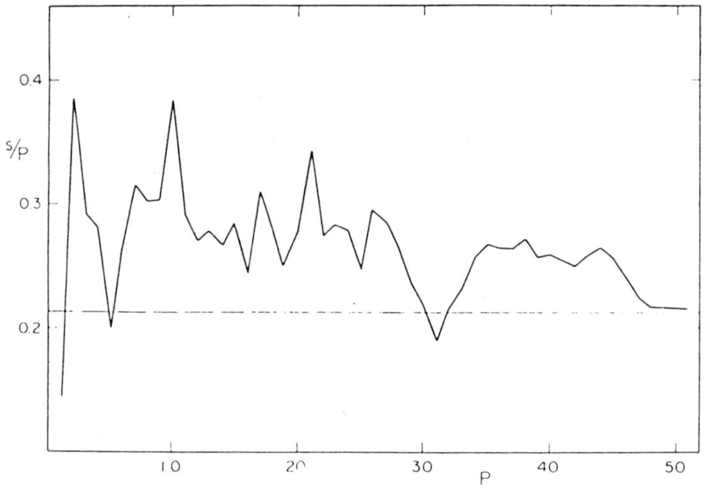

Fig. 2. s/P is plotted against P (Myr). Crater ages taken from Rampino and Stothers (1986). One significant period is seen at $P = 31$ Myr.

given by Equation (2.4) plus a random component. The numerical value of $\overline{s}_m$ depends on the fraction of the periodic part (F_{lp}) and on the standard deviation σ of the sequence given by Equation (2.4). The result is summarized in Table II.

The figures in Table II may be used, as follows, to estimate the values that F_{lp} and σ are likely to have. Assume that we are not making an observation, which is realized only by chance. By this, we mean an event which has a probability of being realized greater than 50%. Then, Table II shows that for $\sigma = 6$ Myr, the likely value of F_{lp} is 0.6 or greater. For $\sigma = 7$ Myr, F_{lp} should be close to 0.8 or greater, and so on. For $\sigma = 10$ Myr, one would never regard the signals (time series) as periodic, whatever the value of F_{lp} (even if there is no sporadic component in the signals).

Inspecting Table II, it seems reasonable to conclude that σ should not be greater than 7 Myr, whatever the fraction of the periodic part in the signals. On the other hand, if σ were as small as 5 Myr, values of F_{lp} as small as 0.4 would be compatible with the periodicity.

5. Probable Value of Spread in Encounter Epochs

If the detected period in the crater formation is real, what can we learn about the structure of the galaxy ? Rampino and Stothers (1986) and Clube and Napier (1986) have argued that the most likely candidate which modulates the impactor flux is molecular clouds. In this section, we consider most probable value for the spread in the otherwise exact periodicity on the $R - S$, $C - N$ hypothesis. Since the solar system oscillates above and below the galactic mid-plane, an exact periodicity

TABLE II
The fraction of cases where artificially generated time series exhibit periodicity. F_{lp} is the fraction of the periodic part in the time series and σ is the standard deviation from an exact periodicity. The number of craters considered is 65 in Table IIa, and 98 in Table IIb. In both cases the ages are $\leq$ 600 Myr

Table IIa	$\sigma = 5$(Myr)	6	7	10
$F_{lp} = 1.0$	100%	95	68	8
$= 0.8$	99%	78	47	11
$= 0.7$	95%	67	29	7
$= 0.5$	58%	39	17	8
$= 0.3$	18%	22	12	8

Table IIb	$\sigma = 5$(Myr)	6	7	10
$F_{lp} = 1.0$	100%	99	79	11
0.8	100%	96	54	14
0.7	100%	82	29	6
0.5	71%	27	12	8
0.3	22%	20	16	11

would be expected if the molecular clouds were concentrated in the mid-plane. In fact, the distribution of molecular coulds perpendicular to the galactic plane has a scale-height of approximately $Z_{1/2} = 26\ R^{0.47}$ pc (Scoville and Sanders, 1986), where R is the distance from the galactic center expressed in Kpc. For the solar system, $R = 10$ Kpc, so that $Z_{1/2} = 77$ pc. On the other-hand, the amplitude of the Z-motion of the solar system ought to be close to 72 pc, if the period of galaxy mid-plane crossing is 30 Myr (Thadeus, 1986). More recent observations of molecular clouds and young stars show that $Z_{1/2} \simeq 60$ pc, somewhat smaller than earlier values (Napier, 1989). Stothers (1985) argued that the amplitude of the solar Z-motion may be as large as 88pc. Accepting that $Z_{1/2} = 60$ pc and the amplitude, A of the solar motion $\simeq 80$ pc, we now ask whether one can detect a periodicity of the kind seen in the crater formation rate.

First, we note that half-width is 0.98 times the standard-deviation assuming a normal distribution. During σ Myr, the sun rises to a distance

$$Z_\sigma = A \sin\left(\frac{\sigma}{T_{1/2}} \frac{\pi}{2}\right)$$

where $T_{1/2}$ is half-period of the solar Z motion and A the amplitude. Since Z_σ ought to be at least $Z_{1/2}$, one would require $A \geq 1.70\ Z_{1/2}$ for $\sigma = 6$ Myr. In other words, the solar motion amplitude would have to be 100pc or so. This is much smaller than 0.6 Kpc, which Wolfendale (1989) requires to account for the noise signal ratio but somewhat greater than the probable amplitude. If one adopts $\sigma = 7$ Myr as the limit on the dispersion, one would have to have $A \simeq 1.5 \cdot Z_{1/2}$ or $A \simeq 90$ pc. Although the adopted amplitude $\simeq 88$ pc may be somewhat greater and half-width $\simeq 60$ pc, smaller than those currently accepted, it is significant that with these values the periodicity can be detected if F_{lp} is close to 0.8 or greater. To summarize, provided that the periodic part in the time series is 0.8 or greater, the periodicity could be detected in the crater formation rate if the amplitude of the solar Z-motion $\simeq 90$ pc and the half-width of the molecular cloud distribution $\simeq 60$ pc.

6. Intensity of Comet Shower

For the crater formation rate to vary periodically, at least the flux of LP comets ought to be such that it varies significantly from one epoch to another. We now discuss how the flux may vary by external disturbances.

The concept of comet shower was first explicitly introduced by Hills (1981). Although a few authors (Everhart, 1979; Yabushita, 1983a) had realized that the currently observed population of long-period comets was not consistent with a steady state picture of the Oort cloud, Hills explicitly stated a possibility that a large number of comets may be injected into the inner solar system by close encounters with passing stars. The period when the flux of comets is much larger than the average is termed "comet shower". The shower would be strong, if the solar system is associated with a heavy (populous) inner extension of the Oort cloud (Bailey, 1986). The flux of comets in and between comet showers is shown in Figure 3.

The flux shown in Figure 3 is calculated on the presumption that stellar and GMC encounters are the major perturbations of the Oort cloud. How strong the comet shower can be has been discussed more closely by a number of authors. Thus, Bailey *et al.* (1987) argued that the flux in the inner solar system may vary by a factor of 3, while Weissman (1988), from the discussion on the distribution of age of craters, argued that there is no geological evidence for very strong showers. This is the point used earlier by Clube and Napier against the existence of massive inner core of the Oort cloud.

All these arguments had been made before the importance of the galactic tide on controlling the dynamics of long-period comets was recognized (Byl, 1986; Heissler and Tremaine, 1986; Torbett, 1986). The distribution of matter in the galactic plane is such that the solar system oscillates above and below the galactic mid-plane with a period close to 60 Myr. Since the gravitational pull towards the galactic centre in such that the solar system makes one rotation with a period close to 200 Myr, the former is stronger than the latter by a factor $\simeq 11$. According to Bailey (M. E. Bailey, private communication 1987), the change in perihelion

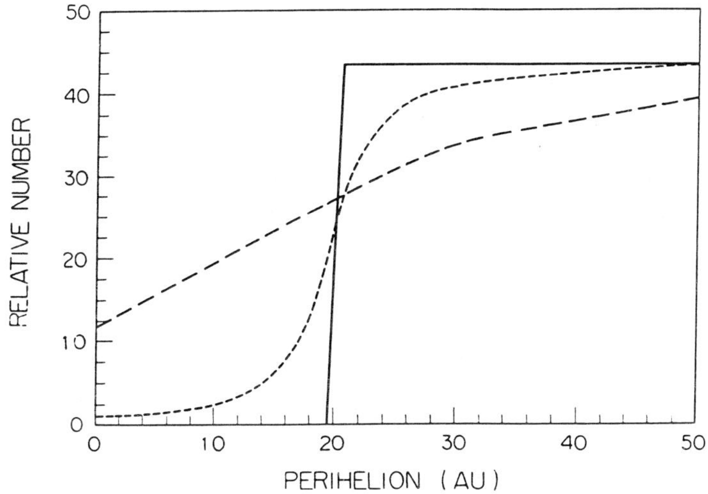

Fig. 3. Flux of LP comets as they are perturbed by the planets, the passing stars and the GMC's. — — — (strong shower), · · · · (moderate shower), ——— (between shower). Taken from Weissman (1986).

distance, q by stellar perturbations, is

$$\delta q \simeq 1.25(a/30000\ \mathrm{AU})^{3.5}\ \mathrm{AU}$$

while, calculations of the orbits of observed 48 LP comets under the galactic tide yields the result that (Yabushita, 1989)

$$\delta q \simeq 12.4\,(25000/a_{\mathrm{orig}})^{-6.3\pm0.2}$$

per revolution, where $1/a_{\mathrm{orig}}$ denotes the original value of $1/a$ (value of $1/a$ before entering the planetary region).

Thus, the motion of LP comets is controlled mainly by the galactic tide. In case of occasional close encounters with stars or GMC's, however, strong perturbation by them is expected. The action of the galactic tide, however, significantly changes the predicted distribution of cometary flux in the solar system. For the detail of calculating the flux, see Tsujii in this volume or Yabushita and Tsujii (1991a).

The cometary flux is typically of the form shown in Figure 4. It differs from the one shown in Figure 3 in that a significant fraction of LP or near parabolic comets penetrate into the inner solar system against the barrier of planetary perturbation. The flux is an increasing function of perihelion distance, q and reaches a maximum at $q \simeq 40$ AU. The flux declines somewhat and approaches a constant level. The ratio of the flux in the inner solar system ($0 < q < 5$ AU) to the outer solar system is designated by f. The ratio is a function of the semi-major axis, a before the comet enters the planetary region.

TABLE III

Ratio of the flux in the outer solar system ($60 < q < 100$ AU) to the one in the inner solar system. This quantity is designated by f. Its value at $a = 30000$ AU is exceptionally small and is probably due to a poor statistic. Its most probable value would be 4.0 as indicated in parenthesis

a (in AU)	15000	20000	25000	30000	35000	40000	45000
f	245	56	7.5	1.3 (4.0)	2.5	2.4	1.3

It is seen immediately that the fraction of comets which penetrate the barrier of planetary perturbation into the inner solar system drastically decreases as a decreases. In other words, only a very small fraction of comets with small a, say, 20000 AU or less can become observable comets.

The ratio f given in Table III can also be used to estimate the strength of comet shower as measured to by the cometary flux. In the absence of the galactic tide, the flux for small values of q is independent of q when the phase-space is uniformly filled (Hills, 1981). In other words, were it not for the galactic tide and planetary perturbation, the flux is independent of q. This means that in a very strong shower, where planetary perturbation has had no time to repel LP comets, the flux in the inner solar system should be the same as in the outer planetary region. Thus, the ratio f listed in Table III gives the factor by which the flux in the inner solar system can increase in the case of very strong external disturbances. Duncan *et al.* (1987) noted that the flux in the inner solar system can increase at most by a factor 20 in the case of very strong showers. The numerical values given in Table III shows, however, that the flux can increase by the factor 50, say, if there is a strong external perturbation such that orbits of comets with $a = 20000$ AU are greatly disturbed. Clube and Napier (1986) derived the equation

$$\delta q \simeq 2Q \frac{\delta V_A}{V_C} \left(\frac{q}{Q}\right)^{1/2}$$

where δq is the perturbation in perihelion distance, δV_A the change in tangential velocity at aphelion, q and Q are the perihelion and aphelion distance, respectively and V_C is the circular velocity. They calculate that for $Q = 4 \times 10^4$ AU and $q = 50$ AU, δq can be as large as 50 AU or greater if δV_A is 2.5×10^{-3} km s^{-1} or greater. Such an impulse is given by an encounter with a GMC with mass $10^{4.5} M_0$ and impact parameter 20 pc and relative velocity $\simeq 20$ km s^{-1}. It is reasonable then to expect that cometary orbits with $a \geq 2 \times 10^4$ AU will be strongly perturbed by such encounters. Such an encounter is expected to occur once in 15 Myr. The increase in the flux in the inner solar system will then be at most 50 times (Table III) that of the steady-state flux. How strong the shower will be depends on the distribution of a among the cometary population. If the cloud is heavily populated by those comets with $a > 20000$ AU, there will be rather mild (moderate) shower, as seen

TABLE IV
Intensity $\overline{f}$ of comet shower by a disturbance such that orbits with $a > a_{\rm min}$ are disturbed. The distribution of a is that of observed comets

$a_{\rm min}$ (in AU)	15000	20000	25000	30000
Shower intensity factor $\overline{f}$	25.8	9.3	3.1	2.1

from Table III. For instance, if the cloud consists mainly of those with $a = 30000$ AU, the shower flux will be some 4 times that of the steady-state flux and so on. By assuming that the currently observed population is in a steady state, it is possible to calculated the strength of the maximum intensity of a comet shower. Let $n(a)da$ be the number of observed comets with semi-major axes between a and $a + da$, and $f(a)$ be the ratio of the flux in the inner solar system to the one outside (namely the factor f given in Table III). The averaged shower intensity $\overline{f}$ is then given by

$$\overline{f} = \frac{\int_{a_{\rm min}}^{a_{\rm max}} n(a)f(a)da}{\int_{a_{\rm min}}^{a_{\rm max}} n(a)da} .$$

On carrying out the numerical integration from $a_{\rm min} = 20000$ AU to $a_{\rm max} = 50000$ AU, it has been found that $\overline{f} = 9.34$. In Table IV we give the shower intensity parameter $\overline{f}$ as a function of the minimum semi-major axis such that all of comets with a greater than $a_{\rm min}$ are disturbed.

It may be noted that from the distribution of $1/a$ which is obtained by evolution from an initial disk surrounding the solar system, Duncan *et al.* (1987) found that the shower intensity is at most 20, however strong the disturbance may be. The shower strength (intensity) given in Table IV has been obtained on the presumption that the solar system is currently not in a comet shower. As discussed in Section 3, the analysis of the crater records indicates, however, that the presumption is probably not correct (The solar system is now probably in an period of comet shower). If what we observe now is a comet shower, or the remnant of it in a recent past, the intensity could be calculated by assuming a model for the distribution of semi-major axes, a.

7. Periodicity in Extinction Rate ?

It is now widely disputed whether the rate of extinction of fauna exhibits periodicity or otherwise. The dispute was triggered by a paper published by Raup and Sepkoski (1984). They derived the rate of extinction of marine fauna as a function of geological age and found peaks which are more or less equally spaced. Indeed,

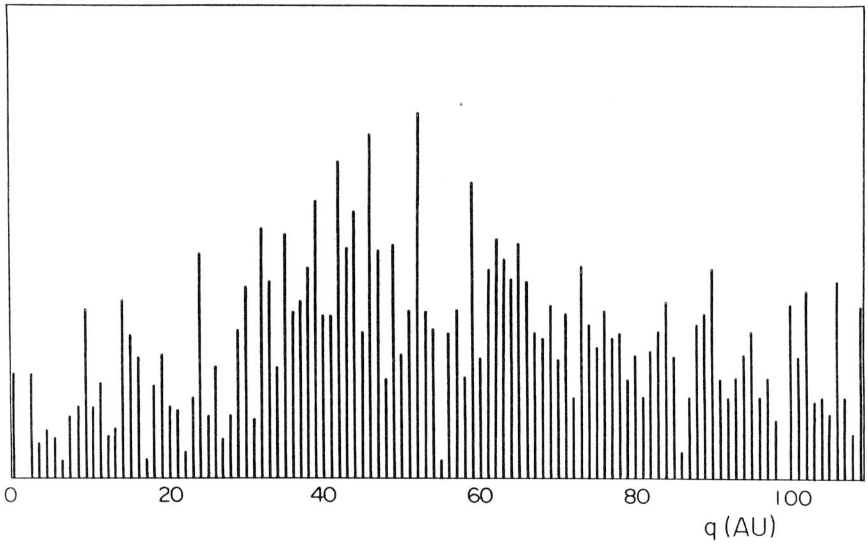

Fig. 4. Flux of LP comets as perturbed by the galactic tide and the perturbations of the four major planets. a (semi-major axes) before entering the planetary region is 30000 AU.

it is this work which gave birth to the work of Alvarez and Muller (1984) where it was argued that the ages of large craters are quantized (periodic). In the opinion of the present author, the period derived by Raup and Sepkoski, as well as that by Alvarez and Muller are not taken to prove periodicity. The times of the peaks of the extinction rate can be analyzed by adopting the criterion proposed by Broadbent. Figure 5 gives the variation of s/P for assumed period, P.

It is seen that s/P lies always above the level required for the periodicity hypothesis to be acceptable. Thus, the claimed periodicity does not appear to be substantiated by statistics. This is the reason why the present author believes the claimed periodicity to be spurious. It has, however, significance in that it motivated the examination of the distribution of crater ages.

8. Hypothesis of LP Comets being the Source of SP Comets

Grieve (1987) has reviewed the relation between the number of large craters ($D \geq 10$ km) and the potential impactors that are currently observed. See also Weissman (1989) for the same topic. In brief, there is so significant discrepancy between the number of known terrestrial craters and the number of impactors which are inferred from the observational data. This situation casts some doubt on the assertion that the present epoch is in a major comet shower. There is, however, an important problem as to the class of impactors. As mentioned in the introduction, a large fraction of impactors are asteroids and short periodic comets, while LP comets play only a minor role in hitting the earth (6% according to Olsson-Steel, 1987).

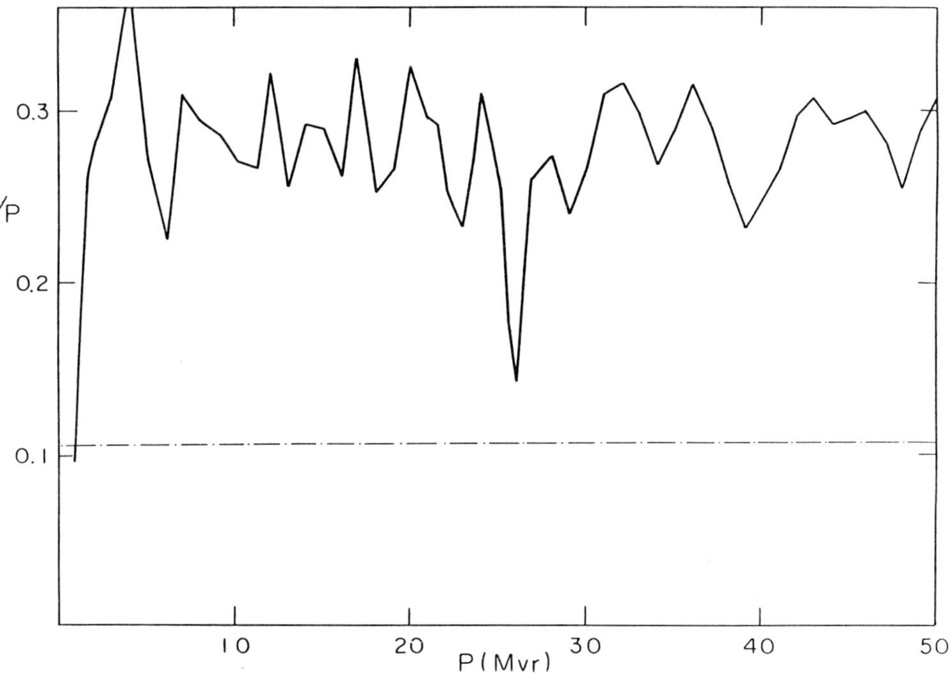

Fig. 5. s/P is plotted for the ages of extinction of marine fauna given by Raup and Sepkoski (1984). A period is seen at 26 Myr but it does not satisfy the adopted criterion for periodicity.

So, even if the flux of LP comets is modulated more on less periodically by the GMC's or by spiral arms, one would not expect to detect periodicity as long as the fluxes of asteroids and of SP comets remain constant. Thus, one is led to speculate if all of the earth crossing asteroids and SP comets originate from the flux of LP comets. Note that Quinn *et al.* (1990) take a different view that SP comets are supplied from a disk-like source known as the Kuiper belt which lies just beyond Neptune. This is the reason why the origin of SP comets plays a significant role in our understanding of fluctuations in the impact rate.

Broadly speaking, SP comets might be captured from the flux of LP comets by the action of the major planets or they may originate, as mentioned, from a source different from the LP flux. Bailey (1986) has argued that SP comets might originate from an inner extension of the observed Oort cloud. Such an extension (inner core) is assumed more heavily populated than the observed cloud. Because of the smaller value of the semi-major axe, they are only weakly perturbed by the galactic tide. Stagg and Bailey (1989) calculated the supply rate of SP comets and obtained a constraint which gives a relation between the inner edge of the assumed extension and an index which specifies the degree of concentration within the cloud.

Duncan *et al.* (1988) calculated the evolution of LP comets as they are perturbed by Jupiter and compared the predicted distributions of inclinations and semi-major axe of comets which now have short orbital periods (< 200 yr). They concluded that the predicted distributions differ from observation, thus favoring the origin from the Kuiper belt.

The origin of SP comets from the flux of LP comets has a long history and is a difficult problem. One difficulty is that it is hard to simulate the evolution from an LP to SP orbit, because it is an accumulation of many small perturbations. Another difficulty is that dynamical evolution is bound to be associated with physical evolution (and variation in absolute magnitude) for which there is very little information. Third, the flux of LP comets for perihelia outside 2 AU from the Sun, which cannot be observed can only be estimated by theoretical models. There has however, been considerable effort to estimate the flux and the probability of capture so that it is now possible to discuss the origin of SP comets with more information than before.

Broadly speaking, the capture is divided into two categories, depending on where the capture initially takes place. The site which was extensively discussed is the Jovian zone, where $q = 4 \sim 6$ AU. There is an extensive literature on the subject starting from the work of Russell (1920). For a review, see Bailey *et al.* (1990) or Stagg and Baily (1989).

First, consider the capture in the Jovian zone. By following the orbits of many hypothetical comets under the action of Jupiter alone, Quinn *et al.* (1990) obtained a value greater than 0.0008 as the capture probability, while Yabushita and Tsujii (1991b) obtained 0.00048 for the probability by following orbits as perturbed by the four major planets. One must note that a large contribution comes from the orbits with large inclinations ($30° < i < 150°$). On the other hand, Everhart (1967) and Hughes (1988) who followed Everhart's work consider capture among the low inclination orbits.

Whether the calculated capture rate can be regarded as consistent with the supply rate of SP comets depends on how long is the lifetime of SP comets and how the brightness decreases with the dynamical evolution. There is a marked difference between absolute magnitudes (H_{10}) of LP and of SP comets. Hughes (1988) derived $\Delta H_{10} = 2.75$, while Yabushita and Hasegawa (1981) found $\Delta H_{10} = 4$. On the presumption that the observed difference is the one which a comet physically evolves with the dynamical evolution, the supply rate of SP comets can be estimated by dividing the number of observed SP comets by the lifetime.

The estimated supply rate varies from 0.07 yr^{-1} (Clube and Napier 1982, lifetime=1500 yr) to 0.006 yr^{-1} for lifetime $= 2 \times 10^4$ yr. For the longer lifetime, the required flux of LP comets is thus 7.5 (=0.006/0.0008) to 12.5 (=0.006/0.00048).

Note that this is the required flux of LP comets in the Jovian zone, regardless of inclinations. In reality, one should adopt smaller value of capture probability, because only low inclination comets can become observed SP comets while SP comets of large inclinations have not been observed, although Olsson-Steel (1989) gives reasons why they might well exist.

So, the required flux is probably 5 times greater. For shorter lifetimes, the required LP flux would have to be even greater in inverse proportion to the lifetime.

Although there is no observation of comets with perihelia in the Jovian zone, a flux as large as $30 \sim 50$ yr^{-1} in $4 < 5 < 6$ AU is not plausible. The difficulty can be surmounted only by assuming a much longer lifetime for SP comets.

Next, we consider the flux in the outer planetary region. Stagg and Bailey (1989) calculated the capture probability for comets in near-parabolic orbits with q up to 34 AU. Then, if the flux of near-parabolic comets is known, it should be possible to calculate the supply rate.

The flux can be calculated by taking into account the galactic tide, as well as the perturbation of the major planets (Yabushita and Tsujii, 1991a,b or Tsujii, 1992). The supply rate of observable SP comets from the flux with $a \geq 25000$ AU is 2×10^{-3} yr^{-1} while the rate from the flux with $a < 20000$ AU is between $1.5 \sim 3 \times 10^{-3}$ yr^{-1}.

Whether the predicted supply rate is consistent with that required by observation depends on how long is the lifetime. As mentioned, if the lifetime were as short as 1500 yr, the predicted supply rate is less than that required by observation by on order of magnitude, while if it were as long as 2×10^4 yr, there will be no significant discrepancy.

At present, it does not appear possible to draw a definite conclusion on this topic, because of the lack of detailed knowledge on the lifetimes.

9. Capture from Interstellar Medium ?

Arguments have been presented from time to time that comets are of interstellar origin and that they were captured from interstellar medium. Astrophysical considerations provide processes whereby cometary nuclei may be formed. Yabushita (1983b) considers globules of low temperature as the site of nuclear formation, while Napier and Humphries (1986) consider stellar radiation playing major roles in forming interstellar planetesimals in dense molecular clouds. Although the hyperbolic character of some of LP comets is commonly ascribed to non-gravitational accelerations, there are reasons to believe that some of the hyperbolic velocities may be genuine (Yabushita and Hasegawa, 1978; Yabushita, 1991). The capture probability of hyperbolic comets under the action of the planets, a hypothetical solar companion and by encounter with a passing star or molecular cloud was investigated by Valtonen (1983) who found that the largest probability is due to the second process, while the first process provides the smallest probability.

Accepting the existence of interstellar comets, Clube and Napier (1984) argued that the solar system captures comets including giant ones in a molecular cloud, and this may give rise to the periodicity.

Here the present author wishes to point out a difficulty in associating the capture to the periodicity. The following formula is used by Clube and Napier to give the number of comets captured dNc ;

$$dNc = 2\sqrt{\pi}x_0^2 \exp(-x_0^2)\left(\frac{GM_\odot}{v_0^2}\right)^2 n\,dR\,,\ x_0 = \frac{v_0}{\sqrt{2}\sigma}$$

where dNc is the number of captured comets while the sun traverses a distance $dR = \sqrt{2}\sigma x_0 dt$; σ is the velocity dispersion of comets in the molecular cloud and v_0 is sun's drift velocity relative to the molecular cloud. It is apparent that most of the comets are captured at large perihelion distances so that when they are captured, they are unobservable comets. For then to become observable, their perihelion distances need be decreased mainly by the perturbation of the galactic tide, which takes a few times 10^7 to 10^8 yr (Byl, 1986) ; the exact time required depends on the orbital configurations relative to the galactic disk. Thus, it is not reasonable to expect that the capture immediately gives rise to periodicity in the flux of impactors. However, as shown by Valtonen (1983), capture of interstellar comets in the presence of the third body (molecular cloud) is much more effective than the capture by the planets and must take place whenever the solar system encounters a GMC, if they contain interstellar planetesimals.

10. Discussion

So far, our argument has been as follows. First, the hypothesis of periodicity of crater formation rate appears to be substantiated by the Broadbent criterion, and the derived period is 30 Myr. Further, the distribution of ages indicates that the current epoch is in a comet shower. Second, the fraction of impactors which are periodic in time was estimated in Section 4. For the periodicity to be detected, the fraction F_{lp} of the periodic component ought to be 40% or more if the standard-deviation from an exact periodicity were 5 Myr or less, while if σ were larger than 7 Myr, we would not detect periodicity, even if all of the impactors were periodic (no sporadic component). Third, as long as the supply rate of SP comets is concerned, it is possible to account for the origin of SP comets by capture from the theoretical flux of LP comets on the presumption that the solar system is not in a comet shower and the lifetimes of SP comets as long as 2×10^4 yr (See, however, Quinn *et al.* (1990) for an alternative view). Then, the modulation of LP comet flux necessarily modulates the flux of SP comets so that a moderate variation in the impactor flux (currently between 12 and 18%) is theoretically possible, while if one supposes that the SP comets are supplied by a source such as the hypothetical Kuiper belt, the variation in the impactor flux will be small (less than 10%). The constraint on the detectability of periodicity would require a periodic component of 60%, which is considerably greater than the one derived from observation (12$\sim$ 18%).

To summarize, an astronomical model which can account for the detected periodicity in the crater formation rate requires ;

(1) Deviation from the exact periodicity characterized by the standard-deviation $\leq 6 \sim 7$ Myr.

(2) Periodic component in the impactor flux which is greater than the observed value by a factor 3 for $\sigma < 6$ Myr and 4 for $\sigma < 7$ Myr.

(3) The amplitude of Z-motion of the solar system would have to be 100 pc ($\sigma < 6$ Myr) or 120 pc ($\sigma < 5$ Myr), while for $\sigma < 7$ Myr, the amplitude $\simeq 90$ pc would be compatible with the periodicity.

The second constraint poses some difficulties. Although short and long-periodic comets occupy only $12 \sim 18\%$ of the earth impactors, some of objects now known as asteroids may be comets, after all. Yeomans (1991) investigated motions of 12 near-Earth asteroids, and found that Appolo and Icarus have non-gravitational components in their motions, thus indicating probable cometary origin. It is thus not unreasonable to suppose that the fraction in the impactor flux which can vary in time is as large as 40%. However, if $\sigma \simeq 7$ Myr and the required periodic component is close to 80% or greater, it would indicate that the majority of earth impactors would have to be of cometary origin, and this certainly would not be accepted by the majority of the workers in the field.

References

Alvarez, W. & Muller, R. A.: 1984, *Nature* **308**, 718

Bailey, M. E: 1985, in Carusi, A. & Valsecchi, G., ed(s)., *Dynamics of Comets*, Reidel, Dordrecht

Bailey, M. E.: 1986, *Nature* **324**, 350

Bailey, M. E., Wilkinson, D. A. & Wolfendale, A. W.: 1987, *Mon. Not. R. Astr. Soc.* **227**, 863

Bailey, M. E., S. M. M. Clube, & Napier, W. M.: 1990, *Origin of Comets*, Pergamon Press, Oxford

Broadbent, S. R.: 1955, *Biometrica* **42**, 45

Broadbent, S. R.: 1956, *Biometrica* **43**, 32

Byl, J.: 1986, *Earth, Moon and Planets* **36**, 263

Clube, S. V. M. & Napier, W. M.: 1982, *Q. Jl. R. Astr. Soc.* **23**, 45

Clube, S. V. M. & Napier, W. M.: 1986, in Smoluchowski, R., Barcall, J. N. & Mathews,M. S., ed(s)., *The Galaxy and the Solar System*, Univ. Arizona Press, Tucson

Duncan, M., Quinn, T. & Tremaine, S.: 1987, *Astron. J.* **94**, 1330

Everhart, E.: 1979, in Duncombe, R. L., ed(s)., *The Dynamics of the Solar System*, Reidel, Dordrecht,

Grieve, R. A. F.: 1987, *Ann. Rev. Earth Planet. Sci.* **15**, 245

Heissler, J. & Tremaine, S.: 1986, *Icarus* **65**, 13

Hills, J. G.: 1981, *Astron. J.* **86**, 1730

Matese, J.J. and Whitmire, D.P.: 1986, *Icarus* **65**, 37

Napier, W. M. & Humphries, C. M.: 1986, *Mon. Not. R. Astr. Soc.* **221**, 105

Napier, W. M.: 1989, in Clube, S.V.M., ed(s)., *In Catastrophes and Evolution*, Oxford Univ. Press, Oxford

Olsson-Steel, D.: 1987, *Mon. Not. R. Astr. Soc.* **227**, 501

Quinn, T., Tremaine, S. & Duncan, M.: 1990, *Astrophys. J.* **355**, 667

Rampino, M. R. & Stothers, R. B.: 1986, in Smoluchowski, R., Barcall, J. N. & Mathews, M. S., ed(s)., *The Galaxy and the Solar System*, Univ. Arizona Press, Tucson

Raup, D. M. & Sepkoski, J. J.: 1984, *Proc. Natl. Acad. Sci.* **301**, 801

Russel, H. N.: 1920, *Astron. J.* **33**, 49

Scoville, N. Z. & Sanders, D. B.: 1986, in Smoluchowski, R., Barcall, J. N. & Mathews, M. S., ed(s)., *The Galaxy and the Solar System*, Univ. Arizona Press, Tucson

Seyfert, C. K. & Sirkin, L. A.: 1979, *Earth History and Plate Tectonics,*, Harper & Row, New York

Shoemaker, E. M. & Wolfe, R. F.: 1986, in Smoluchowski, R., Barcall, J. N. & Mathews, M. S., ed(s)., *The Galaxy and the Solar System*, Univ. Arizona Press, Tucson

Stagg, & Bailey, M. E.: 1989, *Mon. Not. R. Astr. Soc.* **241**, 507

physics and chemistry of the atmosphere, and hence geophysical processes under the controlling influence of climate and biological processes under the controlling influence of primordially synthesised pre-biotic chemicals surviving atmospheric entry. Airbursts and impact craters on the other hand implied a catastrophic history on all timescales, the above being characteristic of (i) geological and biological evolution, (ii) the recent evolution of *homo sapiens,* and (iii) the evolution of civilization, respectively.

Both main aspects of the theory seemed at first to be questionable. Thus significant atmospheric aerosols were generally discounted on the grounds that most directly observed meteoric inputs due to comets were neither sufficient nor adequately preserved. The airbursts and impact craters due to cometary material were also discounted on the grounds that most such effects were generally asteroidal in origin. However (i) the realisation that the meteoroidal input is dominated by the sporadic component which is itself strongly represented by a single broad stream based on the orbit of a known cometary source (Stohl 1984), and (ii) the detection of a previously unobserved but nevertheless substantial population of $\sim 10^6$ g fragile meteoroids undergoing electrostatic disruption during passage of the auroral plasma (e.g. Bigg and Thompson, 1969; Fechtig, 1982; Wetherill and Revelle, 1982) and of such bodies arriving in exceptional swarms following recent cometary disruption in space (Dorman *et al.*, 1978; Kaufmann *et al.*, 1989) fundamentally altered our perception as to the possible dominant source of extraterrestrial input. Furthermore it was widely recognised that some 50% of the flux of kilometre-plus, near-Earth objects were likely to be in the form of defunct comets (Wetherill, 1988). Evidence of an earlier massive meteoroidal swarm e.g. the eleventh century fireball flux in the Taurid meteoroidal complex (Astapovic and Terenteva, 1969) indicated variations in the sub-kilometre flux that were by no means negligible. Contrary to expectation therefore, the evolution of the Earth was quite conceivably under the influence of disintegrating comets (Clube and Napier, 1990).

2. The Oort Cloud and Giant Comets

The involvement of the Oort cloud was a crucial development not previously considered on account of the perceived physical nature of comets.

Thus the presumed low temperatures of primordial comets during their slow accretion in proto-stellar discs favoured an essentially homogenous composition and hence evolutionary products during insolation (dust, gas) which were not expected to have any serious influence upon the Earth. As a basis for investigation however, this put the cart before the horse! Thus variations in temperature among presumed gravitational condensations such as may arise in a rapidly compressed proto-stellar medium, for example, are expected to be associated with various degrees of physical and chemical processing, favouring an essentially inhomogenous composition for primordial comets. It is permissible therefore to envisage comets made up of active and less active regions, with a tendency to split during insolation, and hence additional evolutionary products (meteoroids, asteroids) which can be expected to have a more serious influence upon the Earth. Indeed, large differentiated comets analogous to meteorite 'parent bodies' are not then excluded and these may provide a primordial environment in which considerable water-based processing of simple cometary material occurs. It follows that the additional evolutionary products are then capable of delivering variously processed cometary materials to the Earth in the form of smaller friable bodies producing aerosols and airbursts (sub-kilometre meteoroids) and larger robust bodies producing impact craters (kilometre-plus asteroids). At the same time, whilst crater-forming impacts and low-level airbursts of the Tunguska kind are a well recognised aspect of cometary inputs, the regular introduction of aerosol particles through high level disintegration of cometary meteoroids (see above) seems now to provide a basis for expecting that complex pre-biotic organic molecules synthesized in comets are capable of surviving atmospheric entry (e.g. Marcus and Olsen, 1991), and if such deliveries along with impacts and airbursts are correlated in time, then it is conceivable that the evolution of life is regularly promoted by comets through speciation and extinction, *acting jointly*.

By a straightforward modification of star-formation theory therefore, giving rise to a modest change in our understanding of the formation of comets, and supposing also that the bulk of comets interacting with the Earth originate from the Oort cloud, we may readily envisage centrally robust, differentiated giant comets that undergo splitting during insolation and thereby release cometary asteroids and meteoroids into planet-crossing orbits. Such objects clearly contribute to planetary cratering and zodiacal dust,depending only on the mass function of comets, and it is reassuring that very massive cometary bodies like Chiron are now expected frequently in Jovian and sub-Jovian space (Hahn and Bailey, 1990). However it is through the massive aerosols induced by the potentially unseen flux of disintegrating meteoroids that reaches our planet, initiating and modulating both its biochemistry and its climate, that giant comets may be expected to exercise a controlling influence on life on Earth.

To the extent that the debris of exceptionally large comets have not generally been thought of as making up a significant part of the population of inactive bodies in the near-Earth environment i.e. sub-kilometre meteoroids and kilometre-plus asteroids, giant comets are an overlooked but potentially fundamental key to any galacto-terrestrial relationship of the kind considered here. Indeed to the extent that these recognised minor bodies of the Solar System bear witness to the dynamical and physical evolution of giant comets, such bodies may now be seen as having Galactic and geological implications which were not formerly predicted.

3. Galactic Perspective

Based on an observed large scale expansion of the Galaxy and arising from associated considerations relating to the origin of spiral structure (Clube, 1973; Bailey and Clube, 1978; Clube and Watson, 1979), the general behaviour of rotationally flattened stellar systems is readily characterized by recurring, gravitationally controlled flows to and from their centres. At first this seems a rather remote issue from comets but in dealing with rotationally flattened stellar systems, we draw attention to significant interludes of maximum compression in the commonest large spiral galaxies which may be associated with the growth and decay of dark matter in galactic nuclei, due to the late evolution of successive central superstars. Likewise, we draw attention to an outward phase of the gravitationally controlled flow in these systems which naturally comprises oppositely directed streams of plasma (jets) from the nuclear region, subsequently confined to the rotating disc by the ambient interstellar gas. Outflows of this kind, relating to AGNs (e.g. De Young, 1989), may evidently be thought of as producing spiral arms like those observed since adiabatically expanding plasma of low density, which recombines during confinement to form cold molecular hydrogen, may be rapidly condensed into star-forming regions carried along by the ambient gas. Thus, galaxy observations in general, admitting also the large scale expansion of the Galaxy, allow us to envisage a star formation process involving rapid compression (i.e. the alternative comet forming scenario discussed in Section 2) provided our understanding and formulation of gravitational action (Dicke, 1962; Clube, 1991) are consistent with the intermittent existence of dark matter in regions of high gravitational potential, such as are achieved during the late stages of stellar and superstellar evolution. The explanation of dark matter in the universe is of course unknown, and the subject is introduced here merely to demonstrate the likely interconnection of most astrophysical mysteries. The importance of the galactic perspective resides in the fact that our formulation of gravitational action, which is still a matter for empirical science, also has a bearing on the true nature of star forming regions and hence comets.

Initially, such star-forming regions as are produced in spiral outflows of this kind may be expected to contain proto-stars which separate out individually as aggregates of gravitationally condensing 'parent bodies' with masses that extend over the range of giant comets to proto-planets (Clube, 1988). These aggregates resemble the primordial configurations envisaged by McCrea (1978) as leading to the Hayashi phase of proto-star formation. A well known feature of these configurations is the emergent angular momentum distribution which very naturally conforms to that of the Sun and its planets, incidentally avoiding the seemingly *ad hoc* process of angular momentum redistribution required of the solar nebula theory. However the principal advantage accruing here from these configurations is the accelerated condensation of primordial cometary material resulting in bodies that temporarily retain heat and are differentiated in the manner of meteorite parent bodies generally. Thus we plausibly expect star-forming regions and incipient spiral arms to be populated with differentiated giant comets (and their debris). Indeed the possibility is even raised that the observed Oort cloud may be subsequently replenished from star-forming regions traversed by the Sun, and may not therefore be primordial.

Be that as it may, based on these simple considerations, one readily arrives at a specific proposition concerning the internal constitution of comets which, one expects, may then be verified empirically through comet studies generally. As we have already seen, the understanding of Earth history is certainly relevant but well before this stage is reached in the evolution of comets, we may consider the observed Oort cloud, in particular its orbital energy distribution. Traditionally this was explained by Oort (1950) in terms of a diminishing tendency on the part of insolated comets to fade away rather than disrupt. This proposal has never been adequately reconciled with observations of cometary activity (Yabushita, 1983; Whipple, 1957; see Section 6.1) but from the perspective of a theory in which large differentiated comets are expected, the tendency is a natural consequence of such bodies dominating the mass function and, as a result of splitting during insolation, giving rise to an excess of short-lived comets in Jovian and sub-Jovian space along with many inert (presumably carbonaceous chondritic) bodies in Earth-crossing orbits.Both aspects of Jovian and near-Earth space are in fact observed. Treating the Solar System as a test-particle in the Galaxy therefore, the question is very naturally raised whether the Earth responds in an observable way to cometary asteroids and meteoroids (i.e. to impacts, airbursts and aerosols) due to the unexpected presence of differentiated comets in spiral arms.

4. Terrestrial Perspective

Such considerations led straightforwardly to the notion that galactic modulations of the terrestrial record are inevitable and may be introduced through intermittent replenishment of the Oort cloud from spiral arms, sometimes amounting to its complete overturn, and through regular and irregular fluctuations in the differential gravitational field across the Oort cloud, thereby modulating the cometary flux. The thesis was advanced that the inner Solar System flux of comets, through the agency of inhomogenously structured giant comets, would give rise to massive (short lived) streams of meteoroidal debris which controlled evolution on Earth (Clube and Napier 1982a,b,c, 1984a, 1986, 1990, 1992). To the extent such meteoroidal debris includes short-lived $\geq$ kilometre-sized objects (cometary asteroids), an essentially catastrophic evolution of the Earth was evidently implied, with otherwise unexplained variations on timescales naturally reflecting (i) the Galactic replenishment and perturbation functions, (ii) the arrival rate of giant comets in extreme short-period orbits, and (iii) commensurabilities with disintegration orbits.

It is important to appreciate that this form of "terrestrial catastrophism", based fundamentally on the existence of giant comets, contrasts sharply with the conventional form in which the near-Earth asteroid population is supposedly derived more or less exclusively from the main asteroid belt (Shoemaker, 1983; McLaren, 1983; Chapman and Morrison, 1989). Thus, under the latter circumstances, the near-Earth asteroid population, being meteoritic rather than cometary, is not regarded as being associated with a rapidly disintegrating meteoroidal flux; nor is it associated with a Galactic perturbation function. One therefore discounts any catastrophically induced geological timescales which are due to disintegration orbit commensurabilities, giant comet arrivals and a Galactic perturbation function: the focus of interest is merely on random impacts by $\geq$ kilometre-sized asteroids. The "giant comet" and "conventional" theories also differ so far as the major extinction events are concerned: conventionally these are due to isolated large asteroids whereas, on the alternative theory, a barrage of impactors, with further depositions over $\sim 10^5$ yr, the disintegration time of an exceptional giant comet, is most probably involved.

It should of course be said that stochastically distributed encounters between the Earth and objects deviating from the asteroid belt are not excluded by the giant comet theory of terrestrial catastrophism. Nevertheless it is clear that the emphasis on differentiated giant comets and the rather more frequent role of catastrophism in Earth history due to meteoroidal debris, even extending to the timescale of civilization, demands a careful reappraisal of the roles played by the various minor bodies of the Solar system.

5. Galacto-Terrestrial Relationship

With the discovery at the KT boundary of globally distributed cosmic dust (Alvarez *et al.*, 1980) and impact debris (Bohor *et al.*, 1984), the study of "terrestrial catastrophism" was recognised as a crucially important scientific discipline and the subject advanced several notches on the scale of public awareness. This interest has been maintained, even to the extent of reviving this century an awareness of an important celestial hazard to civilisation (Steel, 1991), and there is now a degree of pressure on the scientific community to resolve the basic issues at stake.

Several explanations of the new findings at the KT boundary were immediately advanced. First, in accordance with conventional theory, supposing near-Earth asteroids are more or less exclusively derived from the asteroid belt, an isolated impact by an $\sim$ 10 km (meteoritic) asteroid was proposed. Then, based on fresh evidence (Raup and Sepkoski, 1984, 1986) for a periodic cycle in the mass extinction record (P $\sim$ 26 Myr,ϕ $\sim$ 10 Myr), close to the previously estimated 30 Myr and including the KT event, the idea was put forward that a binary companion in an extreme elliptical orbit about the Sun (Davis *et al.*, 1984) may be regularly releasing comet showers (Hills, 1981, 1984) from a dense inner core of the Oort cloud (Bailey, 1983). This explanation was not readily sustained however since the proposed companion star had no independent observational support whilst its orbit was both unusual and unstable (Clube and Napier, 1984b). Nevertheless the phase ϕ of the cycle was broadly compatible with the standard Oort cloud as intermediary since the half-period of perturbed comets giving rise to catastrophism is $\sim$ 2-4 Myr, whence it follows that the Galactic plane crossings closely bisect the intervals between the extinction peaks. The new evidence from the extinction record strongly supports the Galactic 'clock' therefore though it is something of a surprise perhaps that the strongest terrestrial signals coincide with the extremities of the Sun's vertical oscillation.

A Galactic hypothesis, in which the Oort cloud is strongly perturbed by molecular clouds close to the plane (Rampino and Stothers, 1984) is evidently not compatible with the extinction cycle phase nor is it clear that the vertical distribution of molecular gas in the Galactic plane is compatible with a detectable cyclic variation in the cometary flux (Thaddeus and Chanan, 1985). On the other hand, the discovery of several likely impact craters at the KT boundary (e.g. Manson, Chicxulub), along with a more extended deposition of cosmic dust (Bhandari 1991) associated with extraterrestrial amino acids (Zhao and Bada, 1989), laid down over $\sim 10^5$ yr, has ultimately focussed attention on the likely role played by a disintegrating giant comet at this mass extinction (Zahnle and Grinspoon, 1990; Shoemaker and Izett, 1992). That the best-studied mass extinction during the Phanerozoic should be in accordance with the more general considerations of previous sections, leading to the giant comet thesis, is clearly a significant advance. Meanwhile it has come to be appreciated that the forces responsible for underlying variations in the cometary flux are best expressed in terms of the local density of perturbers in Galactic phase space, one significant aspect of which is the well known Galactic tide (see Bailey *et al.*, 1990 for a recent review; also Section 6.3).

6. Near-Earth Objects

Near-Earth objects (NEOs) are generally considered to be divided into two broad categories: inactive asteroids ($\sim$ 90%) and active comets ($\sim$ 10%). The term 'asteroid' in this context is something of a misnomer however since this category of NEOs is attributed (Wetherill 1988) to the asteroid belt ($\sim$ 45%) and dead comets ($\sim$ 45%) in roughly equal measures. The active comets are likewise more or less equally distributed between the LP and SP comet populations. Since the observed LP and SP comet fluxes at high inclinations are broadly consistent with the calculated perturbations due to planets (Duncan *et al.*, 1988) but are discrepant by a factor $\sim$ 10-100 at lower inclinations, notwithstanding the more efficient capture and general compatibility of captured orbital parameters (see also Everhart, 1972), some 50% of the present NEO flux, due to comets, is apparently supplied from an additional low inclination source. This source is commonly identified as the Kuiper belt (cf. Bailey, 1992), presumably on the general grounds that dead comets therefrom may be active comets with inert crusts. The other possibility, as we have seen, is that the LP population, which preferentially experiences prograde captures close to the ecliptic , includes large differentiated comets which split during insolation, thereby intermittently giving rise to an excess of low inclination SP comets and inert bodies in Earth-crossing orbits.

6.1. Giant Comets

The generally uniform activity of observed cometary material and the initially strong fading of LP comets are no longer considered compatible (Whipple 1992). This suggests the LP and SP comet populations may now be best understood in terms of cometary splitting (e.g. Kresak, 1991), even to the point of total disruption of many smaller comets (an observed process) and a tendency during insolation, based on an evolving partition between active and inactive cells in otherwise broadly uniform cometary material, towards the formation of increasingly robust cometary asteroids with declining pockets of activity. Not only then do we understand the extreme SP comet population excess in terms of a large differentiated source but it seems likely that conspicuous members of this excess are generally smaller in size than observed members of the LP comet population.

In practice, the cumulative absolute magnitude distributions of LP (P> 200 yr) and extreme SP (P< 15 yr) comets both possess prominent "knees" corresponding to the sharp fall-off in the detection rate at distances from the Sun where the water-ice sublimation rate is substantially reduced (Hughes, 1989). In the presence of uniformly partitioned material, the ~ 5 magnitudes' difference between these knees would indicate a factor ~ 10 difference in the linear size of LP and extreme SP comets. In a steady state, we might then expect a proportional excess of the latter as great as $\sim 10^3$ compared with the observed excess, based on dynamical arguments, of ~ 10–10^2. An additional subliminal population of extreme SP comets is indeed implied both by the lower frequency of observed comets beyond the knee and by the rapid interchanges between visible and invisible comets in SP orbits (e.g. Nakamura and Yoshikawa, 1992). Thus it is to be expected that the current sub-Jovian flux of comets includes a substantial, relatively short-lived population of cometary asteroids in near-Earth orbits, some of which survive to be regular members of the NEO population. This is to be expected even more if the splitting of a single, large, differentiated source occurs in a sub-Jovian orbit. Since Comet P/Encke is currently observed in a sub-Jovian orbit in association with an extended meteoroid complex (Steel *et al.*, 1991), a substantial population of disintegrating cometary asteroids within this complex is likely to be replenishing the current shortfall in meteors producing the zodiacal dust (Grün *et al.*, 1986). The evidence, in other words, now points to a time-dependent dust production process in the inner Solar System, and the circumstances under which the likely population of cometary asteroids in this complex erodes and fragments are clearly of interest (see Section 7).

The difference between the standard and giant comet theories of minor bodies in the Solar System, then, is that standard theory traces Oort cloud comets to only a small fraction of the SP population on a more or less one-to-one basis without any substantial growth in numbers due to splitting. Expressing this another way, the observed mass function of Oort cloud comets and the existence of such bodies as Chiron and 1992 AD in the more distant SP population are explicitly set aside in favour of an additional cometary reservoir, e.g. the Kuiper Belt, and since there is no significant splitting, there is no *a priori* basis for expecting inactive cells to separate out in the form of cometary asteroids and meteoroids in Earth-crossing orbits, rather it is supposed that inert crusts frequently form around otherwise active comets. According to giant comet theory on the other hand, the SP population is traced in principle to individual, more massive members of the Oort cloud and we expect a substantial unseen population of inactive kilometre-plus and sub-kilometre bodies in near-Earth orbits. Such inactive cometary material has a tendency to occupy orbits accessible to the NEO population as a whole (Milani *et al.*, 1989). Thus to the extent that the present distribution of minor bodies is representative of the Solar System for a significant fraction of its lifetime, we expect catastrophic processes on both long and short timescales, due to the Galactic perturbation function and the disintegration of successive individual giant comets.

6.2. Unitary Events

The fundamental perception then is of massive meteoroid streams lasting $\sim 10^4$ yr, due to each successive giant comet, followed by a somewhat longer interval, say $\sim 10^5$ yr, during which the core of a giant comet undergoes more intermittent decay. We may describe the random intervals between giant comets (say, with diameters $\geq$ 100 km) as CHRONs, whilst assuming that these objects arrive in batches associated with spiral arm and interarm passages (see later), giving rise to episodically sustained fluxes and absences of duration $\sim$ 50-100 Myr, described as SUPERCHRONs. This terminology corresponds with that used to describe the known episodicity of the geological record, recognising also that the initially concentrated and subsequently dispersed phases of the evolution of a massive meteoroid stream may then be associated with glacials and interglacials respectively. This represents a departure from the conventional explanation of extreme glacial/interglacial conditions on Earth, according to which there is a substantial albedo feedback amplification of the weakest Milankovitch perturbation. Thus we envisage an episodic galacto-terrestrial relationship in which disintegrating giant comets represent unitary events in Earth history lasting $\sim 10^4$ yr, separated by intervals $\geq 10^5$ yr.

The glacials corresponding to unitary events, through the generation of polar ice-caps, perturb the Earth's equilibrium figure, causing mantle currents to flow on a global scale, thereby simultaneously enhancing both plate tectonics and core dynamics. The unitary events are therefore represented by enhanced geological activity (orogeny, sea-floor spreading, flood basalt eruptions etc.) and geomagnetic disturbances (including reversals). Moreover, since we deal with comets whose origin allows conditions under which complex pre-biotic organic molecules are synthesised, the near contemporary incidence of impacts and aerosols due to massive meteoroid streams is a potential source of extinction-speciation events. It should be noted that impacts alone are less likely to produce major punctuations in the evolutionary record since any prior balance of species which is locally disturbed through such impacts is liable to be recovered through the normal process of natural selection. It follows that the predicted turnover in biological mutations is greater in the giant comet than in the conventional theory of terrestrial catastrophism. Thus, we expect a high degree of correlation between extinction-speciation and geomagnetic reversal events, each closely following the giant comet flux.

6.3. Galactic Perturbation Function

The initial major peaks in the extinction record (working back from the present) occur at $\sim$ 10,38 and 65 Myr with evidence for an overall $\sim$ 26 Myr cycle based on a phase of $\sim$ 10 Myr (Raup and Sepkoski, *loc cit*). More or less identical major peaks occur in the geomagnetic reversal frequency (Creer and Pal, 1989) with a similar overall sequence through preceding epochs. This correlation indicates a probable association with giant comet events provided the galactic perturbation function due to the vertical oscillation of the Sun has a half-period P/2 $\sim$ 26 Myr and is out of phase with galactic plane crossings by $\phi \sim$ P/4 : we recall that there is a lag of $\sim$ 3 Myr between Oort cloud pertubation and its reflection in the SP cometary flux due to the travel time of LP comets. Rampino and Caldeira (1992) conclude from their study of a more extensive array of geological events (extinctions, ocean anoxic and black shale events, major stratographic sequence boundaries, major evaporite deposits, continental flood basalt eruptions, sea-floor spreading events and major orogenies) that there is a clear clustering of events during the past 240 Myr with an underlying periodicity and phase of 26.6 ± 0.4 Myr and 9.1 ± 0.7 Myr respectively.

Since the geomagnetic reversal record is characterized by interpulses between the more conspicuous peaks, giving a significant periodicity of $P/4$ (e.g. Creer and Pal, *loc cit*), the terrestrial evidence apparently points to an underlying cycle of period P due to the Sun's vertical motion in the Galaxy, associated in turn with strong Oort cloud perturbations close to the extremities of the Sun's oscillation and somewhat weaker perturbations as the Sun crosses the plane. This uniquely galactic signature in the terrestrial record seems to indicate that account should be taken of both the spatial and the velocity distribution of perturbers in galactic phase space giving rise to Oort cloud perturbation. Thus if the cometary flux F_c arises chiefly through tidal perturbations due to random encounters in the Galactic environment, we obtain (Heisler and Tremaine, 1986)

$$F_c \propto \rho(z)\frac{\Phi(x)}{x}$$

where $\rho(z)$ reflects the spatial distribution of material in the galactic plane and $\Phi(x)$ is the error function based on $x = V_{\odot}/\sqrt{2}\sigma$ reflecting the solar speed $V_{\odot}$ and stellar velocity dispersion σ, identically oriented, at the relevant instant of perturbation. It is evident that $\rho(z)$ and $\Phi(x)/x$ maximise in and out of the plane respectively, giving rise to a $P/4$ cycle in F_c of alternating strength, depending on the precise phase-space distribution of material close to the galactic plane. In general, it is not to be expected that this cycle will be prominent in the stellar disc since σ is usually at least as large as $V_{\odot}$. However the situation is substantially different for low velocity material concentrated in spiral arms (Clube and Napier, 1992) and we cannot then exclude the possibility that $\rho(z)$ and $\Phi(x)/x$ undergo significant variations with height z from the plane. Thus, in addition to the CHRON-SUPERCHRON structure of the geomagnetic reversal frequency record, characterising the action of giant comets, we expect a '15 Myr' cycle, also a uniquely cometary signature attributable to spiral arm passages.

One final point: while the galacto-terrestrial relationship formulated here is developed in terms of the action of giant comets, the question is obviously left open to what extent the flux of NEOs from the asteroid belt also influences the terrestrial record. This is a fundamental question answered in principle by the extent to which the observed 15 Myr cycle is reproduced by the cratering record. The cratering record is still very sparse (e.g. Grieve 1989) and does not as yet provide a satisfactory basis for periodicity studies (however, see Yabushita, 1991).

7. The Unitary Event Corresponding to the Late Pleistocene-Holocene

We consider now the unitary event corresponding to the most recent major glaciation experienced by the Earth at the end of the Pleistocene (i.e. $\sim$ 25,000 to 10,000 yr BP), commonly attributed at present to the latest weak Milankovitch perturbation. This attribution involves an albedo feedback mechanism which is not easily quantified. Furthermore it takes no account of recent modulations of the terrestrial record which are readily related to the dynamical and physical evolution of a separate astronomically observed system.

It is well known that the more conspicuous meteor streams due to relatively high velocity particles are misleading as to the total meteoroid mass impinging upon the Earth. Thus the incoming mass is dominated by an observed population of relatively low velocity meteoroids in substantially prograde, low inclination orbits known as the sporadic meteor flux. The sporadic meteors have a primary component comprising the so-called helion (HE) and anti-helion (AH) flux at each longitude due to material in fairly eccentric orbits, and a secondary component in near-circular orbits which corresponds to the zodiacal cloud. The secondary is a derivative of the primary component, arising through the action of Poynting-Robertson drag. The orbital aphelia associated with the sporadic flux are closely matched by the population of large meteoroids producing the occasional spectacular fireballs, and tend to lie in the sub-Jovian realm beyond the main concentration of the asteroid belt (Dohnanyi, 1978). Many of these larger meteoroids also break up at great altitudes above the Earth's surface, as we have seen, contributing a total mass of cometary material comparable to the influx of particles from the zodiacal cloud. These latter are recognized amongst the particles collected from the stratosphere by their chondritic composition and tell-tale tracks due to cosmic rays and solar wind, indicating their prior independent existence as interplanetary dust. The cometary material arriving in bulk on the other hand, which remains relatively shielded in space, apparently reduces to micron and submicron particles capable of residing for decades in the atmosphere, which are essentially undetected in the stratosphere.

Some 50% of the AH/HE sporadic flux is now known to be concentrated in a very broad eccentric stream with a similar range of aphelia resulting in pre- and post-perihelion intersections with the Earth's orbit in November and June respectively, each intersection lasting several months (Stohl, 1984). This broad eccentric stream is closely aligned with the so-called Taurid meteor complex which has apparently experienced a history of hierarchical splitting (at perihelion) and fragmentation (elsewhere) during the last $\sim 10^4$ yr (Steel *et al.*, 1991), and it seems very likely now, in view of the known timescales of orbital precession and Poynting-Robertson drag, that these various meteoroidal components of the near-Earth environment are the evolutionary products of a single large progenitor in a Taurid orbit which has disintegrated during these $\sim 10^4$ yr. Depending on the rapidity of this object's descent from the Oort cloud to a sub-Jovian orbit and on the dynamical process that determines its final orbit, it is plausible also to suppose that there was a prior period of perhaps similar duration but of even greater activity which was associated more with the release of volatiles and dust than with the disintegration into meteoroids, and during which there would therefore have been an extremely intense meteor stream. The intersections of this stream with the Earth can be expected to have had a significant influence on the Earth's atmosphere for extended periods of time (Clube and Napier, 1984), and it is reasonable on the basis of observed dust etc. to suppose that the meteoroidal mass present during the last $\sim 2 \times 10^4$ yr is that of a single giant comet of diameter ~ 100 km. The astronomical evidence, then, is broadly consistent with a current glacial/interglacial, that of the late Pleistocene/Holocene, due to a single large comet.

Our interplanetary environment already includes among its present complement of minor bodies at least two large objects in cometary orbits, namely Chiron and 1992 AD with diameters ~ 300 and ~ 100 km respectively. Orbital integrations suggest a survival time for Chiron prior to ejection of $\sim 2 \times 10^5$ yr, during which it makes several transitions to the extreme SP population where it spends a total of about 5000 yr (Hahn and Bailey, 1990). This observation places a sensible upper limit on the size of giant comets remaining in sub-Jovian orbits, since the observed mass function of Oort cloud comets implies a comet of about 100 km diameter is then always present in the extreme SP population in some evolutionary state. It follows that among these candidates, we may plausibly expect a deflection from time to time, due to extreme cometary activity (i.e. non-gravitational forces), into a more stable orbital configuration where the insolated giant comet completes its physical evolution. The existence of the Taurid meteoroid complex in a sub-Jovian orbit, possibly derived from a single large comet, is consistent therefore with the observation of Chiron *and* with the proposition that such comets represent unitary events in Earth history.

At least half-a-dozen asteroids, in addition to Comet P/Encke, are now likely members of the Taurid Complex (Asher and Clube, 1992). These seem likely to have split away from the progenitor in the past whilst the latter may only exist now as a cluster of smaller cometary asteroids and meteoroids. Whether or not a Taurid Complex progenitor still exists is not of course established but there is specific evidence for a considerable meteoroidal swarm associated with a 7:2 Jovian mean motion resonance which seems to have been at the heart of the Complex for some 10^4 yr. This raises the interesting question whether P/Encke is a reactivated cometary asteroid and whether it was released by its librating source (either a large cometary asteroid or cluster of smaller asteroids) in unison with P/Encke's disjointed trail (observed by IRAS in 1983), due to a fragmentation event in the resonance towards the end of the eighteenth century. Thus, much of the observed trail lies within the resonance whilst P/Encke is permanently outside. Modelling constraints indicate this scenario is possible for a libration period of about 400 yr, though such a libration is not particularly stable to planetary perturbations on timescales $\sim$ 500-1000 yr. However, at least one orbit arising from the above constraints has been found to avoid close encounters during the nodal intersection period centred on $\sim$ 500 AD, suggesting that a longer term protected state might exist which is dictated by the dynamical conditions under which the progenitor orbit arose. Such a state would allow the possibility that this particular librating source, through regular encounters with the resonant swarm, has generated further meteoroids and dust every $\sim$ 200 yr. A dust generation process of this kind would certainly be of interest for the values of both its period and its phase since the contemporary climate and climate-proxy cycles on Earth have very similar values i.e. the proposed dust and meteoroid enhancements have coincided with global coolings ca 1900, 1700, 1500 etc. AD. We may be in the presence, it seems, of an erratic-cum-regular dust-making machine which not only continues to feed the zodiacal cloud but which may also be responsible for occasional multi-Tunguska bombardments during the Holocene (Clube and Napier, 1990).

These conclusions are of a provisional nature at present, and stand to fall dependent on the detection of a Taurid Complex source and the dynamical validation of its capture into a resonant orbit at least 5000 yr ago, but it may well be significant that Chinese fireball observations during the last two millennia indicate substantial Taurid stream enhancements during the 5th, 11th and 15th centuries (Hasegawa, 1992) which have not otherwise been explained. The first of these coincides with nodal intersections of the Earth's orbit by the librating source, and it may also be relevant that this coincides with the European Dark Age. Thus it is striking, to say the least, that a period of sustained eschatological concern and known recession in civilization should now coincide with a period of likely close encounters between the Earth and a multi-Tunguska swarm. Further progress along these lines is clearly important so far as unitary events are concerned since misunderstandings as to the true nature of human history and the cometary environment may impede our acceptance of a general galacto-terrestrial relationship. The ancient fear of comets in Europe and broom-stars in China (i.e. guest stars which sweep the Earth clean at the start of a new age) is of course well known but has been rather casually set aside during the scientific era as being of no material significance!

Acknowledgements

I would like to thank Drs. D.J. Asher, M.E. Bailey, W.M. Napier and D.I. Steel for helpful discussions; also Dr. S. Yabushita, Professor W.H. McCrea and the Royal Society for logistical support.

References

Alvarez, L.W., Alvarez, W., Asaro, F. and Mitchell, H.V.: 1980, *Science* **208,** 1095
Asher, D.J. and Clube, S.V.M.:1992, in preparation
Astapovic, I.S. and Terenteva, A.K.: 1968, *IAU Symp* **33,** 308
Bailey, M.E.: 1983, *ACM I (*ed Lagerkvist, C.I. and Rickman, H., Uppsala), 383
Bailey, M.E.: 1992, *Celest. Mech.,* this issue, p. 49
Bailey, M.E. and Clube, S.V.M.: 1978, *Nature* **275,** 278
Bailey, M.E. Clube, S.V.M. and Napier, W.M.: 1990, *The Origin of Comets (*Pergamon)
Bhandari, N.: 1991, *Current Science* **61,**2, 97
Bigg, E.K. and Thompson, W.J.: 1969, *Nature* **222,** 156
Bohor, B.F., Foord, E.E., Modreski, P.J. and Triplehorn, D.M.: 1984 *Science* **224,** 867
Chapman, C.R. and Morrison, D.: 1989, *Cosmic Catastrophes (*Plenum)
Clube, S.V.M.: 1973, *Mon. Not. R. Astr. Soc.* **161,** 445
Clube, S.V.M.: 1978, *Vistas Astron.* **22,** 77
Clube, S.V.M.: 1988, *Dust in the Universe (*ed Bailey, M.E. and Williams, D.A., CUP), 331
Clube, S.V.M.: 1990, *Apeiron* **9,** 30
Clube, S.V.M. and Napier, W.M.: 1982a, *Q. Jl. R. Astr. Soc.* **23,** 45
Clube, S.V.M. and Napier, W.M.: 1982b, *Earth Planet Sci. Lett.* **57,** 251
Clube, S.V.M. and Napier, W.M.: 1982c, *The Cosmic Serpent (*Faber)
Clube, S.V.M. and Napier, W.M.: 1984a, *Mon. Not. R. Astr. Soc.* **211,** 953
Clube, S.V.M. and Napier, W.M.: 1984b, *Nature* **311,** 635
Clube, S.V.M. and Napier, W.M.: 1986, *The Galaxy and the Solar System (*ed Smolochowski, R., Bahcall, J.N. and Matthews, M.S., Arizona), 260
Clube, S.V.M. and Napier, W.M.: 1990, *The Cosmic Winter (*Blackwell)
Clube, S.V.M. and Napier, W.M.: 1992, *Mon. not. R. Astr. Soc.,* submitted
Clube, S.V.M. and Watson, F.G.: 1979, *Mon. Not. R. Astr. Soc.* **187,** 863
Creer, K.M. and Pal, P.C.: 1989, *Catastrophes and Evolution (*ed Clube, S.V.M., CUP), 113
Davis, M., Hut, P. and Muller, R.A.: 1984, *Nature* **308,** 715
De Young, D.S.: 1989, *Astrophys. J.* **342,** L59
Dicke, R.H.: 1962, *Enrico Fermi Course XX (*ed Möller, C., Academic), 1
Dohnanyi, J.S.: 1978, *Cosmic Dust (*ed McDonnell, J.A.M., Wiley), 527
Duncan,M., Quinn, T. and Tremaine, S.: 1988, *Astrophys. J. Lett.* **328,** L69
Everhart, E.: 1972, *Astrophys. J. Lett.* **10,** 131
Fechtig, H.: 1982, *Comets (*ed Wilkening, L., Arizona), 370
Grieve, R.A.F.: 1989, *Catastrophes and Evolution (*ed Clube, S.V.M., CUP), 57
Grun, E., Zook, H.A., Fechtig, H. and Giese, R.H.: 1986, *Icarus* **62,** 244
Hahn, G. and Bailey, M.E.: 1990, *Nature* **348** 132
Hasegawa, I.: 1992, *Celest. Mech.,* this issue, p. 129
Heisler, J. and Tremaine, S.D.: 1986, *Icarus* **65,** 13
Hills, J.G.: 1981, *Astron. J.* **86,** 1730
Hills, J.G.: 1984, *Nature* **311,** 636
Holmes, A.: 1927, *The age of the Earth (*Benn)
Hughes, D.W.:1989, *ACM III (*ed Lagerkvist C.I. *et al,* Uppsala)

Kauffman, P., Kuntz, V.L.R., Pase Leme, N.M., Piazza, L.R., Vilas Boas, J.W.S., Brecher, K. and Crouchley, J.: 1989, *Science* **246,** 787
Kresak, L.: 1991, *Comets in the Post-Halley era (*eds R.L. Newburn *et al,* Kluwer), I, 607
Marcus, J.N. and Olsen, M.A.: 1991, *Comets in the Post-Halley era (*eds R.L. Newburn *et al,* Kluwer), I, 439
McCrea, W.H.: 1978, *The Origin of the Solar System (*ed Dermott, S.F., Wiley), 75
McCrea, W.H.: 1981, *Proc. R. Soc., Ser. A***375,** 1
McLaren, D.J.: 1983, *Geol. Soc. Am. Bull.* **94,** 313
Milani, A., Carpino, M., Hahn, G. and Nobili, A.M.: 1989, *Icarus* **78,** 212
Nakamura, T. and Yoshikawa,: 1991, *Publ. Nat. Astron. Obs. Japan* **2,** 293
Napier, W.M. and Clube, S.V.M.: 1979, *Nature* **282,** 455
Oort, J.H.: 1927, *Bull. Astron. Inst. Neth.* **4,** 79
Oort, J.H.: 1932, *Bull. Astron. Inst. Neth.***6,** 249
Oort, J.H.: 1950, *Bull. Astron. Inst. Neth.* **11,** 91
Rampino, M.R. and Caldeira, K.: 1992, *Celest. Mech.,* this issue, p. 143
Rampino, M.R. and Stothers, R.B.: 1984, *Nature* **308,** 709
Raup, D.M. and Sepkoski, J.J.: 1984, *Proc. Natl. Acad. Sci. (USA),* **81,** 801
Raup, D.M. and Sepkoski, J.J.:1986, *Science* **231,** 833
Shoemaker, E.M.: 1983, *Rev. Earth Planet Sci.* **11,** 461
Shoemaker, E.M. and Izett, G.A.: 1992, *Lunar Planet Sci. Conf.* **23,** in press
Steel, D.I.: 1991, *Nature* **354,** 265
Steel, D.I. Asher, D.J. and Clube, S.V.M.: 1991, *Mon. Not. R. Astr. Soc.* **251,** 632
Stohl, J.: 1984, *ACM I (*ed Lagerkvist, C.I. and Rickman, H., Uppsala), 419
Thaddeus, P. and Chanan, G.A.: 1985, *Nature* **314,** 73
Wetherill, G.W.: 1988, *Icarus* **76,** 1
Wetherill, G.W. and ReVelle, D.O.: 1982, *Comets (*ed Wilkening, L., Arizona), 297
Whipple, F.L.: 1957, *Trans IAU* **9,** 249
Whipple, F.L.: 1967, *NASA SP-* **150,** 409
Whipple, F.L.: 1992, *Celest. Mech.,* this issue, p. 1
Yabushita, S.: 1983, *Q. Jl. Astr. Soc.* **24,** 430
Yabushita, S.: 1991, *Mon. Not. R. Astr. Soc.* **250,**481
Zahnle, K. and Grinspoon, D.: 1990, *Nature* **348,** 157
Zhao, M. and Bada, J.L.: 1989, *Nature* **339,** 463

PROGRESS IN THE ELLIPTIC RESTRICTED 3-BODY PROBLEM: ASTEROIDS IN THE 2/1, 3/1 AND 1/1 RESONANCE

R. DVORAK
Institute of Astronomy, University of Vienna, Türkenschanzstrasse 17, A-1180 Vienna
e-mail: DVORAK@AVIA.UNA.AC.AT

Abstract. We report on recent research of the elliptic restricted three-body-problem concerning the motion of Asteroids in the 2/1, 3/1 and 1/1 mean motion resonances with Jupiter. In a recent review (J.Henrard, 1988) one can find a detailed description of the problem of the Kirkwood gaps. Therefore we concentrate on the new results of the last three years derived numerically and analytically. It is interesting that some of these asteroids originally in a gap can reach such high eccentricities, that they can become Mars crossers, Jupiter crossers and even, in exceptional cases Earth crossers. Such a close approach with a major planet can throw out the body from the main belt or can cause a collision! We also present new results of the largeness of stable zones around the Lagrangean equilibrium points L_4 and L_5 and compare them with existing ones.

Key words: Elliptic restricted problem, Kirkwood gaps, Trojans.

1. Introduction

Twenty years ago V. Szebehely (1970) finished his book "Theory of Orbits", with the subtitle: "The restricted problem of three bodies". He claims there in the foreword, that: "The subject of this treatise occupies a central place in analytical dynamics, celestial mechanics, and space dynamics". This statement may be also true for the more complicated, but more realistic elliptic restricted problem of three bodies. In the former mentioned book only a small chapter is devoted to it; this fact is understandable as there was still so much to do on the circular problem. The last decades have brought an incredibly fast development of the capacity of the computers in memory and speed, and therefore it is now possible to treat the more difficult elliptic restricted problem (=EP) with the aid of numerical experiments. Many problems of the dynamics of celestial bodies can only be explained within the framework of the EP. The circular problem (=CP) fails e.g. to explain the Kirkwood gaps in the asteroidal main belt and is definitely not the right model to describe the difficult capture process of a comet with a hyperbolic orbit to a short periodic comet. So scientists in the last two decades started to study this model more and more and succeeded in explaining an unsolved riddle of the last 150 years: We have now strong evidence, that the asteroids in the 3:1 mean motion resonance with Jupiter are thrown out because of the chaoticity of their orbit and the following close approach with Mars (or the Earth or Jupiter). This was one of the first discoveries of chaotic motion in the Solar System. Some evidence for this chaoticity of the motion was already found by Froeschlé and Scholl (1981,1982), the final discovery is due to Wisdom's papers (1982, 1983). But nevertheless it

Celestial Mechanics and Dynamical Astronomy **54:** 195–205, 1992.

is not yet clear whether the 2:1 resonance gap can also be explained through the appearance of chaotic motion.

In this sense the aim of this paper is to present the recent results on the motion of Asteroids in the 2/1 and 3/1 resonance. Some new results concerning stability regions around the triangular equilibrium points will also be discussed in the framework of the EP. Although analytical work exists on the stability of the equilibrium points up to high orders, no estimation of the largeness of a stable zone around was determined up to now.

2. Description of the Model

Let us first describe the elliptic restricted three body problem, which is the basic physical model of all the results presented in this article:

It is the general 3-body problem for point masses with the following restrictions:

(1) the 3rd body is supposed not to influence gravitationally the two primary bodies ($m_3 = 0$)

(2) it follows that m_1 and m_2 have a simple Keplerian orbit

(3) m_3 is moving in the same plane as m_1 and m_2

If the Keplerian orbit of the primaries is a circle we speak of the (circular) restricted problem; if m_3 does not move in the plane of m_1 and m_2 we call it the three dimensional restricted problem.

The equations of motion in a rotating pulsating system x and y, where the two primary bodies have fixed positions at the x-axis are the following ones (e.g. Szebehely, 1970):

$$\ddot{x} - 2\dot{y} = \frac{\partial\Omega}{\partial x} \qquad \ddot{y} + 2\dot{x} = \frac{\partial\Omega}{\partial y} \tag{1}$$

with the potential Ω of the dynamical system in the rotating axes.

$$\Omega = \frac{1}{2}[(1-\mu)r_1^2 + \mu r_2^2] + \frac{1-\mu}{r_1} + \frac{\mu}{r_2} \tag{2}$$

where $m_1 + m_2 = 1$ and one of the masses is μ; r_1 and r_2 are the respective distances of the 3rd body from the primaries. It is well known that one can derive the Jacobian constant by multiplying the former equations by $\dot{y}$ respectively $\dot{x}$, summing up and integrating:

$$v^2 = 2\Omega + C_1 \tag{3}$$

This expression leads to the zero velocity curves already derived by Hill some 100 years ago. It is important to note, that with their aid one can define forbidden regions of motion of the third body (e.g. V. Szebehely, 1970)

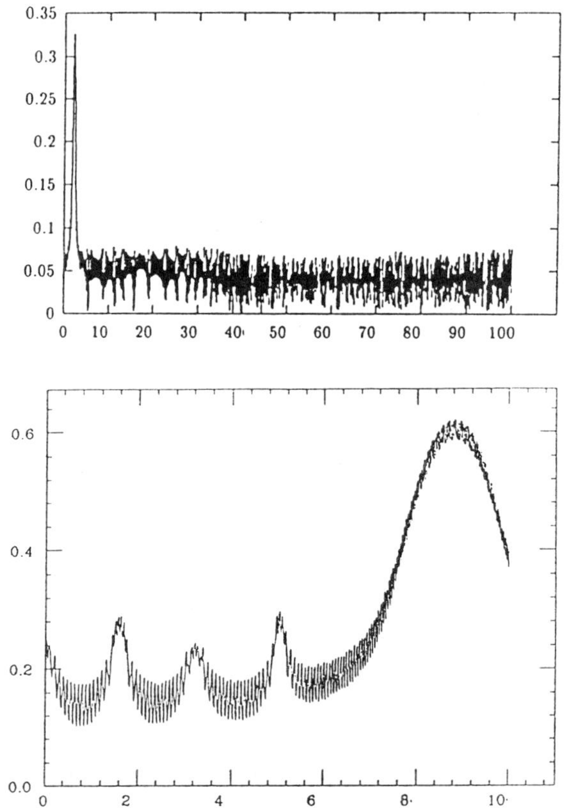

Fig. 1. Two typical chaotic orbits in the 3/1 resonance, a possible Mars crosser (upper graph) and a possible Earth crosser(lower graph); time is measured in 10^4 years (x-axis); eccentricity (y-axis)

When taking into account the eccentricity of the primaries' orbit, a rotating and pulsating coordinate systems may be used, where the equations of motion for the third body can be written as

$$x'' - 2y' = \frac{\partial \omega}{\partial x} \tag{4}$$

$$y'' - 2x' = \frac{\partial \omega}{\partial y} \tag{5}$$

with $' := \frac{d}{df}$. The potential ω varies with the period of the primaries:

$$\omega(x, y, f) = \frac{\Omega(x, y)}{1 + e \cos f} \tag{6}$$

with Ω given from the circular problem.

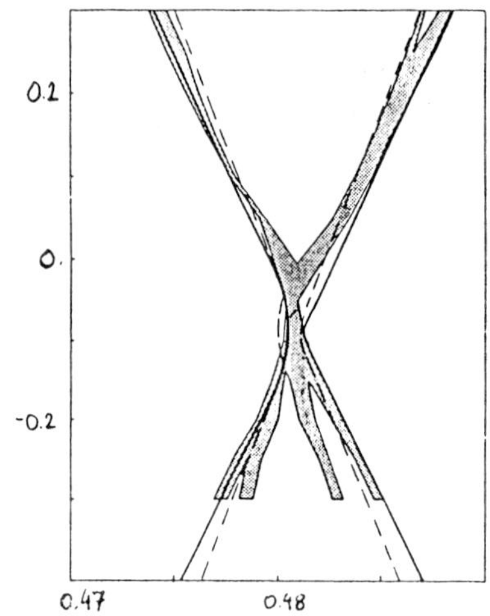

Fig. 2. Hyperbola like level curves derived with a semi-numerical perturbation method by Henrard and Caranicolas (1990) in the representative plane semi-major-axis versus $e.\sin\sigma$ with σ the resonant angle. The shaded area shows the superimposed chaotic region derived by Wisdom (1983).

In the EP no 1st integral can be derived, that means that no zero velocity curves can be used for a qualitative analysis of the motion. However the system is much closer to many problems in Celestial Mechanics and Dynamical Astronomy. The EP is especially useful when one wants to treat the motions of Asteroids under the gravitational attraction of the Sun and one massive planet (e.g. Jupiter) on an eccentric orbit around the Sun.

3. The 3:1 Resonance

We show in Figure 1 two typical chaotic orbits in the 3/1 resonance: the first one is a fictitious asteroid, which we started with a low initial eccentricity ($e = 0.05$). Its orbit suffers from an immediate increase of the eccentricity up to $e = 0.32$, then the graph shows only small variations in the eccentricity ($0.0 \leq e \leq 0.075$) for very long time (1 Myr) (Figure 1a). The second one, which was started with an initial eccentricity of 0.25, suffered from relatively small variations at first and then one recognizes a sudden increase up to 0.6 in the eccentricity! Note that here the time scale is one order smaller, the intergration time was only 10^5 years (Figure 1b). While the first fictitious Asteroid would be (in a full planetary model) a crosser of the orbit of Mars, the second one would even cross the Earth's orbit.

Recently Henrard and Caranicolas (1990) could explain that the chaoticity of the motion in this gap comes from two different sources: (1) due to the the separatrices of secondary resonances which leads to a chaotic layer and (2) due to a slow crossing of the separatrices. The 1st source seems to be the more important because it is responsible for the random jumps between high eccentric and low eccentric

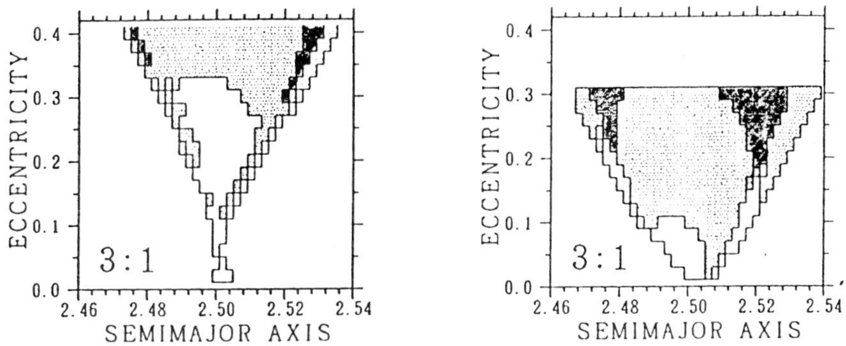

Fig. 3. Initial values of a and e for orbits of Asteroids in the 3/1 resonance with Jupiter which can become planet crossers during their dynamical life times. The light gray region indicate possible mars crossers, the dark region possible Jupiter and Earth crossers. The right picture corresponds to the lower part of Figure 2 upside down (after Yoshikawa, 1989).

modes. The latter behaviour can quite well be seen from Figure 1a.

The structure of the gap itself was also very well explained by Henrard and Caranicolas (1990): In the middle of the resonance no chaotic motion appears (Figure 2). It is dominant on the X structured region in the a versus $e.sin\sigma$ (σ is the resonant angle) initial condition diagram. The same shape was found by Wisdom (1983) and by Yoshikawa (1991) numerically (Figure 3). In a recent numerical approach Dvorak (1991) determined also the structure of the gap with the aid of the maximum eccentricities of the Asteroids' orbits during the whole integration time. One can see the results on a contour plot (Figure 4); there exists a relatively sharp line of possible high eccentric orbits up to an initial eccentricity $e = 0.2$. From that value on a bifurcation is visible which coincidents quite well with the other mentioned results (Figures 2 and 3).

4. The 2:1 Resonance

The key for understanding the structure of the gap is also due to Henrard and Lemaître (1986 und 1987) where they developed a semianalytical method for the 2:1 resonance. Although they concluded that their results do "not support the theory of formation of the 2/1 Kirkwood gap by removal of the asteroids through close encounters with Mars" they put in evidence the important role of chaoticity. This former statement agrees with Yokishawa (1989, 1990 and 1991), who doubted that pure gravitational forces are responsible for the formation of this gap. A new analytical treatment based on the stability of periodic orbits by Morbidelli and Giorgilli (1991) could find for the 2/1 Kirkwood gap "no mechanical explanation in the framework of the complete restricted problem of three bodies".

In the next three figures we show the structure of the 2/1 resonance derived by a semianalytical approach (Figure 5), numerical integrations by Yoshikawa, 1991

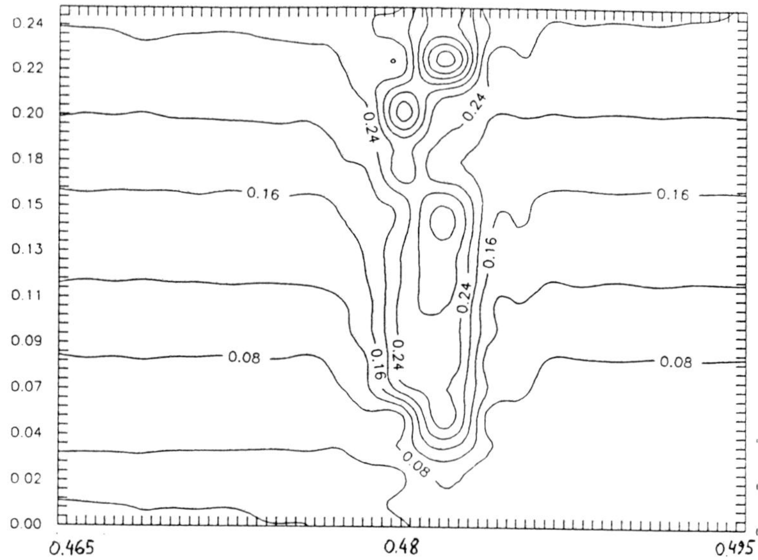

Fig. 4. Contour plot of the initial condition diagram for the 3/1 resonance, semi major axis *versus* eccentricity. Contours are plotted for the maximum value of the eccentricity during the whole integration time (10^4 periods of the Jupiter)

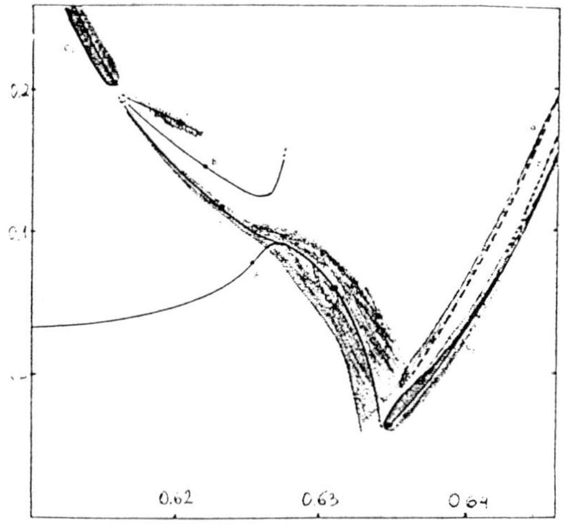

Fig. 5. Determination of regions in the initial a versus initial e diagram where chaotic motion (shaded area) is expected after Henrard and Lemaître (1987).

(Figure 6) and Dvorak 1992 (Figure 7). Chaotic motion arises on the edges of the resonance, but not in the inner part of the 2/1 resonance with Jupiter. All graphs show the same overall picture: no region of chaoticity appears in the middle of the resonance also for initially low eccentric orbits.

A typical orbit is shown in Figure 8, where an orbit looks quite regular. Then,

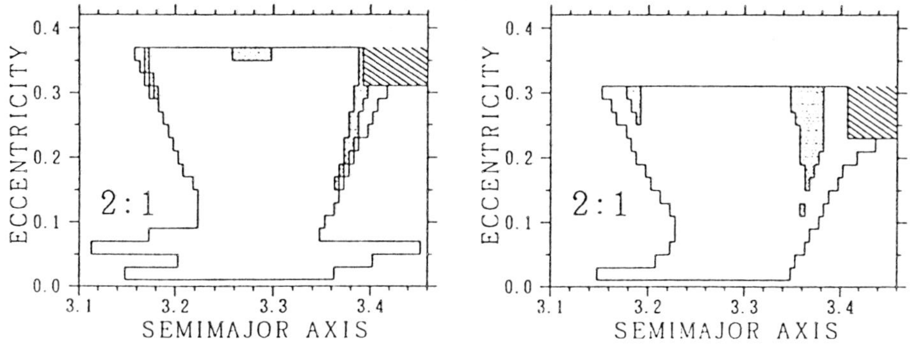

Fig. 6. Initial values of a and e for orbits of Asteroids in the 2/1 resonance with Jupiter which can become planet crossers during their dynamical life time. The light gray region indicate possible mars crossers, the dark region possible Jupiter and Earth crossers.

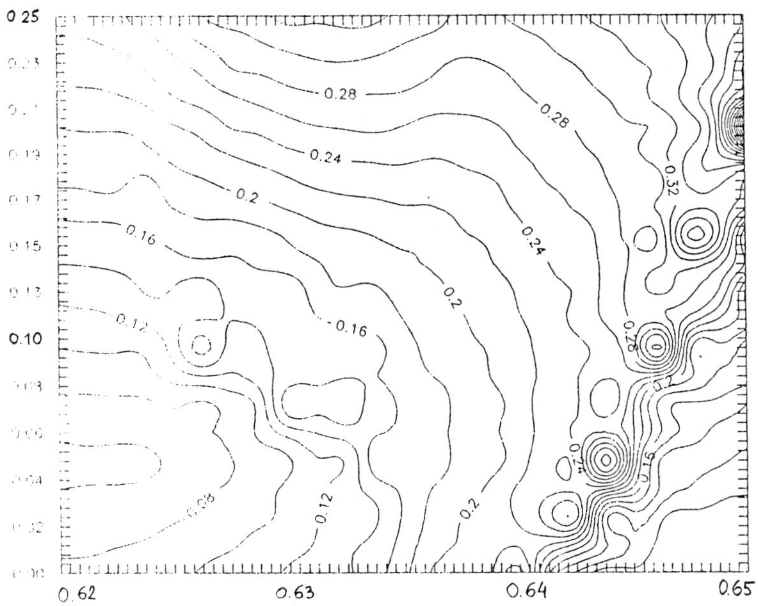

Fig. 7. Contour plot of the initial condition diagram for the 2/1 resonance, initial semi major axis versus initial eccentricity. Contours are plotted for the maximum value of the eccentricity during the whole integration time (10^4 periods of the Jupiter).

after a sudden close approach to the second primary body (Jupiter), it starts to be very irregular and the body will finally be ejected from the Planetary System (=hyperbolic orbit). It has to be emphasized that for the same calculations (same initial conditions for the fictitious asteroid) in the circular model the orbit stays quasiperiodic with a much lower maximum value of the eccentricity e. It seems that the existence of the Jacobi-Integral does not allow high increases of the eccentricity. From this it is again evident, that the eccentricity of Jupiter's orbit (e=0.048 during

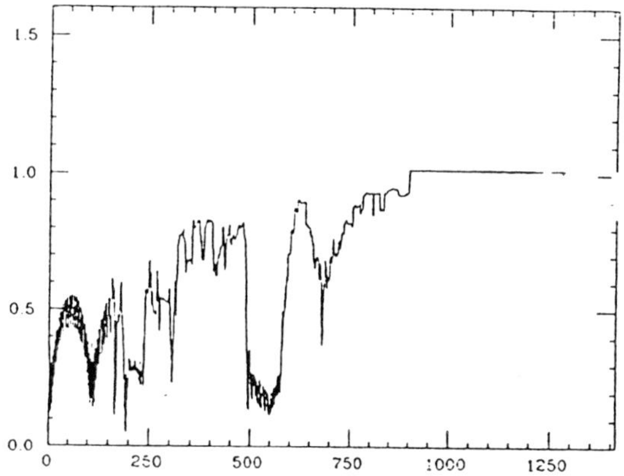

Fig. 8. A typically chaotic orbit in the 2/1 resonance, time is measured in 100 years.

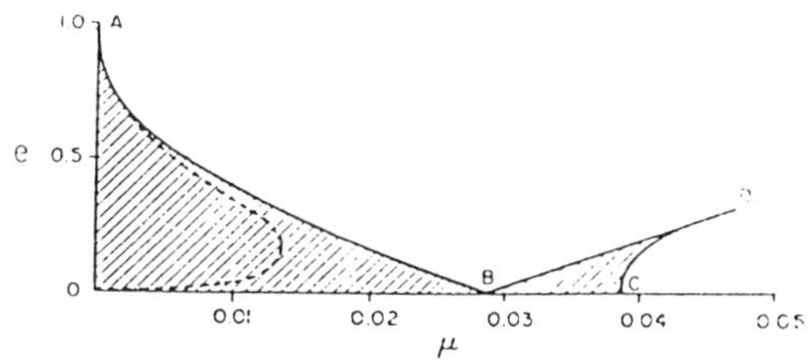

Fig. 9. Linear stability of the triangular points (after Danby, 1964)

all the integrations by Dvorak and also by Yoshikawa and Wisdom) leads to a drastic change of the orbit.

5. The 1:1 Resonance

Apart from the motion of asteroids in the main belt, there is another interesting question namely asteroids in the 1:1 mean motion resonance with Jupiter. Since long time we know that in fact both Lagrangean triangular points are populated with the so called Trojan Asteroids. They move in very interesting orbits around these equilibrium points (e.g. the horseshoe orbits). The knowledge of such orbits is also connected with the determination of the stability of the triangular stability points itself. There exist many studies of this kind for the CP, and some for the EP. It is well established since long time, that in the CP from the value $\mu_{\text{crit}} = 0.040064$ (e.g. Roy, 1988) on these point are stable: that means for all the planets in the Solar System (always regarded as a separate system Sun - Planet - massless body) the points L_4 and L_5 are stable. To determine the stability of these points one can proceed in the EP as one has to do it for the CP: the equations of motion (1) can be linearized around the libration points. But in the EP the result of the linearized stability analysis is not solvable in closed form. Nevertheless Danby (1964) and

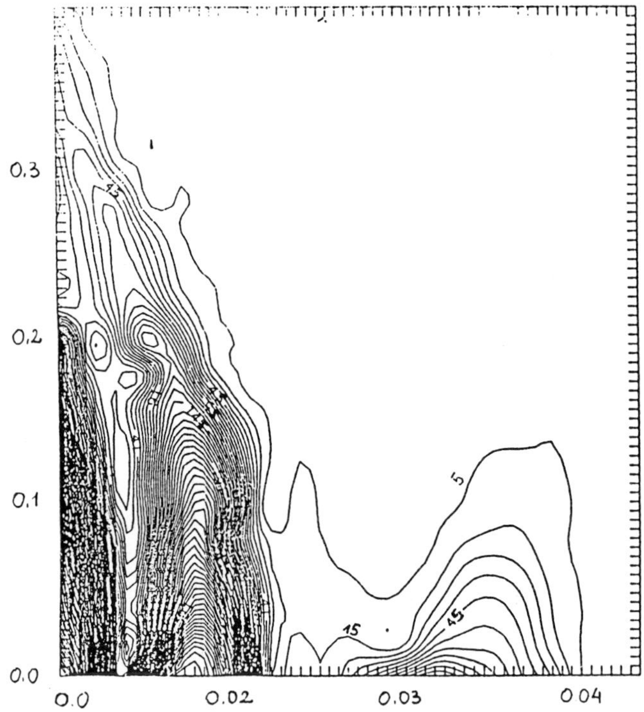

Fig. 10. Contour Plot of the Largeness of the stability zones around L_4 in the EP. x-axis is the mass ration, y-axis is the eccentricity, contours are plotted for the number of stable orbits found around the equilibrium point (Lohinger and Dvorak, to be published)

Bennett (1965) determined the stability characteristics for them in dependence of the mass ratio and the eccentricity.

The shaded area (Figure 9) is the linear stable one. The x-axis corresponds to the CP, where one recognizes 2 critical values of μ. $\mu^* = 0.02859548$ is lying in a region of stable Lagrange points and μ_0 is identical with the former mentioned μ_{crit}. From the figure one can see, that in general higher eccentric orbits of the primaries lead to less stable equilibrium point for the other parameter μ. This is not true for values between $\mu^* = 0.02859548$ and μ_0 where the eccentricity of the primaries stabilizes the equilibrium points.

It is expected, that a stable equilibrium point is surrounded by a region of stable (regular) motion, that means that a body close to such a point stays there. One has to keep in mind that a body moving exactly in L_4 or in L_5 corresponds to a periodic orbit. It is well known that a stable periodic orbit is surrounded by regular orbits laying on a torus around it.

The question arises how large is this region of stable motion around the Lagrange points. Up to now no analytical treatment of such problem has been undertaken. Once more the numerical analysis can be used to produce such interesting results.

There exist already a study of this kind for the CP, where zones around the triangular points have been determined for all mass ratios (Dvorak and Lohinger, 1990). The results fit quite well to the former discussed appearance of one unstable point in the stable region. This work has now been extended to the whole parameter space eccentricity versus μ and the largeness of such regions has been determined.

For our numerical investigation we had to choose an efficient integration method and to define the stability of the motion. The stability conditions were chosen in such a way to be "precise" enough to give a qualitatively correct picture of motion. For the study here we adapted the following stability definition:

An orbit is called stable when its distance to L_4 remains smaller than 0.86 units of the distance of the primaries during the whole integration time of 200 periods of the primaries.

Parallelly we used the Lie-Integrator (Hanslmeier and Dvorak, 1984) and the Bulirsch-Stoer adapted by J. Kribbel (1987). As a grid size we took 0.005 units of the distance of primaries in the rotating x-y frame.

On the x-axis we plotted the mass ratio μ, on the y-axis we plotted the eccentricity of the primaries and the contours are plotted according the number of stable orbits, (this is in our definition equivalent to the largeness of the stable region around the equilibrium point). As the mass ratio tends to 0.0 (in the 2-body problem all motions are stable) the stable zones increase. It is interesting to note, that the semianalytical work (compare Figures 9 and 10) is very well confirmed. As a new result the largeness of the stability zone indicates the zone where stable motion for a massless body can still be expected.

6. Conclusions

What can we conclude from the recent results of the motion of Asteroids in resonances? Although many old problems seem to be answered in a satisfactory way we are still not able to understand the structure of the Asteroid belt as a whole. The motion of the Trojans seems quite well be understood; the cause of the gap in the 3/1 mean resonance with Jupiter is also clear. However the 2/1 and 3/2 resonance are worth to be studied again analytically and numerically in spite of very important contributions to the topic (e.g. Schubart, 1982). But a lot of work in this domain is on its way. So we can hope to give a "final" explanation of this riddle of Celestial Mechanics within the next few years.

References

Benett, A.: 1965, "Characteristic exponents of the five equilibrium solutions in the elliptically restricted problem", Icarus 1, 177

Danby, J.M.A.: 1964, "Stability of the triangular points in the elliptic restricted problem of three bodies", Astron. J. 68, 159

Dvorak, R., Lohinger, E.:1991, "Stability Zones around the Triangular Lagrangean Points", in Roy, A.E. (ed) *Predictability, Stability, and Chaos in n-body Dynamical Systems*, Plenum Publishing Corporation (in press)

Dvorak, R.:1992, "New Results on the Motion of Asteroids in Resonances", in Ferraz-Mello, S. (ed), IAU Symposium 152 "Chaos, Resonance and Collective Dynamical Phenomena in the Solar System", (in press)

Froeschlé, C., and Scholl, H.: 1981, "The Stochasticity of Peculiar Orbits in the 2/1 Kirkwood Gap", Astron.Astrophys., 93, p. 62

Froeschlé, C., and Scholl, H.: 1982, "A Systematic Exploration of Three-dimensional Asteroidal Motion at the 2/1 Resonance", Astron. Astrophys. 111 p. 346

Hanslmeier, A., and Dvorak, R.:1984, "Numerical Integration with Lie-Series", Astron. Astrophy. 132, 203

Henrard, J.: 1988, "Resonances in the Planar Elliptic restricted Problem", in Roy, A. (ed), *Long Term Behaviour of Natural and Artificial N-Body Systems*.

Henrard, J., and Caranicolas, N.D.: 1990, "Motion near the 3/1 Resonance of the planar elliptic restricted three body problem", Cel. Mech. 47, p. 99

Kribbel, J.: 1987, private communication

Lemaître, A., and Henrard, J.: 1986, "A perturbation method for problems with two critical arguments", Cel. Mech. 39, 213

Lemaître, A., and Henrard, J.: 1987, "A perturbative treatment of the 2/1 Jovian resonance", Icarus 69, 266

Lemaître, A., and Henrard, J.: 1990, "On the Origin of Chaotic Behaviour in the 2/1 Kirkwood Gap", Icarus, 83, p. 391

Morbidelli, A. and Giorgilli, A.:1990, "On the dynamics in the asteroids belt. Part II: Detailed study of the main resonances" Cel. Mech., 47, p. 173

Roy, A.E.: 1988, "Orbital Motion", Adam Hilger, Bristol and Philadelphia, 3rd edition,

Schubart, J.: 1982, "Three Characteristic Parameters of Orbits of Hilda-type Asteroids", Astron. Astrophys. 114, 200

Szebehely, V.: 1970: "Theory of Orbits", Academic Press, Inc.

Wisdom, J. : 1982, "The origin of the Kirkwood gaps: A mapping for Asteroidal motion near the 3/1 commensurability", Astron. J. 87, p. 577

Wisdom, J. : 1983, " Chaotic behaviour and the origin of the 3/1 Kirkwood gap", Icarus, 56, p. 51

Yoshikawa, M.: 1989, "A Survey of the Motions of Asteroids in the Commensurabilities with Jupiter", Astron. Astrophys. 213, p. 436

Yoshikawa, M.: 1990, "Motions of Asteroids at the Kirkwood Gaps I. On the 3:1 Resonance with Jupiter", Icarus, 87, p. 78

Yoshikawa, M.: 1991, "Motions of Asteroids at the Kirkwood Gaps II. On the 5:2, 7:3, and 2:1 Resonances with Jupiter", Icarus 92, p. 94

ASTEROID FAMILIES:
RECENT RESULTS AND PRESENT SCENARIO

V. ZAPPALÀ and A. CELLINO
Osservatorio Astronomico di Torino, I-10025 Pino Torinese (TO), Italy

Abstract. After several decades of frustrating results showing a generally poor agreement among different asteroid family classifications, recent studies based on high accuracy proper elements, as well as on objective statistical methods of cluster analysis have largely improved the situation. Now, a number of asteroid families have been recognized on the basis of different methods of cluster analysis, using asteroid proper elements data sets computed by means of different theories. For these reasons, they should be considered of very high reliability. Moreover, spectroscopic observations confirm in some cases these results, indicating surface compositions of the family members in agreement with a geochemically plausible parent body. However, in particular zones of the belt, like the Flora region, further efforts should be performed in order to establish the real consistence of the resulting clusterings of objects. In addition, the size distribution and the taxonomic types of some well established families seem to indicate particular features of the family sample when compared with the field objects. We recall that asteroid families, in the framework of asteroid collisional evolution, are of the highest importance for understanding the mechanisms of injection of fragments into the Earth-crossing zone through mean-motion and secular resonances and, as a consequence, for evaluating the impact rate on Earth of asteroidal objects.

Key words: Asteroid, asteroid families, proper elements, cluster analysis.

1. Introduction

Asteroid dynamical families have long been a subject of research for planetological science during the present century. Their discovery dates back to 1918, when Hirayama noticed the existence of big concentrations in the distribution of the asteroid orbital elements a, e and i. In some cases the observed condensations appeared too conspicuous to be due to chance. Hirayama devoted to this subject a long series of papers (Hirayama, 1918, 1919, 1920, 1923, 1928, 1933), and succeeded in evidencing a number of clusterings, which he called *families*. The best examples of these are the three well known Eos, Themis and Koronis families.

The basic and today generally accepted idea for explaining the existence of asteroid families is also due to Hirayama. According with this hypothesis, families were formed as the results of collisional events. They originated from the catastrophic disruption of single parent bodies, which were destroyed by high-velocity impacts with other objects of asteroidal size. In other words, the present family members should be collisional fragments which were ejected with comparatively low relative velocities, so that their orbital elements did not change too much with respect to each other and to the original parent body. This fact makes them still recognizable today.

Celestial Mechanics and Dynamical Astronomy **54:** 207–227, 1992.

According with this scenario, the name "family" should be assigned only to groupings composed by members genetically related, rather than to any defined clustering of objects in the space of orbital elements. As a consequence, groupings that are recognizable only because they occupy particular zones of the belt which are isolated as a result of the presence of secular and mean motion resonances, should not be termed families. A well known example in this sense is given by the objects of the so-called Phocaea zone.

A very important point concerning asteroid dynamical families is that, as was already recognized by Hirayama, the time variations of the orbital osculating elements of asteroids due to the gravitational perturbations of the planets, mainly Jupiter, make these elements not suitable for the purposes of families identification. Instead, the *proper elements* a', e', i', defined as constants of integration of the differential equations giving the variation of the orbital elements in the framework of the secular perturbation theory, provide a set of elements which in principle could be considered constant. For this reason proper elements have been computed and analyzed by the authors who have studied the problem of asteroid family identification during the last decades.

Among them, and after the work of Hirayama, the most important studies have been published by Brouwer (1951), Arnold (1969), Lindblad and Southworth (1971), Carusi and Massaro (1978), Williams (1979, 1989), Kozai (1979). In addition, Van Houten *et al.* (1970) carried out a search for asteroid families using solely the data set provided by the Palomar Leiden Survey (PLS).

Of course, the different searches for asteroid families quoted above, differ largely in some important respects; in particular, different authors used various data sets of asteroid proper elements, different theories of proper elements computation, as well as different criteria of clustering identification. Moreover, various statistical tests of the results were applied by some authors, while they were not attempted at all in other studies. The general situation is shown in Table I. For a general review on this subject, see Valsecchi *et al.* (1989).

As can be seen in Table I, in spite of the noticeable effort produced by different researchers, the situation concerning asteroid families identification has been far from satisfactory until recently. In fact, a general agreement among the various proposed family lists has not long been achieved neither for what concerns the number nor the consistency of the families, with the only exception of the three major cases of Eos, Themis and Koronis.

On the other hand, the interest about asteroid families has been growing recently, mainly as a consequence of the improvements of our knowledge of the general process of collisional evolution of the asteroid belt (Chapman *et al.*, 1989; Davis *et al.*, 1989). In fact, it is evident that in this respect asteroid families have a crucial importance: their number can be considered as a significant constraint for the models aimed to describe the overall collisional evolution of asteroids, while the observed physical properties of their members can provide important information on the processes of catastrophic breakup of bodies of asteroidal size, as well as

TABLE I
Results of the asteroid families searches published until 1989. The general disagreement is evident for what concerns the number of recognized families. The results of the two most recent attempts of families identification (Zappalà *et al.*, 1990; Bendjoya *et al.*, 1991) are presented and discussed in the text

Author(s)	Numbered Asteroids	PLS Asteroids	Families found
Hirayama (1933)	1223	no	5
Brouwer (1951)	1563	no	28
Arnold (1969)	1735	no	37
Van Houten *et al.* (1970)	—	yes	14
Lindblad and Southworth (1971)	1735	yes	34
Carusi and Massaro (1978)	1861	yes	15
Williams (1979)	1796	yes	104
Kozai (1979)	2125	no	72
Williams (1989)	2065	yes	117

on the internal composition of asteroids. In this sense, the family members could give a unique possibility for "looking inside" asteroidal bodies, since their present exposed surfaces come from the interior of their parent bodies.

For these reasons, it is evident that the discrepancies among the different existing family lists are particularly frustrating, since they do not allow us to take profit of the potentially important physical information contained in the presently existing asteroid families. These discrepancies are due to various reasons. As already mentioned, they are basically a consequence of: (1) the different data bases considered by the different authors; (2) the different methods used to compute the orbital proper elements; (3) the different identification procedures used.

Among these causes, (1) is probably less important. While it is evident that in general the larger the data set, the better are the results, we can expect that the lack of a number of (mostly small) objects can mainly influence the ability to identify some families not easily recognizable and largely composed by small members. On the other hand, we do not think that the differences on the adopted data sets listed in Table I can explain the big discrepancies of the results. As an example, it is hard to believe that a difference of 61 objects between the Lindblad and Southworth (1971) and the Williams (1979) data sets can justify the big discrepancy on the number of families found by these authors (34 and 104, respectively).

More important appear to be the differences among both the adopted proper elements computation theories, and the methods used for identifying families out of the background in the space of proper elements. For what concerns the former problem, since it is often hard to quantify precisely the accuracies of the different theories of proper elements computation, as well as the influence that these can have on the overall results of asteroid families identification, some numerical and empirical approaches are probably more suitable. For instance, tests devoted to study, for selected asteroids chosen in different zones of the belt, the stability of the proper elements by means of numerical integrations spanning over a relatively large period time have been performed by Milani and Kneževič (1990). On the other hand, a good test could be also to apply a given method of families identification to a data set of asteroids whose proper elements are computed by means of different techniques. This can in principle allow us to infer what is the importance of the proper elements computation for the purposes of families identification. Such an exercise has been recently carried out (Zappalà *et al.*, 1992), and the results will be reviewed in Section 2.2.

On the other hand, the adopted identification method has a crucial importance. So far, most of the procedures applied by the various authors have been eminently visual. Moreover, in several cases no statistical test for the significance of the resulting clusterings found in the space of proper elements was performed. Instead, what we need is in principle a method which should satisfy some properties of *objectivity* and *reproducibility*. For this reason, some automated procedure managed by computer should be preferred, in order to avoid well known biases affecting the response of human eye in this kind of applications. In addition, the reliability of the resulting clusterings should be quantified on the basis of statistics, and taking into account the uncertainties of the adopted values of proper elements. In the next Section we present the results of two new searches for asteroid families which have been recently carried out (Zappalà *et al.*, 1990; Bendjoya *et al.*, 1991), taking into account the above requirements.

2. Families Reliability Tests

2.1. Different Methods Applied to the Same Data Set

Recently, two new independent methods of families identification have been applied to the same data set of asteroid proper elements (Zappalà *et al.*, 1990; Bendjoya *et al.*, 1991). This fact allows for the first time a direct comparison between different procedures, since the possible discrepancies in this case do not depend on anything but the methods themselves. Moreover, both methods are "objective", in the sense that they do not use any visual inspection of the data, but they largely employ mathematical algorithms executed by computer.

The data set used in both researches has been the 4.2 version of the list of asteroid proper elements computed by Milani and Knežević. Such a sample, including about 4100 numbered asteroids, has been the largest ever used for families identification purposes. The proper elements of this data set have been computed by means of

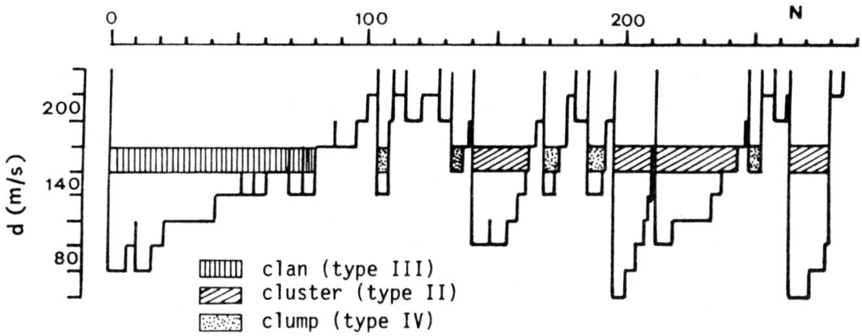

Fig. 1. An example of stalactite diagram referring to the sample of asteroids whose proper semi-major axis ranges between 2.5 and 2.825 AU (3/1 and 5/2 mean motion resonances with Jupiter). Examples of different nomenclatures for the resulting groupings are also indicated (see Section 4).

a second-order (in the planetary masses), fourth-degree (in the eccentricities and inclinations) secular perturbation theory, based on the Lie series technique. The accuracies of these proper elements depends in a complicated way on the location in the asteroid belt, and have been tested by means of numerical integrations of orbits (see Milani and Knežević, 1990).

The first method is the so-called Hierarchical Clustering Method (HCM), and has been described by Zappalà *et al.* (1990). Basically, this method uses well known techniques of Multivariate Data Analysis in order to analyze the structures of the clusterings present in the space of proper elements. The basic idea is to introduce a definition of distance in this space; the mutual distances between the objects of the considered sample are then computed, and agglomerations are performed in such a way that it is possible to build a *dendrogram* containing all necessary information for families identification purposes. In particular, for any given value d of the distance, it is possible to derive immediately what are the existing clusterings composed by objects whose mutual distances are less than d. The results are expressed under the form of *stalactite diagrams*, an example of which is given in Figure 1. It has to be noticed that the adopted distance definition in the space of proper elements is connected with the assumption of a collisional origin for families, thus it exploits the known relationship (Gauss equations) between the ejection velocities of the fragments of a catastrophic impact and the corresponding differences of the orbital elements. With this assumption, the distances between the objects are actually expressed in m sec^{-1}. In order to evidence the presence of families, one has to derive a confidence level of distance, for which no clusterings can be found having a non-negligible probability to be due to chance. Such a critical value of the distance can be found by generating, in the same portion of the proper elements space occupied by a given sample of real asteroids, a fictitious population of objects randomly chosen, but with the constraint to separately fit the a', e', and i' distributions of the real sample. In this way, the critical distance can be assumed to be the deepest level at which it is still possible to find sizeable agglomerations of of objects of the fictitious population (5 objects according to Zappalà *et al.*, 1990). In this way, the families present in the analyzed sample are evidenced, even if it is clear that a true collisional origin of these clusters cannot be assessed without a detailed analysis of the physical properties of their members (taxonomic types, observed mass distribution, etc.).

More recently, Bendjoya *et al.* (1991) have applied another procedure for asteroid families identification, namely the so-called Wavelet Analysis Method (WAM). The basic concept is that of *wavelet transform*, a mathematical tool which has been recently applied to the analysis of signals; its main advantage is that it is able to provide information on both the localization and the size of an event in a signal. If $s(x)$ is a one-dimension space signal in $L^2(\Re)$, the wavelet transform of this signal is a two dimension function $S(a, b)$, where a is a strictly positive scale-variable (size) and b a space-variable (localization). $S(a, b)$ is the decomposition of $s(x)$ onto a base of functions obtained from the dilatations and the translations of a unique function $\psi(x)$ of $L^2(\Re)$.

$$S(a, b) = a^{-\frac{1}{2}} \int_{-\infty}^{+\infty} s(x)\psi^*(\frac{x-b}{a})dx \tag{1}$$

$\psi(x)$ is called the analyzing wavelet. $a^{-\frac{1}{2}}$ is a $L^2(\Re)$ normalization factor and * denotes the complex conjugate.

The wavelet transform acts like a sort of mathematical zoom and $S(a, b)$ can be seen as a representation of the a-sized details of $s(x)$ around the location b. As an analyzing wavelet, it can be shown that a good choice is given by the so-called mexican hat function (see Figure 2), given by:

$$\psi_\sigma(x) = (n - \frac{r^2}{\sigma^2})\exp(-\frac{r^2}{\sigma^2}) \tag{2}$$

n is the dimension of the space in which the signal is defined and r is the distance defined from the metric chosen in the considered space. For the application of wavelet analysis to the problem of asteroid families identification, each asteroid is represented by a point in the (a', e', i') space. The signal to be analyzed in this case is thus a set of Dirac functions $\delta(\mathbf{x} - \boldsymbol{\eta}_j)$, $\boldsymbol{\eta}_j$ being the position of the jth point. In computer form, the wavelet transform is derived in a discrete way by computing the wavelet coefficients of the signal on each node of a frame superimposed to it. Let us consider a plane of data points over which a network is superimposed. The wavelet coefficient for the node $\mathbf{b}_i$ is obtained by substituting in Equation (1) $s(x)$ by the set of Dirac functions $\delta(\mathbf{x} - \boldsymbol{\eta}_j)$:

$$C(\sigma, \mathbf{b}_i) = \sum_{j=1}^{N} \psi\Big(\frac{\boldsymbol{\eta}_j - \mathbf{b}_i}{\sigma}\Big) \tag{3}$$

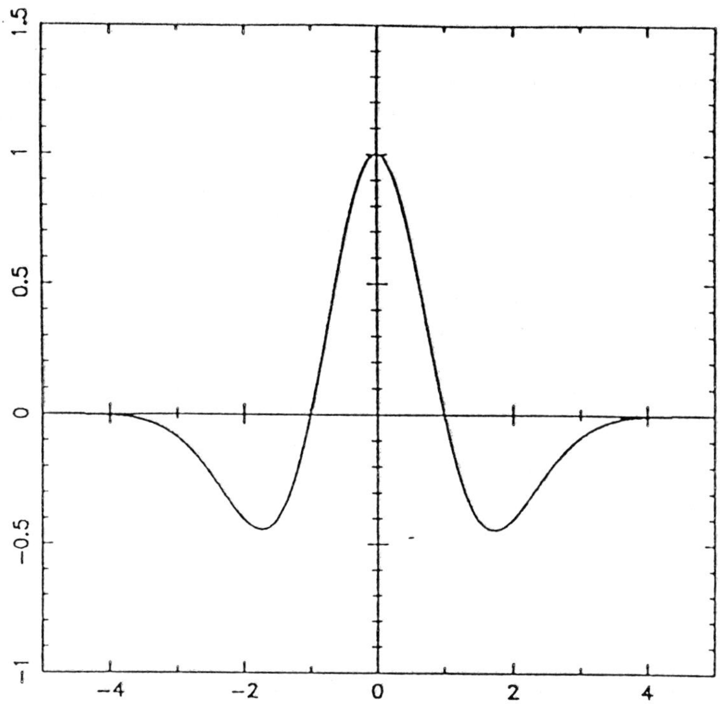

Fig. 2. Plot of the $1D$ mexican hat function. (Adapted from Bendjoya and Cellino, 1992).

where N is the number of data points. $C(\sigma, \mathbf{b}_i)$ is the wavelet coefficient at the node $\mathbf{b}_i$ for the scale σ. With the above choice of the mexican hat function as the analyzing wavelet, it follows that:

$$C(\sigma, \mathbf{b}_i) = \sum_{j=1}^{N} w_j$$

where

$$w_j = \left(2 - \frac{r_{ij}^2}{\sigma^2}\right)\exp\left(-\frac{r_{ij}^2}{2\sigma^2}\right) \tag{4}$$

r_{ij} being the distance between the ith node $\mathbf{b}_i$ and the jth point $\boldsymbol{\eta}_j$. This distance is expressed on the basis of the chosen metric, which in the case of the Bendjoya *et al.* (1991) paper has been chosen to be identical to the metric used by Zappalà *et al.* (1990). It is easy to understand that a node centered on a uniformly populated area (see Figure 3, cases a and b) will have a wavelet coefficient close to zero, since the positive weights will be compensated by the negative ones, provided that the size of this area is big compared to the studied scale. On the opposite, for a node centered just on a structure (or on a hole) at a given scale, the wavelet coefficient

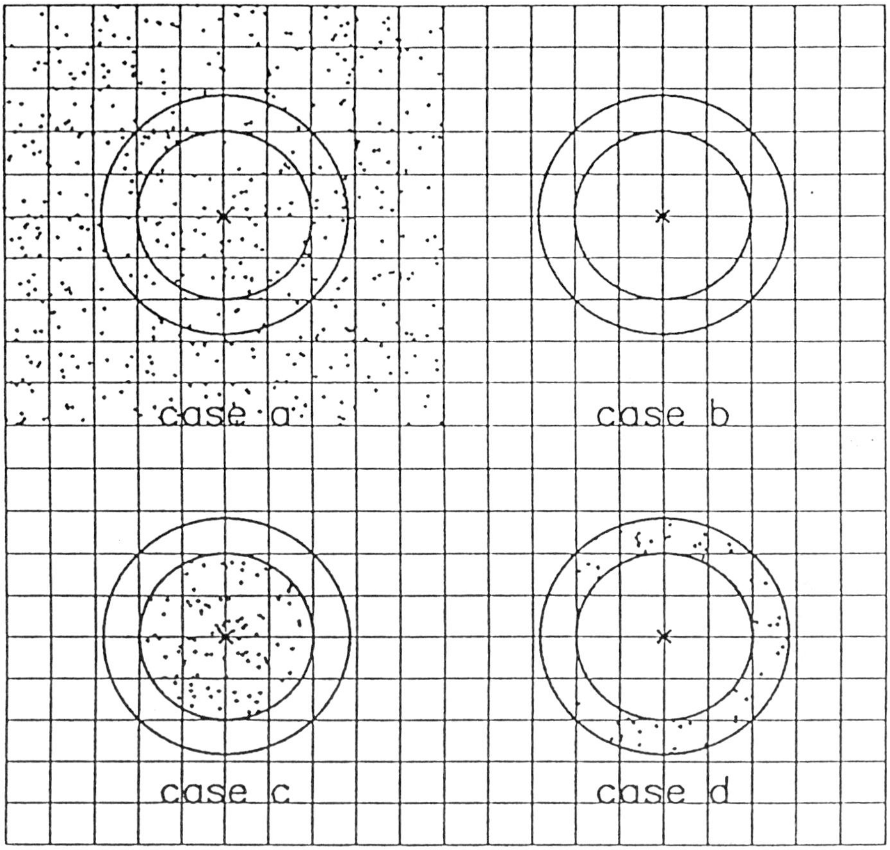

Fig. 3. Structure detection in the framework of Wavelet Cluster Analysis. (Adapted from Bendjoya and Cellino, 1992).

(see Figure 3, cases c and d) will be strongly positive (or negative). In this way, a map of wavelet coefficients can be built, at different scales, in which all the positive coefficients are indicators of the presence of structures. Also in this case, there is the problem of identifying which clusterings cannot be due to chance. The solution is similar to that found in the case of the HCM: a quasi-random population is created, the wavelet analysis is applied to this fictitious sample, and a threshold is found for the wavelet coefficients, which gives the level of confidence for which the structures having larger coefficients can be accepted as "families". Finally, it has to be noticed the importance of the adopted scale for the family definition. As a matter of fact, at scales sufficiently large different structures merge together, while they disappear at scales too small (as en effect of "overzooming"). Thus, families are defined at the largest scale for which they do not grow appreciably, whereas they grow drastically at a scale slightly larger. In other words, the chosen scale is the largest for which a well-defined structure still exists, before merging with other

TABLE II
The families found by Zappalà *et al.* (1990) (HCM) and by Bendjoya *et al.* (1991) (WAM), and percentage of the common objects (with respect to the number of members found for each WAM family). For a definition of different zones of the asteroid belt, see also Table IV

Zone of the belt	HCM family	WAM family	Least-numbered asteroid		Intersection (%)
2	21a	–	525	Adelaide	–
2	21b	–	883	Matterania	–
2	22	–	763	Cupido	–
2	23	–	1047	Geisha	–
2	–	21	281	Lucretia	–
2	–	22	422	Berolina	–
2	–	23	700	Auravictrix	–
2	–	24	963	Iduberga	–
3	31	31a	4	Vesta	71
3	32	32	650	Amalasuntha	75
3	33	35	878	Mildred	85
3	34	33	1378	Leonce	100
3	35	31b	1933	Tinchen	50
3	–	34	142	Polana	–
4	41a	41a	15	Eunomia	96
4	42	43	110	Lydia	100
4	43	–	141	Lumen	–
4	44	41b	145	Adeona	100
4	45	44	170	Maria	90
4	46	45	668	Dora	100
4	47	46	847	Agnia	86
4	48	42	1272	Gefion	100
5	51	51	158	Koronis	76
6	61	61	221	Eos	85
7	71	71	24	Themis	51
7	72	–	137	Moelibea	–
7	73	72	490	Veritas	86

structures of the same region at larger scales.

As was mentioned at the beginning of this Section, the fact that both HCM and WAM have been applied to the same data set of asteroid proper elements, and also using the same definition of a metric in this space, allows us to perform a comparison particularly fruitful between their respective results. Since these two methods are mutually independent, it is clear that *a priori* a noticeable agreement between the results should be considered as a serious clue of the reality of the proposed families. In this sense, Table II shows that the situation appears to be particularly satisfactory. With the exception of the Flora region, the agreement is noticeably good for what concerns both the number of resulting families, and the identities of their members. In addition to the well known Eos, Themis and Koronis families, a dozen of other families are found by the two methods, and should be considered as highly reliable. Among them, we have to notice the existence of a small family having as the largest body the asteroid 4 Vesta. This large body is known to exhibit a peculiar spectrum dominated by the bands of basaltic material at 1 and 2 μm, a fact that makes Vesta also a promising possible parent body for a peculiar class of meteorites: the eucrites (see also Section 5). On the other hand, the apparent disagreement of the results in the zone of Flora is not surprising, due to both the particularly crowded background present in this zone, and the relatively high uncertainties in the proper elements of the asteroids belonging to this zone of the belt (see also Section 5).

2.2. Same Identification Method Applied to Different Proper Elements

Another interesting possibility to check the families reliability and to better understand what is the origin of the observed discrepancies among different identification attempts could be to apply a given identification method to data bases whose proper elements are computed by means of different procedures. In this respect, a study has been performed by Zappalà *et al.* (1992). These authors have selected the most recent data bases of asteroid proper elements, i.e. those of Williams (1989) and of Milani and Kneževič (1991, version 5.7), both based on high-order perturbation theories. Unfortunately they contain different samples of asteroids; the former refers to the first 2065 numbered objects with the addition of about 1200 PLS asteroids, the second contains 4258 numbered objects. In spite of the obvious problems deriving from the use of different samples, it can be interesting to take profit of this "disadvantage" by looking at the number of members belonging to the common families both in the region of their intersection (till asteroid 2065) and in the two other regions (PLS and high-numbered asteroids). The reliability of a given family can be improved if it is found to contain, in addition to a number of objects belonging to the intersection region of the two data sets, also a considerable amount of objects coming from both of the other regions.

The adopted families identification method has been the HCM (Zappalà *et al.*, 1990). We recall that by means of his analysis Williams (1989) found 117 families, while Zappalà *et al.* (1990) recognized 21 families only.

According with the Zappalà *et al.* (1992) analysis, this large discrepancy is mainly due to the method used for family identification rather than to the method of proper elements computation. In fact, considering groupings having at least 5 members, they have found 26 families using the Williams' data set against 20 families recognized when the elements of Milani and Kneževič have been considered;

TABLE III
A comparison between the families found by applying the Hierarchical Clustering Method to both the asteroid proper elements data set of Williams and the MK57 data set of Milani and Knežević. An asterisk (*) indicates families not found in the Zappalà *et al.* (1990) paper, due to the smaller data set used

Family list	Williams		MK 57		Intersection	remarks
Zappalà *et al.* (1990)]	Num.	PLS	Num	H.Num.		
31 + 35	4	5	5	12	4	Vesta
32	5	3	9	32	5	F-type
33	4	29	2	21	2	
41	17	9	25	77	16	Eunomia
44	8	7	8	12	7	Adeona
45	14	2	17	21	13	Maria
46	7	5	6	14	6	
48	5	5	3	20	3	
51	54	29	54	101	54	Koronis
61	89	9	80	127	79	Eos
71	75	38	72	165	71	Themis
1183 (*)	3	10	2	12	1	
135 (*)	1	18	1	16	1	Hertha
808 (*)	4	7	3	4	3	

the "intersection clusterings" were found to be 16. The agreement was highly improved when only families containing at least 10 members were taken into account: from Williams' data set they obtained 15 clusterings, 14 of which represent also the complete sample recognized using the Milani and Kneževiĉ data set. Table III shows in a more detailed way the results of the latter analysis. The columns list the family identification number (that adopted by Zappalà *et al.*, 1990 or, in the last three cases, the number of the least-numbered asteroid of the grouping), the members belonging to the Williams numbered and PLS objects, the members belonging to the Milani and Kneževiĉ numbered asteroids (up to 2065 and beyond 2065), and the common objects present in both classifications. As already noticed the agreement is quite satisfactory, showing that the results are almost independent on the methods used for proper elements determination. In our opinion this result is important and encouraging for future studies on families, since it indicates that the most critical point is the choice of an objective statistical method for clustering identification. Moreover, that analysis shows that, even in the case of smaller permitted clusterings, less than 30 families can be identified using the Williams' data set. This small number has to be compared with the much larger amount of families found by Williams (117) using the same data set. The discrepancy can be easily explained if we admit that the Williams' criterion of family identification was too liberal or, alternatively, that the Zappalà *et al.* method was too conservative. However, the latter possibility seems in our opinion questionable, due to the reliability tests performed and to the previously discussed agreement with the results of Bendjoya *et al.* (1991). On the other hand, Williams used a quite subjective method, mainly based on a "visual" definition of clusterings. We suggest that he assigned an acceptable reliability also to groupings whose statistical significance was poor, often of the same level of the background groupings due to chance only. This argument can be understood better by looking at a typical *stalactite diagram* used by Zappalà *et al.* (1990) for their family representation. That of Figure 1 refers to their Zone 4 of the Main Belt, which is limited by the 1/3 and the 2/5 mean motion resonances with Jupiter. As already explained, the stalactite diagram shows, for different levels of the *distance*, the resulting clusterings. The width of a stalactite at each level is given by the number of objects belonging to the clustering at that distance level. The 160 m sec^{-1} level is the *quasi-random level*, i.e. the distance level under which no clusterings are found in the case of fictitious populations of quasi-randomly generated objects. It is easy to see that there are basically three kinds of stalactites. The so called type II is sharp and generally deep, and corresponds to families which are well defined and are found to have a large intersection of common objects both in the Williams and the Milani and Kneževič data bases. Type III is less sharp and its members grow gradually as long as the distance level increases; it follows that the level at which one "cuts" the stalactite in order to define the family membership is very crucial: a statistically significant grouping surely exists, but it is very hard to decide which are its real members. Numerical simulations performed by Bendjoya *et al.* (1992)(see the next Section for a more detailed presentation of the results) show that in such cases one has to make a choice between two possibilities: (a) to cut the stalactite at a low level, losing probably a significant amount of family members; (b) to be more liberal, cutting the stalactite at higher levels, taking into account that in this way a large percentage of interlopers can be included. This choice will probably remain a "philosophical" one, as long as observations can solve the compositional "puzzle" of the original parent body, allowing to eliminate objects whose physical and chemical features cannot be genetically related to the parent body. Finally, type IV stalactites are those that do not reach at present - or marginally reach - the rank of possible families on the basis of the Zappalà *et al.* (1990) statistical criterion. Most of them have been considered as real families by Williams (1989), and are therefore the main responsible for the large discrepancy found comparing Williams and Zappalà *et al.* classifications. Later in the text we will present a suggested new terminology for naming asteroid groupings identified in different ways (Farinella *et al.*, 1992), mainly based on the points addressed before.

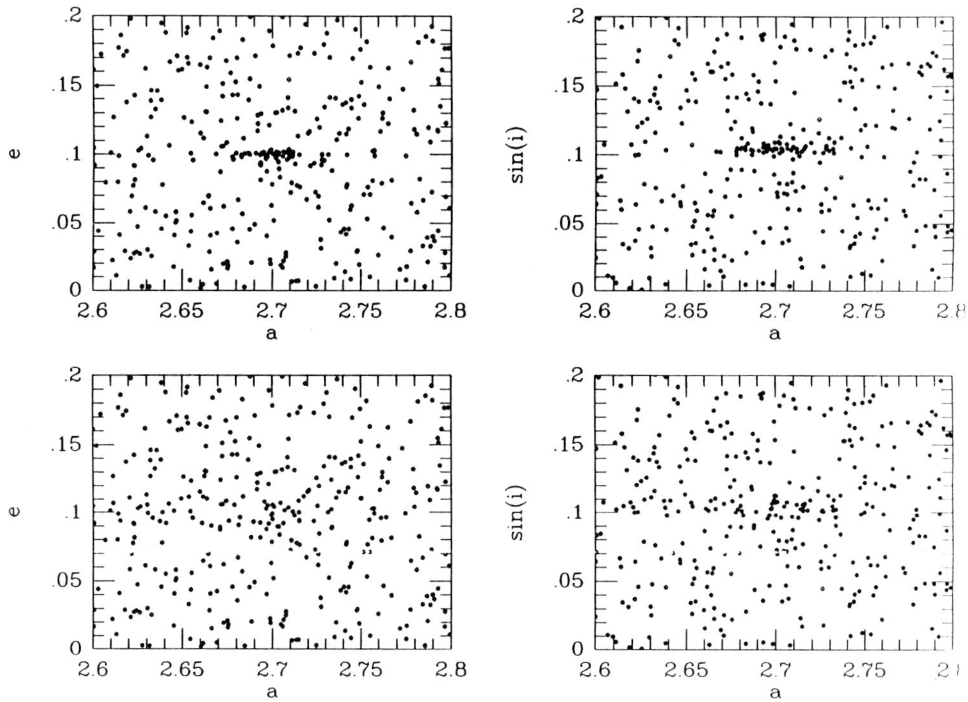

Fig. 4a. Projections on the a–e and a–$\sin i$ planes of the simulated family with a background of 300 objects, and with $V_{min} = 50$ m sec^{-1} (top), and $V_{min} = 150$ m sec^{-1} (bottom).

TABLE IV
Number densities for the different zones of the asteroid belt and for the three cases analyzed in the simulations

	Density
Zone 2 ($2.065 \leq a < 2.300$ AU)	0.16
Zone 3 ($2.300 \leq a < 2.501$ AU)	0.08
Zone 4 ($2.501 \leq a < 2.825$ AU)	0.04
Zone 5 ($2.825 \leq a < 2.958$ AU)	0.03
Zone 6 ($2.958 \leq a < 3.030$ AU)	0.05
Zone 7 ($3.030 \leq a < 3.278$ AU)	0.04
Simulations Background 300	0.04
Simulations Background 600	0.08
Simulations Background 900	0.12

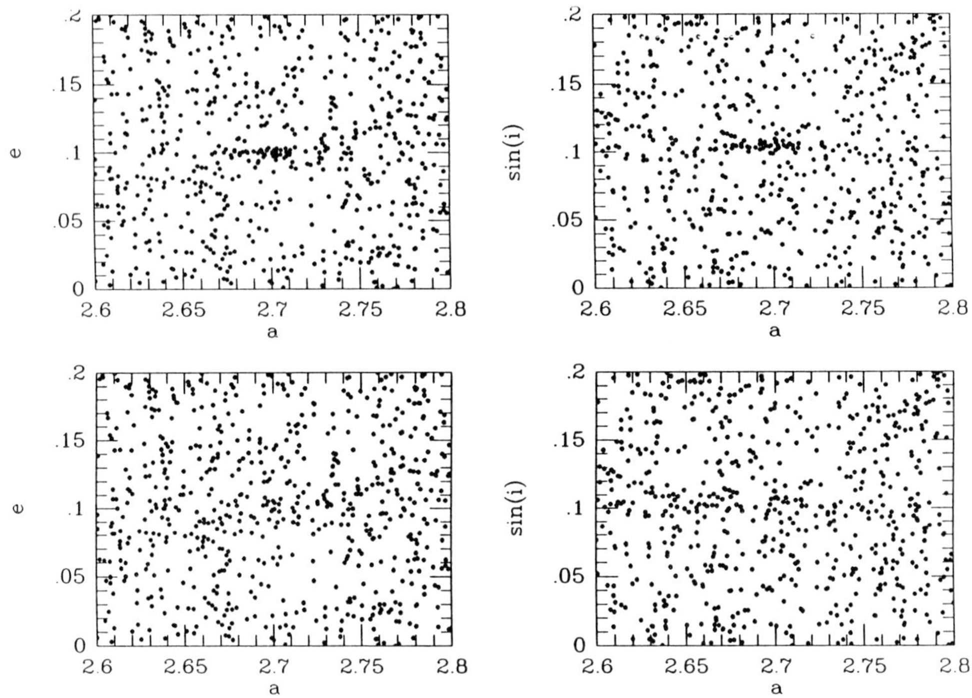

Fig. 4b. Projections on the a–e and a–$\sin i$ planes of the simulated family with a background of 600 objects, and with $V_{\rm min} = 50$ m sec^{-1} (top), and $V_{\rm min} = 150$ m sec^{-1} (bottom).

3. Simulated Families

In order to check the efficiency of the two most recent methods of families identification (Zappalà *et al.*, 1990, Bendjoya *et al.*, 1991), Bendjoya *et al.* (1992) have applied them to a family scenario which is *a priori* known. Admitting that asteroid families are generated in a collisional event, a sample of simulated families was artificially built, by assigning to each family member a typical ejection velocity according with a power-law distribution, justified by laboratory fragmentation experiments. The number dN of fragments having a mean ejection velocity between v and v+dv is given by:

$$dN = 0 \qquad \text{if } V < V_{\rm min}$$
$$dN = CV^{-\alpha}dV \qquad \text{if } V > V_{\rm min}$$

where $V_{\rm min}$ is the minimum ejection velocity and C is a constant. The exponent α has been taken equal to 3.25 in order to follow the results coming from laboratory experiments, (see, e.g., Fujiwara and Tsukamoto, 1980) while the diameter of the parent body has been chosen equal to 100 km. A family of 50 members has been simulated for $V_{\rm min}$ equal to 50, 100 and 150 m sec^{-1}, and then it has been plunged in a background of 300, 600 and 900 fictitious asteroids. 9 cases have been obtained, corresponding to different densities and to different dispersions

TABLE V
Comparison of the results of simulated families identification by means of Hierarchical Clustering Method (HCM) and Wavelet Analysis Method (WAM)

Background		$V_{\min}$ (m sec^{-1})					
		50		100		150	
		$N_{\text{Fam.}}$	$N_{\text{Int.}}$	$N_{\text{Fam.}}$	$N_{\text{Int.}}$	$N_{\text{Fam.}}$	$N_{\text{Int.}}$
300	HCM	50	2	46	1	45	5
	WAM	50	1	47	1	41	3
600	HCM	50	3	46	4	36	11
	WAM	49	4	46	4	40	9
900	HCM	50	2	46	2	29	4
	WAM	49	4	46	7	36	11

of the family members; Figures 4a and 4b show some of these cases in the *a*-*e* and in the *a*-sin*(i)* planes. Table IV gives the number densities for each case, which can be compared with the real number densities of the different zones of the asteroid belt (as defined by Zappalà *et al.*, 1990). It appears that, except for zone 2 which represents the so called "Flora region", most of the real families have to be compared with the simulations corresponding to backgrounds of 300 and 600 asteroids. The results are quite encouraging. In the case of a background of 300 objects the family was always recognized (with at least 41 members), for every ejection velocity and by both methods. The number of interlopers never exceeded 5 objects. Similar results have been also obtained when a background of 600 asteroids was considered. However, in this case the number of interlopers increased till 11 for the highest ejection velocity (150 m sec^{-1}). The cases related to a background of 900 objects show an increase of this trend (however, at least 29 family members have been detected by both methods), pointing out the problems connected with the number of interlopers already evidenced for type III groupings. In particular, in a simulation obtained using a very high minimum ejection velocity (200 m sec^{-1}) and a background of 900 asteroids, the method of Zappalà *et al.* (1990) recognized only 8 asteroids to be part of the family, with the addition of only 2 interlopers, while the method of Bendjoya *et al.* (1991) was able to identify 29 family objects but with the addition of 27 interlopers. On the other hand, we have to notice that the quite "conservative" result obtained by applying the method of Zappalà *et al.* (1990) can be in principle improved, in the sense of being able to recognize a much larger family, if one decides to scale with the density of the background the critical distance level for cutting the stalactites; further studies on this subject are in progress, even if they cannot really overcome the "interlopers" problem connected with type III clusterings. Table V presents in details the results obtained by Bendjoya *et al.* (1992), evidencing for each case the number of "real" family members as well as the number of the interlopers.

That work permits to believe *a posteriori* in the reliability of the previous analyses performed by Zappalà *et al.* (1990) and Bendjoya *et al.* (1991) on real families, since, as already noticed, the number densities of the zones of the real asteroid belt are mostly comparable with those of the easier cases of the simulations (50 objects in a background of 300 asteroids). For the most difficult cases (such as for the Flora region, where, on the other hand, the two classifications have shown a very poor agreement) the study permits to point out the need for making a distinction between different kind of clusterings and for having a new nomenclature.

4. A New Nomenclature for Asteroid Family Identification

From the results outlined in Sections 2 and 3, it seems that confusion and ambiguities about family classification arise mainly as a consequence of using the single word "family" for naming asteroid groupings identified in different ways and subjected to different significance tests. This is particularly true for groupings identified "by eye"(i.e., through visual inspection of the asteroid distribution) when they are compared with groupings found by almost automated clustering algorithms. Farinella *et al.* (1992) suggested a "family scheme" by recognizing four types of groupings in proper elements space, distinguished both by the techniques used for their identification and by the adopted tests of statistical significance. Figure 5 reports this scheme. Farinella *et al.* (1992) call the type IV families, introduced in Section 2, **clumps**, and they can be defined as groupings recognized by visual inspection but not subjected to rigorous statistical tests. Groupings which have shown to be statistically significant are divided in two categories: **clusters** (or type II families) and **clans** (or type III families). Clusters are groupings for which an unequivocal membership definition is possible through a clearcut separation from the "random" background and from other groupings; clans are statistically significant also, but lie within a background so dense and/or are so close to each other that they cannot be separated in a clearcut manner. Obviously, new data could in the future allow clumps to become clusters or clans, as well as clans can become clusters or vice versa. Finally, Farinella *et al.* (1992) proposed to call **families** only those clusters or clans for which some physical evidence indicates that their members have a genetic relationship (and not just a dynamical one, like the asteroids of the Phocaea region). This evidence can be based on collisional physics (e.g., mass and velocity distributions should be in agreement with the results of laboratory experiments or at least physically reasonable); or it could be based on taxonomy, from which a given family shows to be cosmochemically plausible.

5. Open Problems and Future Work

In the framework outlined before, it is hard to insert the grouping(s) found in the region of Flora (zone 2 in Zappalà *et al.*, 1990). If we accept the idea that the region

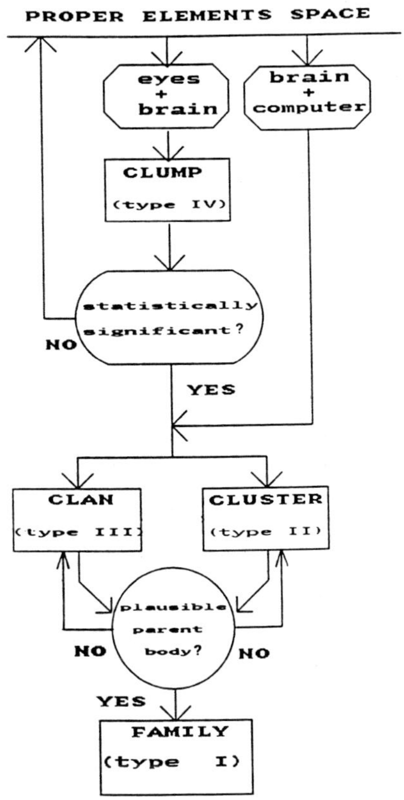

Fig. 5. Scheme of the proposed new nomenclature for asteroid groupings. (Adapted from Farinella *et al.*, 1992).

of Flora is actually a region very highly populated with respect to the other regions of the belt (after having taken into account the obvious problems of observational selection effects) and therefore dominated by a very dense background, it follows that we have to admit only the presence of several clumps, but not that of clusters, clans or families. On the other hand, if we believe that the background cannot be very different from that of the adjacent zone 3, a very large clan can be recognized. The large dispersion of its members can be possibly due to the presence of strong secular and mean-motion resonances. Further studies should be performed on this problem, mainly aimed to improve the stability of the proper elements, which, even using high-order perturbation theories, seems to be not yet completely satisfactory.

A support to the hypothesis of a big clan is given by the analysis of the size distribution of the asteroid belt. Cellino *et al.* (1991) have shown, through an analysis of the IRAS data, that families have a slope of the power-law describing their size distribution much steeper than that of field asteroids. The same feature is present when all the asteroids belonging to the zone of Flora are taken into account.

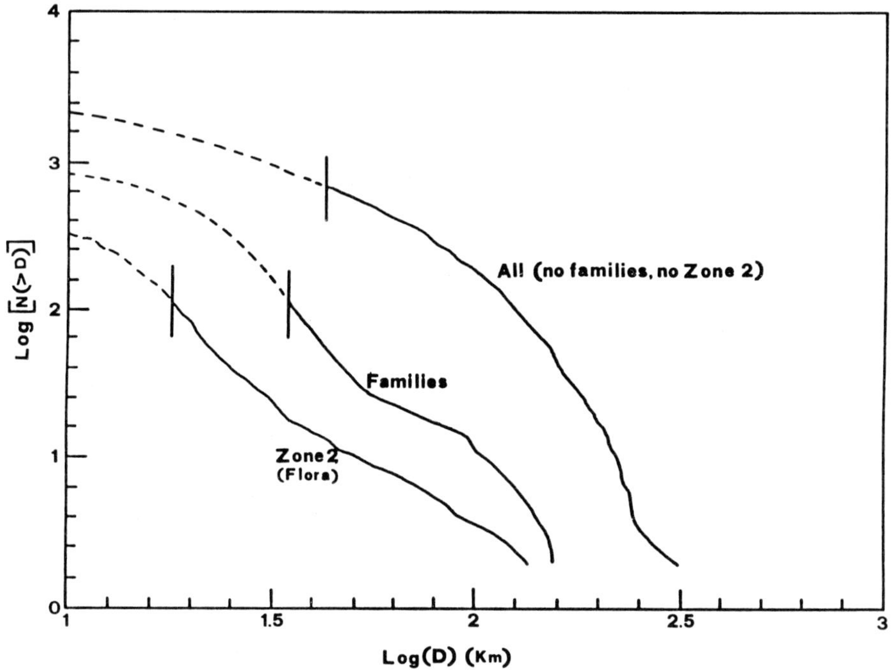

Fig. 6. The cumulative size distributions of family asteroids, Flora region objects, and asteroids not belonging to either of the above samples.

Figure 6 shows the distributions of non-family, family and zone 2 asteroids. One can conclude that the similarity of the distributions of family and Flora region asteroids plays in favor of the idea of a single very large clan in zone 2. However, it remains still to investigate why the families show such a different size distribution. Is this fact due to the relatively young age of the families, i.e. to the fact that families still represent the outcome of a single impact and not a fully relaxed collisional population ? Simulations performed for testing the dispersion of family members due to their subsequent collisional evolution can partially solve this problem (Davis, private communication).

Another interesting question about families is the following: could the impact event which has originated a family modify the taxonomic type of the fragments with respect to the parent body? The reply is surely "yes" if we consider a differentiated parent body, since the fragmentation should expose fragments coming from the mantle and from the core; on the other hand, the reply should be "no" for an homogeneous parent body. However, it is interesting to notice that two of the most reliable families, i.e., Koronis and Eos, show members whose taxonomic types are well defined and similar (and therefore consistent with an homogeneous parent body), but significantly different from the "normal" types of the zones of the belt they belong to. In particular, Eos family members have a very strange tax-

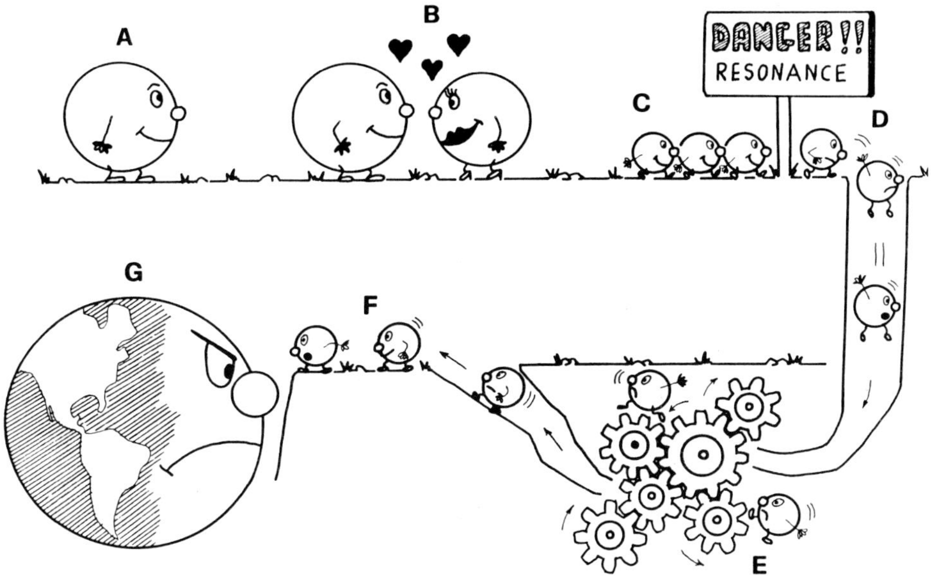

Fig. 7. The geological implications of the asteroidal families are described in this drawing. During the solar system lifetime a main belt asteroid (A) is expected to have a "catastrophic" encounter with another object (B). As a consequence of this collisional event, several fragments can be produced forming an asteroid family (C). However, if the catastrophic impact takes place not too far from some mean-motion or secular resonance, some fragments can be injected into them (D) and their orbital evolution can be controlled by a complex mechanism (E) able to increase the eccentricity well enough to transfer the fragments into the inner-planets region (F). It follows that highly energetic impacts with the Earth have to be considered as relatively frequent events (G).

onomic type, which is defined as K type in Tholen's classification (1984). Should we believe that the Eos parent body, the only one object of relatively large size to be disrupted in the Eos region (zone 6 in Zappalà *et al.* (1990)), was a "unique" asteroid in the belt, or will we think about a modification of a common taxonomic type (maybe S) after the catastrophic fragmentation which originated the family? Moreover, we have also to recall that the S-type associated to the objects of the zone of Flora seem significantly different from the S-type of the other regions (Dermott *et al.*, 1985). Again, one could connect this fact with the relatively young age of some families.

Finally, let us spend some words about the family of Vesta. Very recent observations of some asteroids belonging to the Vesta's family defined by Zappalà *et al.* (1990) have shown spectra which allow to classify these objects as belonging to a taxonomic type which seems intermediate between S and V (Tholen, private communication), and R and V (Binzel, private communication), and to call them "vestoids". Can the Vesta's family be the source of the Earth Approaching Asteroids of class V and, more generally, the source of the eucrite meteorites ? Preliminary simulations performed by Rosa *et al.* (1992) seem to show that a not negligible percentage of fragments coming from a crater-like impact on Vesta could reach the 3/1 resonance with Jupiter, known to represent one of the most efficient sources of injection of objects in the inner planets region (Wisdom, 1985).

As a final remark let us say that the introduction of automated clustering algorithms for families identification seems finally to have solved most of the ambiguities and confusions related to this old but fundamental problem. This should permit to start further and more detailed physical studies of the families in the general framework of the asteroid collisional evolution.

Acknowledgements

We are grateful to Claude Froeschlé, Philippe Bendjoya, Paolo Farinella and Mario Carpino for their helpful discussions.

References

Arnold J.A. (1969) *Astron. J.*, **74**, 1235–1242.
Bendjoya Ph. and Cellino A. (1992). In *Comptes rendus de l'ecole de Goutelas 1991*, in press.
Bendjoya Ph., Slezak E., Froeschlé Cl. (1991) *Astron. Astrophys.*, **251**, 312–330.
Bendjoya Ph., Cellino A., Froeschlé Cl., Zappalà V. (1992). In *Asteroids Comets Meteors 1991*, in press, Flagstaff.
Brouwer D. (1951) *Astron. J.*, **56**, 9–32.
Carusi A., Massaro E. (1978) *Astron. Astrophys. Suppl.*, **34**, 81–90.
Cellino A., Zappalà V., Farinella P. (1991). *M.N.R.A.S.*, in press.
Chapman, C.R., Paolicchi, P., Zappalà V., Binzel R.P., Bell J.F. (1989) In *Asteroids II* (R.P. Binzel, T. Gehrels, M.S. Matthews, Eds.), pp. 386–415, University of Arizona Press, Tucson.
Davis D.R., Weidenschilling S.J., Farinella P., Paolicchi P., Binzel R.P. (1989) In *Asteroids II* (R.P. Binzel, T. Gehrels, M.S. Matthews, Eds.), pp. 805–826, University of Arizona Press, Tucson.
Dermott S.F., Gradie J., Murray C.D. (1985). *Icarus*, **62**, 289–297.
Farinella P., Davis D.R., Cellino A., Zappalà V. (1992) In *Asteroids Comets Meteors 1991*, in press, Flagstaff.
Fujiwara A., Tsukamoto A. (1980). *Icarus*, **44**, 142–153.
Hirayama K. (1918) *Astron. J.*, **31**, 185–188.
Hirayama K. (1919) *Proc. Phys.-Math. Soc. Japan*, **3(1)**, 52–59.
Hirayama K. (1920) *Proc. Phys.-Math. Soc. Japan*, **3(2)**, 236–240.
Hirayama K. (1923) *Japan J. Astron. Geophys.*, **1**, 55–93.
Hirayama K. (1928) *Japan J. Astron. Geophys.*, **6**, 137–162.
Hirayama K. (1933) *Proc. Imp. Acad. Japan*, **9**, 482–485.
Kozai Y. (1979) In *Asteroids* (T. Gehrels, Ed.), Univ. of Arizona Press, Tucson, pp. 334–358.
Lindblad B.A., Southworth R.B.: 1971, In *Physical Studies of Minor Planets* (T. Gehrels, Ed.), NASA SP-267, pp. 337–352.
Milani A., Knežević Z. (1990) *Celestial Mechanics*, **49**, 247–411.
Rosa F., Zappalà V., Cellino A. (1992). In preparation.
Tholen D.J. (1984). Ph. D. Thesis, University of Arizona.
Valsecchi G.B., Carusi A., Knežević Z., Kresák Ľ., Williams J.G. (1989) In *Asteroids II* (R.P. Binzel, T. Gehrels, M.S. Matthews, Eds.), pp. 368–385, University of Arizona Press, Tucson.
Van Houten C.J., Van Houten-Groeneveld I., Herget P., Gehrels T. (1970). *Astron. Astrophys. Suppl.*, **2**, 339–448.
Williams J.G. (1979) In *Asteroids* (T. Gehrels, Ed.), pp. 1040–1063, University of Arizona Press, Tucson.

Williams, J.G. (1989) In *Asteroids II* (R.P. Binzel, T. Gehrels, M.S. Matthews, Eds.), pp. 1034–1072, University of Arizona Press, Tucson.
Wisdom J. (1985). *Nature*, **315**, 731–733.
Zappalà V., Cellino A., Farinella P., Kneževič Z. (1990) *Astr. J.*, **100**, 2030–2046.
Zappalà V., Cellino A., Farinella P. (1992). In *Asteroids Comets Meteors 1991*, in press, Flagstaff.

THE ULTRAVIOLET SPECTRA OF CARBONACEOUS CHONDRITES

T. INAGAKI
Department of Physics, Osaka Kyoiku University, Tennoji, Osaka 543, Japan

S. YABUSHITA
Department of Applied Mathematics and Physics, Kyoto University, Kyoto 606, Japan

and

K. WADA
Department of Nuclear Engineering, Kyoto University, Kyoto 606, Japan

Optical spectroscopy of meteorites is an important laboratory tool for the study of the origin and evolution of solar and galactic interstellar minor bodies. In the region from the visible to the ultraviolet, the absorption spectra have hitherto been measured on the Bruderheim (Egan and Hilgeman, 1975), the Orgueil (Huffman, 1977), the Murchison (Sakata *et al.*, 1977) and the Yamato (Yabushita *et al.*, 1987) chondrites. These measurements were performed with pulverized specimens mixed with KBr powder or suspended in a transparent liquid. Due to these methods of sample preparation, the spectra observed were not very clear and the origin of the observed absorption features was not identified. In the present study, we have studied carbonaceous chondrites in the from of a thin film prepared by vacuum evaporation.

A tiny fragment of the Belgica carbonaceous chondrite B-7904 (type CM2) was allotted to us by the National Institute of Polar Research, Tokyo. The vacuum evaporations were made by using an ordinary vacuum evaporator with a tantalum boat as a resistive heater. The vaporized chondrite inclusions were recondensed on a quartz substrate kept at room temperature. The evaporation temperature was carefully controlled at around 1050 ° C. Below this temperature, no appreciable condensation took place on the substrate, although we observed a significant pressure rise caused presumably by various gases released from the chondrite specimen, including the thermally degraded organic compounds (Shimoyama *et al.*, 1991).

The absorption spectrum obtained for an evaporated film with mass density 4.3 μg cm^{-2} is presented in Figure 1. The spectrum exhibits a sharp peak at 226 nm and a broad shoulder around 270 nm. These features are considerably clearer than those observed previously for the pulverized samples (Sakata *et al.*, 1977; Yabushita *et al.*, 1987). In order to identify the inclusion responsible for the spectral features observed here, we have performed a series of compositional analyses both on the raw meteorite specimen and the evaporated films. When we combined the results from these analyses all together, we came to the conclusion that the most plausible inclusion giving rise to the observed absorption features must

Celestial Mechanics and Dynamical Astronomy **54:** 229–231, 1992.

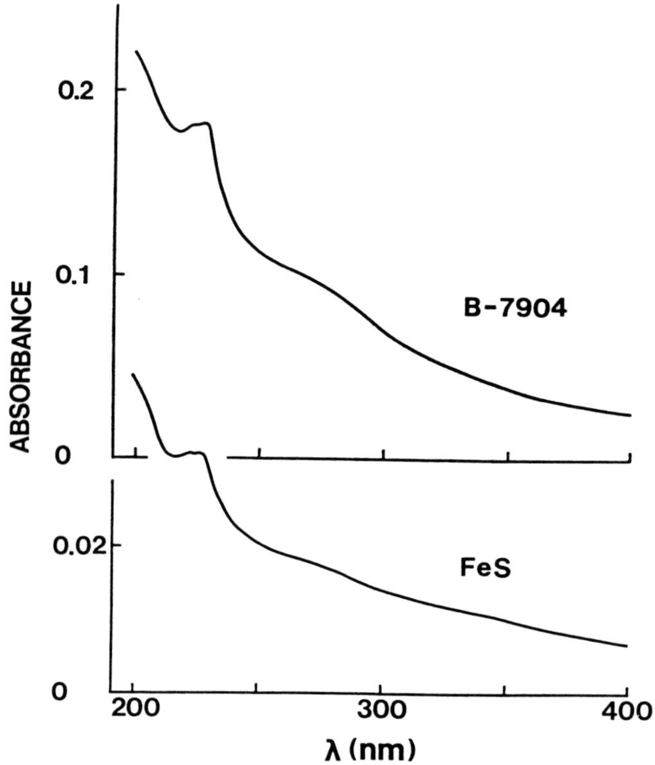

Fig. 1. Absorption spectra of the evaporated thin films of the Belgica carbonaceous chondrite B-7904 and troilite Fe*S*.

be troilite Fe*S*. This mineral is one of the major inclusions common to Antarctic chondrites (Haranura *et al.*, 1983; Tomeoka, 1990). To confirm this, we prepared a thin film sample of commercially available synthetic Fe*S* by vacuum evaporation. The absorption spectrum observed is given in Figure 1 and compared with one obtained for the chondrite film. Similarity between the two data is remarkable, demonstrating convincingly that the component evaporated from the chondrite inclusions is troilite Fe*S*.

The present results provide important information concerning the origin of the absorption features observed previously with a liquid suspension of pulverized carbonaceous chondrites. Differing from the evaporated thin films, the pulverized samples are thermally intact and should preserve all the components of the original inclusions. Nevertheless, the observed absorption curves feature two absorption bumps around 226 and 270 nm, resembling closely those observed here with evaporated films (Sakata *et al.*, 1977; Yabushita *et al.*, 1987). This suggests that the absorption features of the pulverized samples are also caused by troilite Fe*S*. We have confirmed this by measuring Fe*S* pulverized and suspended in hexafluoroisopropanol. The absorption features observed for the liquid suspensions of the pulverized chondrites have long been thought to originate from the organic extract

(Sakata *et al.*, 1977; Yabushita *et al.*, 1987), providing strong support for the organic model of the 220 nm feature of interstellar extinction (for example, Draine, 1989).

So far, very little attention has been paid to troilite FeS regarding its connection with interstellar extinction even though it is a rather common meteoritic mineral. Now that FeS is found to yield a striking absorption at 226 nm, it seems to deserve more attention.

Acknowledgements

The authors would like to thank Drs. T. Hagihara and H. Yamaguchi for performing skillful analyses on the meteorite sample.

References

Draine, B. T.: 1989, in *Interstellar Dust*, eds. L. J. Allamandola and A. G. G. M. Tielens, Kluwer, Dordrecht, pp. 313–327.

Egan, W. G. and Hilgeman, T.: 1975, *Astrophys. J.* **80**, 587.

Haramura, H., Kushiro, I., and Yanai, K.: 1983, *Mem. Natl. Inst. Polar Res.*, Spec. Issue **30**, 109.

Huffman, D.: 1977, *Adv. Phys.*, **26**, 129.

Sakata, A., Nakagawa, N., Iguchi, T., Isobe, S., Morimoto, M., Hoyle, F. and Wickramasinghe, N. C.: 1977, *Nature*, **266**, 241.

Shimoyama, A., Komiya, M., and Harada, K.: 1991, *Proc. NIPR Symp. Antarc. Meteorites*, **4**, 247.

Tomeoka, K.: 1990, *Proc. NIPR Symp. Antarc. Meteorites*, **3**, 40.

Yabushita, S., Inagaki, T., Arakawa, E. T., and Wada, K.: 1987, *Mon. Not. R. astr. Soc.*, **229**, 45.

SOLAR DUST RING OBSERVATION AT THE TOTAL SOLAR ECLIPSE IN MEXICO

SYUZO ISOBE
National Astronomical Observatory Mitaka, Tokyo 181, Japan

TOSHIHIKO TANABE
Faculty of Science, University of Tokyo, Hongo, Tokyo, Japan

and

MOTOKAZU NOGUCHI
National Astronomical Observatory Mitaka, Tokyo 181, Japan

There are dust grains in interplanetary space. Zodiacal light observations show a distribution of dust grains in the ecliptic plane, but cannot show how close to the Sun dust grains survive. Dust grains close to the Sun are heated by solar radiation and are evaporated. Thus, a dust free zone is formed. There is a question as to how large this dust free zone may be.

Peterson (1967) made infrared observations of the outer corona during a total solar eclipse and found an enhancement of infrared radiation at 4 $R_\odot$ ($R_\odot$ is the solar radius measured as an angle). In 1983, Mizutani *et al.* (1984) confirmed this enhancement with infrared observations during the total solar eclipse in Indonesia. Infrared observations show a clear edge to the dust free zone. However, this infrared radiation is thermal radiation from dust grains heated by solar radiation, and the intensity distribution depending on distance from the Sun does not match the density distribution because the intensity of thermal radiation depends strongly on dust grain temperature.

To obtain information about the dust grain distribution and its composition, we need optical data. The dust grains surrounding the Sun scatter solar radiation and this scattered radiation is known as the F-corona. At some values of R, all the dust grains contribute to scattered solar radiation along the line of sight and in most cases forward scattering is dominant and gives zero polarization. However, if there is a high concentration of dust grains just outside the dust free zone, we expect to observe high polarization amongst the radiation scattered by dust grains with a scattering angle around 90^0.

Mukai and Yamamoto (1979) calculated the orbital motion of dust grains surrounding the Sun taking into account thermal evaporation effects and showed that some types of dust grain evaporate at about 4 $r_\odot$ ($r_\odot$ is the solar radius measured in linear distance) where the dust density is high.

In 1983, Isobe *et al.* (1987) made optical observation at 4 visual wavelengths during the total solar eclipse in Indonesia. However, the observed images at 3

Celestial Mechanics and Dynamical Astronomy **54:** 233–235, 1992.

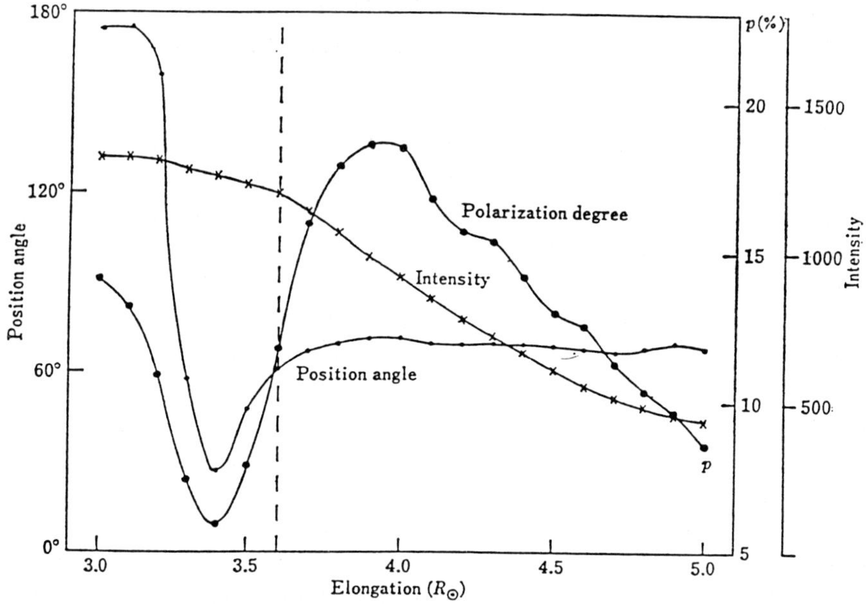

Fig. 1. Polarization distribution along the ecliptic plane obtained by Isobe *et al.* (1987) during the total solar eclipse, Indonesia, in 1983.

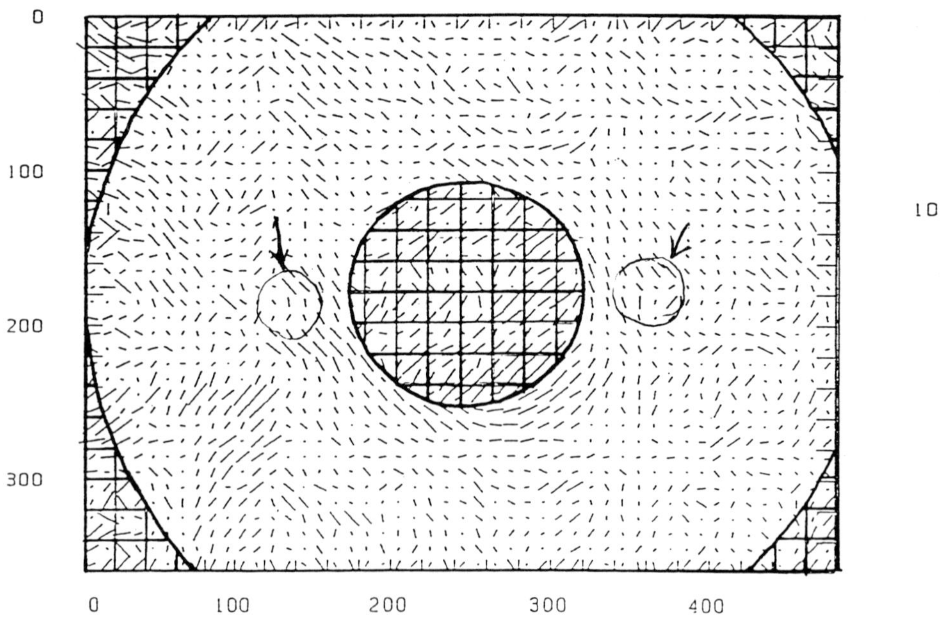

Fig. 2. Polarization distribution surrounding the Sun. The bar at the right side of the figure shows a magnitude of 10% polarization. The Sun is at the center of the inner circle which has a radius of 3 $R_\odot$. Two areas with arrows have high polarization as expected for radiation scattered by dust grains surrounding the Sun.

wavelengths were damaged by saturation effects in the inner corona because of pointing errors with the telescope, and at the wavelength of 8015 A we obtained measurements of the two dimensional polarization distribution and its enhancement at 4 $R_\odot$ in the ecliptic plane. We claimed this high polarization at 4 $r_\odot$ was due to scattered radiation by dust grains surrounding the Sun, but could not confirm the result because of the lack of observations at the other 3 wavelengths.

In 1991, we repeated this observation during the total solar eclipse in Mexico. Since coronal radiation at 4 $R_\odot$ is very faint and is comparable with that of the sky background brightness even during the total solar eclipse, our observation should be carried out at high altitude to reduce sky background brightness. In Indonesia, a balloon gondola was launched, but in Mexico we climbed Popocatepetl to an altitude of 5250 m.

It was a fine day during the test, but unfortunately the corona was covered by very thin cloud. Therefore, we could observe the total solar eclipse, but only obtained noisy data. After the reduction of these data making some allowances for noise effects, we obtained a preliminary polarization distribution map. This is still noisy but shows high polarization at 4 $R_\odot$ in the ecliptic plane.

At the moment, we have reduced our data at only one wavelength, namely 8015 A, but still have to reduce those at the other 3 wavelengths of 5325A, 5965 A, and 7200 A. The wavelength dependence of the polarization at each point in our observed field will then be obtained, with the possibility of new constraints relating to dust grains surrounding the Sun.

This project was supported by a grant from the Ministry of Education, Science, and Culture.

References

Isobe, S., Hirayama. T., Baba, N., and Miura, N.: 1987, *Publs. Astron, Soc. Japan* **39**, 667–677
Mizutani, K., Maihara, T., Hiromoto, N., and Takami, H.: 1984, *Nature* **312**, 134–136
Mukai, T. and Yamamoto, T.: 1979, *Publs. Astron. Soc Japan* **31**, 585–596
Petorson, A.W.: 1967, *Astrophys. J. Letters* **148**, L37-139

SHORT-PERIOD COMETS AND APOLLO-AMOR-ATEN TYPE ASTEROIDS IN VIEW OF TISSERAND INVARIANT

H. KOSAI
National Astronomical Observatory of Japan, Mitaka, Tokyo 181, Japan

Tisserand invariant is a powerful mean of identifying comets. The invariants have been calculated for 154 short-period comets and for 185 Apollo-Amor-Aten type asteroids, using orbital elements given by Marsden (1989) for short-period comets, and Batrakov *et al.* (1990), IAUCs, and MPCs for Apollo-Amor-Aten type asteroids. Note that Apollo, Amor, and Aten type asteroids are constrained by their perihelion and aphelion distances:

$$\begin{aligned} a > 1.0\,\text{AU},\quad & q < 1.017\,\text{AU},\ \text{Apollo (Earth Crosser) type},\\ 1.017\,\text{AU} < \ & q < 1.3\,\text{AU},\ \text{Amor (Mars Crosser) type},\\ a < 1.0\,\text{AU},\quad & Q > 0.983\,\text{AU},\ \text{Aten type}. \end{aligned}$$

Here we adopt the Tisserand invariant as defined by the following equation:

$$J = \frac{1}{a} + \frac{2\sqrt{a - e^2}}{a'\sqrt{a'}} \times \cos I$$

where, a = semi-major axis of comet or asteroid, a' = semi-major axis of Jupiter, e = eccentricity of comet or asteroid, and I = inclination of orbit for comet or asteroid with Jovian orbit.

In Figure 1 a histogram distribution of the Tisserand invariant for short-period comets is given, while in Figure 2 the distribution for Apollo-Amor-Aten asteroids is given.

The short-period comets have J-values less than 0.6, while AAA-asteroids have J-values between 0.3 and 1.5. There is a region where short-period comets and AAA-asteroids are distributed. It is found that asteroids 3552, 5025 P-L, and 1982 YA have small values of the Tisserand invariant, 0.4448, 0.3910, and 0.4612, respectively. It is also found that comets Oterma, Helin-Roman-Crocket, and Gehrels 3 have J-values of 0.5835, 0.5820, and 0.5823, respectively. The well-known P/Encke has the J-value of 0.5811.

It is thus possible that AAA-asteroids given above could be of cometary origin. In this sense, they ought to be observed with greater attention.

Celestial Mechanics and Dynamical Astronomy **54:** 237–240, 1992.

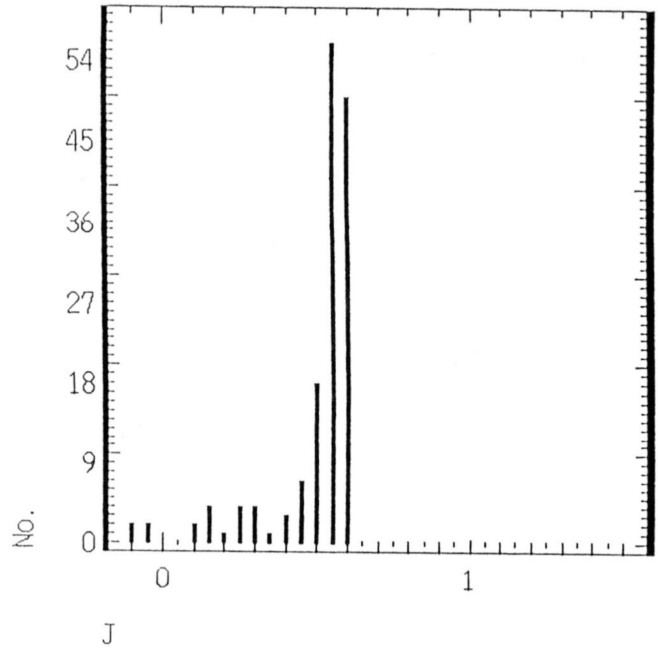

Fig. 1. Histogram distribution of the Tisserand invariant for short-period comets.

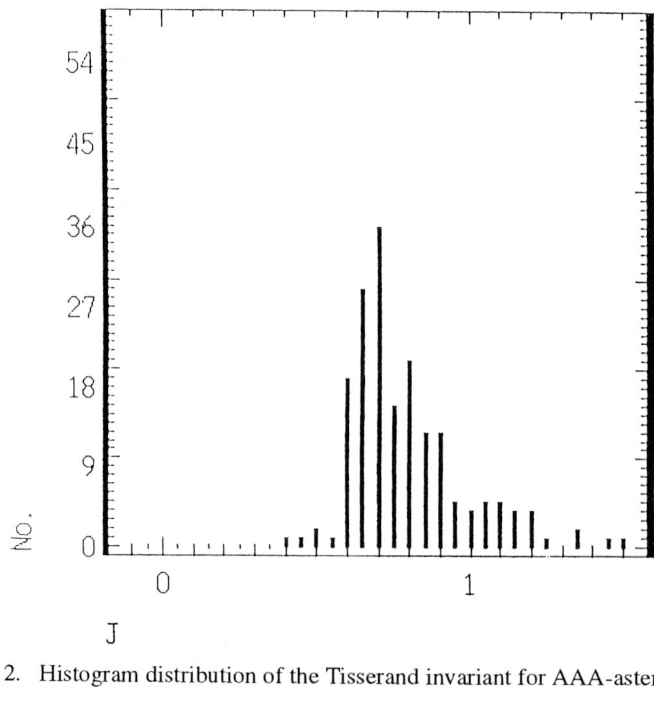

Fig. 2. Histogram distribution of the Tisserand invariant for AAA-asteroids.

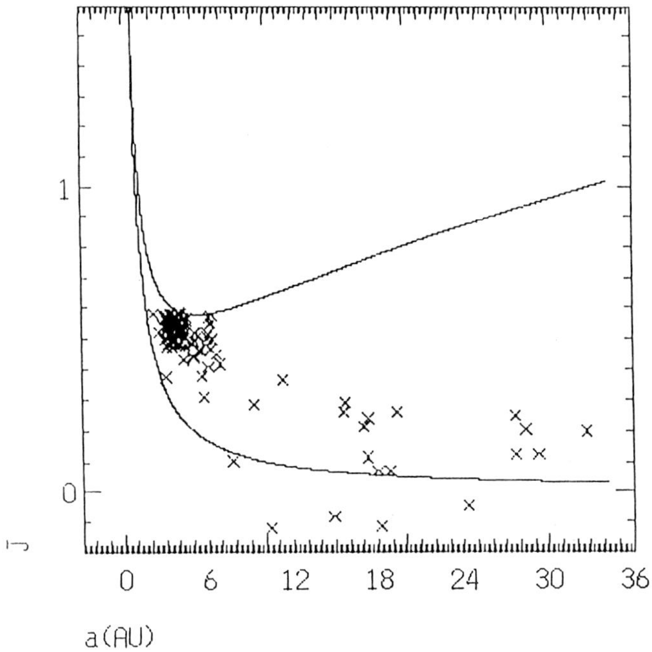

Fig. 3. Constant (J) *vs* semi-major axes (a) for short-period comets. Upper line: $e = 0.0$, $I = 0.0$, lower line: $e = 1.0$, $I = 0.0$.

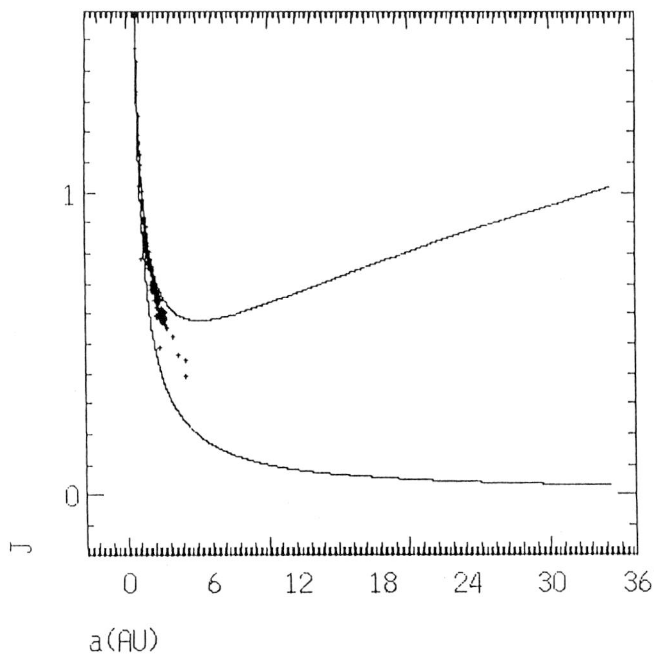

Fig. 4. Constant (J) *vs* semi-major axes (a) for AAA-asteroids. Upper line: $e = 0.0$, $I = 0.0$, lower line: $e = 1.0$, $I = 0.0$.

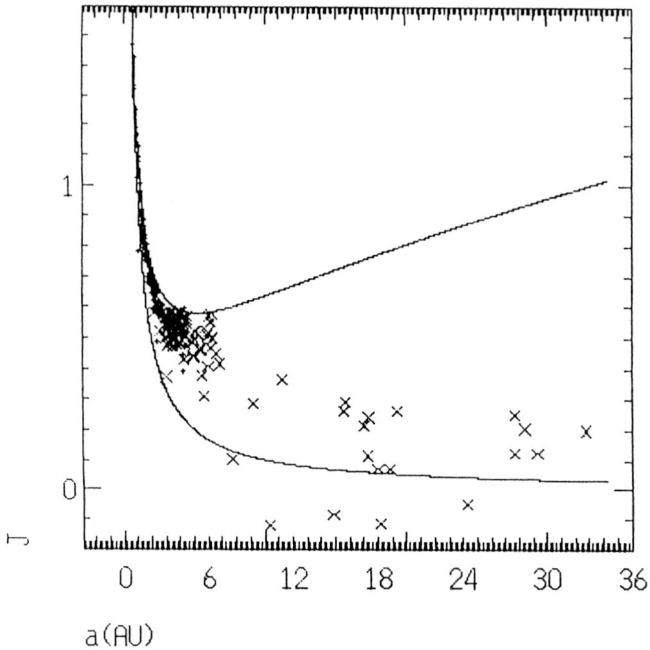

Fig. 5. Constant (J) *vs* semi-major axes (a) for short-period comets and AAA-asteroids. Upper line: $e = 0.0$, $I = 0.0$, lower line: $e = 1.0$, $I = 0.0$.

References

Batrakov, Yu. V., Vinogradova, T. A., Sumzina, N. K., Filenko, L. L., Filinnova, I. A., Chernetenko, Yu. A., and Shor, V. A.: 1990, *Ephemerides of Minor Planets for 1991*, Institute of Theoretical Astronomy, Leningrad

IAUCs: *International Astronomical Union Circulars*, Central Bureau for Astronomical Telegrams

Marsden, B. G.: 1989, *Catalogue of Cometary Orbits, Sixth Edition*, International Astronomical Union, Central Bureau for Astronomical Telegrams/Minor Planet Center

MPCs: *The Minor Planet Circular/Minor Planets and Comets*, Commission 20 of the International Astronomical Union.

THE CHARGING PROCESSES OF DUST PARTICLES AND THE EFFECTS OF LORENTZ SCATTERING IN THE CIRCUM-SOLAR DUST BAND

A. SATHEESH KUMAR and SYUZO ISOBE
National Astronomical Observatory, Mitaka, Tokyo 181, Japan

1. Introduction

The charging processes for the dust particles in the circum-solar dust band at 4 $R_\odot$, and the effects of the interactions between these charged particles and the magnetized ambient solar wind plasma on the evolution of their orbits are analysed. The orbital elements and the distribution in space of the charged dust particles undergo changes due to the effect of additional forces, especially the Lorentz force, acting on such particles, and, hence, the distributions of these particles differ very much from that predicted by taking into account only the effects of gravitational and Poynting-Robertson drag forces.

The acquisition of charges by the dust particles in the interplanetary space due to the incident solar wind and radiations causes these particles to interact with the magnetic field carried by the ambient solar wind plasma through the Lorentz force. The Lorentz force affects the orbital elements and, hence, the orbital distribution of the charged dust particles. Since the Lorentz force is directly proportional to the charge, it is essential to have an accurate estimate of the charge on the dust particle.

The charge q on a dust particle with a mean effective radius s is:

$$q = (\Phi/300)s \quad \text{(esu)} ,$$

where, Φ is the electrostatic potential of the grain in volts.

At 4 $R_\odot$, the physical processes that make the most significant contributions to the equilibrium potential are:

(1) The sticking of charged particles in the solar wind onto the dust grain.

(2) The secondary electron emission due to the impact of solar wind particles on the dust grain. This process becomes very important in the energy ranges of electrons in the average and the maximum phases (about 60 eV and 290 eV, respectively) of the solar cycle.

(3) Photoelectron emission due to the incident solar quanta.

(4) Thermionic emission of electrons from dust particles.

At 4 $R_\odot$, the thermionic emission becomes more effective than even the photo-emission due to the very high temperatures attained by the dust particles in the

Celestial Mechanics and Dynamical Astronomy **54:** 241–244, 1992.

near-solar region ($T > 2000$ K for graphite grains (Mukai and Giese, 1984)). In Millet *et al.* (1980), an estimate for carbon grains in the near-solar orbits yields $\Phi \sim 2$ V, which is a very low value (probably due to the neglect of the secondary electron emission process). Typically, at 1 AU, the potential on graphite grains are estimated to be about +10 V (see Wyatt, 1969). However, due to the additional effects of thermionic and secondary electron emission processes, one should expect higher potentials on the dust particles in the near-solar regions.

2. The Effects of the Lorentz Scattering Force

The Lorentz force arises due to the interaction of the charged dust particle, moving with a velocity $\mathbf{v}$, with the magnetic field $\mathbf{B}$ carried by the solar wind plasma moving with a velocity $\mathbf{v}_{\rm sw}$. This force can be obtained by noting that the electric field $\mathbf{E}$ must vanish in the frame of reference moving with the plasma. Therefore, the force acting on a dust particle possessing a charge q is:

$$\mathbf{F}_{\rm L} = \pm(q/c)\mathbf{V} \times \mathbf{B},$$

where $\mathbf{V} = \mathbf{v} + \mathbf{v}_{\rm sw}$.

The major forces that affect the dust particle orbit are:

the gravitational force: $$F_{\rm g} = -GM_{\odot}/r^2(4/3)\pi\rho s^3$$

the radiation pressure force: $$F_{\rm r} = \pi s^2/cS_0(r_0/r)^2 Q_{\rm pr}$$

the Poynting-Robertson drag: $$F_{\rm PR} = -\pi s^2/cS_0(r_0/r)^2\left[\frac{v/c}{1-v^2/c^2}\right]Q_{\rm pr}$$

where G is the gravitational constant, $M_{\odot}$ is the mass of the Sun, r is the radial distance of the dust particle from the Sun, $S_0 = 1.36 \times 10^6$ erg cm^{-2} s^{-1} is the solar constant, $r_0 = 1$ AU, s and ρ are the radius and the density of the dust particle ($\rho = 2.26$ g cm^{-3} for graphite grains), $Q_{\rm pr}$ is the radiation pressure coefficient, and v and c are the velocity of the dust particle and the velocity of light, respectively.

A comparison between the magnitudes of these forces at 1 AU and at 4 $R_{\odot}$ for different particle radii are shown in Table I. It can be seen that $F_{\rm r} > F_{\rm g}$ for $s \leq 0.3$ μm, and so these particles will be forced out of the solar system by the radiation pressure force. The comparison between the Poynting-Robertson drag force (which causes the dust particles to spiral inward into the inner orbits) and the Lorentz force shows that except for very large particles ($s \gg 10$ μm), $F_{\rm L} \gg F_{\rm PR}$. This result remains valid at 4 $R_{\odot}$ also; but the magnitude of the Lorentz force decreases if the potential Φ is kept the same as that at 1 AU. But, since one can expect a higher potential on the dust particles at 4 $R_{\odot}$, the Lorentz force becomes higher, as seen from Table I, where the ratio $F_{\rm L}/F_{\rm PR}$ is shown for two higher values of Φ. Hence, the Lorentz force should affect even larger particles at 4 $R_{\odot}$ than at 1 AU.

TABLE I
Comparison of the forces acting on a dust particle at 1 AU and at 4 $R_\odot$.

Nature of force (dyn)	$r = 1$ AU $\Phi = 10$ V Grain size (μm)				
	0.1	0.3	0.5	1.0	10.0
F_g	-5.6×10^{-15}	-1.5×10^{-13}	-7.0×10^{-13}	-5.6×10^{-12}	-5.6×10^{-9}
F_r	3.4×10^{-14}	1.9×10^{-13}	3.9×10^{-13}	1.4×10^{-12}	1.4×10^{-10}
F_r/F_g	6.1	1.23	**0.55**	0.25	0.025
F_{PR}	-4.5×10^{-17}	-2.5×10^{-16}	-5.2×10^{-16}	-1.9×10^{-15}	-1.9×10^{-13}
F_L	$\pm 1.3 \times 10^{-14}$	$\pm 4.0 \times 10^{-14}$	$\pm 6.7 \times 10^{-14}$	$\pm 1.3 \times 10^{-13}$	$\pm 1.3 \times 10^{-12}$
F_L/F_{PR}	2.93×10^{2}	1.60×10^{2}	1.29×10^{2}	7.02×10^{1}	**7.02**
t_L/t_{PR}	2.52×10^{-7}	4.15×10^{-6}	1.44×10^{-5}	1.05×10^{-4}	0.105
Force	$r = 4 R_\odot$ $\Phi = 10$ V Grain size (μm)				
	0.1	0.3	0.5	1.0	10.0
F_g	-1.6×10^{-11}	-4.3×10^{-10}	-2.0×10^{-9}	-1.6×10^{-8}	-1.6×10^{-5}
F_r	9.7×10^{-11}	5.3×10^{-10}	1.1×10^{-9}	4.1×10^{-9}	4.1×10^{-7}
F_r/F_g	6.05	1.23	0.55	0.25	0.025
F_{PR}	-6.5×10^{-14}	-3.6×10^{-13}	-7.4×10^{-13}	-2.7×10^{-12}	-2.7×10^{-10}
F_L	$\pm 1.4 \times 10^{-11}$	$\pm 4.2 \times 10^{-11}$	$\pm 6.9 \times 10^{-11}$	$\pm 1.4 \times 10^{-10}$	$\pm 1.4 \times 10^{-9}$
F_L/F_{PR}	2.14×10^{2}	1.17×10^{2}	9.40×10^{1}	5.12×10^{1}	**5.12**
	$r = 4 R_\odot$ $\Phi = 15$ V				
F_L/F_{PR}	3.22×10^{2}	1.76×10^{2}	1.41×10^{2}	7.69×10^{1}	**7.69**
	$r = 4 R_\odot$ $\Phi = 20$ V				
F_L/F_{PR}	4.29×10^{2}	2.34×10^{2}	1.88×10^{2}	1.03×10^{2}	**10.3**

A comparison between the time scales for the Lorentz force and the Poynting-Robertson force (calculated using the relations given in Millet *et al.*, 1980) is also shown in Table I. The characteristic time for the Poynting-Robertson drag is found to be much higher than that for the Lorentz force for most particles at 1 AU. Since the value of the magnetic field in the near-solar region is much higher than that at the earth's orbit, this result should hold true at 4 $R_\odot$. Further, the Lorentz force can affect all the orbital elements of the dust particle (Consolmagno, 1979). Hence, the spatial distribution of these particles should differ from that predicted by taking into account only the Poynting-Robertson effect. This was confirmed for dust particles at 1 AU (Mukai and Giese, 1984), where the inclusion of the Lorentz effect gave a distribution, $n(r) \sim r^{-1.3}$ (reproduced here in Figure 1), satisfying the Helios probes measurements for dust particles in the ecliptic plane.

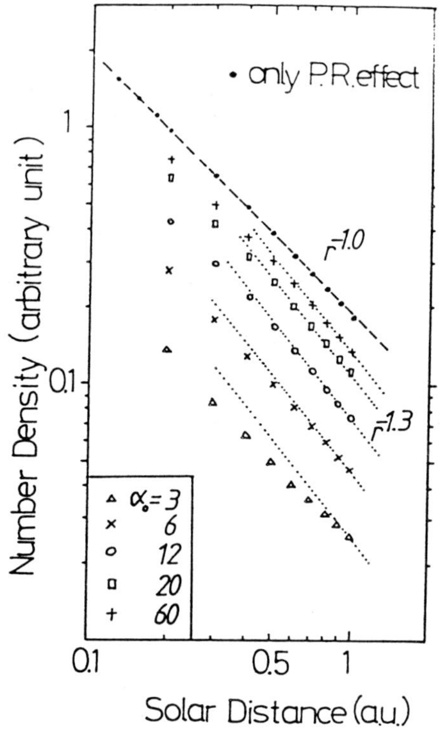

Fig. 1. Spatial distribution of grains under both the Pointing-Robertson and the Lorentz effects as a function of solar distance which is derived by injection function f_2. Dotted lines: $r^{-1.3}$. Mukai and Giese (1984).

3. Conclusions

(1) Due to the higher values of the potential on the dust particle and the ambient solar wind magnetic field, the Lorentz force affects much wider size range of particles in the near-solar regions.

(2) The magnitude of the Lorentz force being much higher and its characteristic time to affect the particle's orbit being much lower than those for the Poynting-Robertson drag force, the Lorentz force is a major perturbing force for dust particles in the circum-solar dust band at 4 $R_{\odot}$.

References

Consolmagno, G.: 1979, *Icarus* **38**, 398
Millet, J., Lafon, J. P. L. and Lamy, Ph. L.: 1980, *Astron. Astrophys.* **92**, 6
Mukai, T. and Giese, R. H.: 1984, *Astron. Astrophys.* **131**, 355
Wyatt, S. P.: 1969, *Planetary Space Sci.* **17**, 155

SHOCK METAMORPHISM ON THE MOON

YASUNORI MIURA and TOSHIO KATO
Faculty of Science, Yamaguchi University, Yamaguchi 753, Japan

Abstract. Shock metamorphism of the lunar samples is discussed. All types of lunar glasses formed by various-size collision-type impact are found as impact glass, ropy glass and agglutinates. The agglutinates bonded by crystal and glassy materials contain hydrogen and helium from the solar wind components. Lunar shocked minerals of plagioclase and silica show anomalous compositions and densities. There are typical two formation processes on planetary materials formed by shock events; that is (1) shocked quartz formed by silica-rich target rocks (esp. on evolved planets of the Earth and Mars), and (2) shocked silica with minor Al contents formed from plagioclase-rich primordial crusts of the Moon. The both shocked silica grows to coarse-grain normal crystals after high-temperature metamorphism which cannot distinguish the original main formation event of impact process.

Key words: Shock metamorphism, the Moon, collision type, agglutinates, formation process, shocked silica.

1. Introduction

Evidence of collision shocks in the cosmic area has been studied mainly through shock metamorphism which is defined by physical, chemical and mineralogical changes due to a high pressure shock wave within a few microseconds.

Shock waves can be produced in the earth's atmosphere and near vacuum of space as "collision" and "collisionless" shock events, respectively. Collision shocks occur when particles of a neutral gas (i.e. atoms and molecules) in the earth's atmosphere collide. By contrast, collisionless shocks occur when the "plasma" from the Sun (i.e. electrically charged particles) encounters the magnetic fields of planets and comets (Tidman and Krall, 1971). Thus, shock waves are considered to play a significant role in space environments.

The purpose of this study is to elucidate the evidence for shock metamorphism on the Moon.

2. Material Evidences of Collision Shocks

The main cause of collision shock waves on the Moon is meteorite. This forms lunar regolith soils and smaller rocks (Miura and Kato, 1989; Miura, 1989).

Quantitative treatments of shock metamorphism using definite chemical and structural data of amorphous glass phases are usually difficult to interpret due to the metastable states. Thus, among three material features of shock metamorphism (i.e. planar, diaplectic and melt glass features), the structural, chemical and textural data of the planar and diaplectic features may be discussed quantitatively, as in this study (French, 1968).

Celestial Mechanics and Dynamical Astronomy **54:** 245–248, 1992.

TABLE I
Diaplectic glasses formed by collision shocks in the lunar regolith

Diaplectic glass	Constituent materials	Remarks
Impact glass	Only glass	Large impact melt by large meteorites
Ropy glass	Glass, broken pieces	By medium-size meteorite impact
Agglutinates	Glass, Fe-metal droplet, mineral grains, solar-winds (H_2, He)	Many bombardments by micrometeoroids and solar winds

TABLE II
Geological processes involving shocked minerals on evolved and primordial planetary surfaces

(1) On evolved planets and meteoroids:	
Quartz → (shock waves)→ High pressure silica	→ (glass coating) → Stishovite/Coesite → (rapid depression) → Shocked quartz (with high ρ) ↓ (high temperature metamorphism) ↓ Quartz (normal)
(2) On primordial planetary and lunar surfaces:	
Plagioclase→ (shock waves)→ Diaplectic plagioclases	→ (vaporization) → (quenching) → Silica minerals Anomalous quartz (with high ρ) ↓ (high temperature metamorphism) ↓ Quartz (normal)

3. Shock Features of the Lunar Regolith

Specific shock-metamorphic effects for feldspar and quartz grains are produced by shock pressure; that is, planar features ($10 \sim 30$ GPa), diaplectic glass ($30 \sim 50$

GPa), melting (50 ~ 90 GPa) and vaporization (~ 90 GPa) (French, 1968; Stöffler, 1971; Miura, 1989, 1991a,b). The shock metamorphic effects are very difficult to find in randomly distributed samples of lunar impact rocks. Lunar hard rocks randomly distributed by meteoritic impact contain little high-pressure silica, though impact and diaplectic glasses formed by high-pressure can be found in all types of lunar glassy materials (i.e. impact glass by large impact, ropy glass by medium-size impact, and agglutinates by micrometeoroids impact as listed in Table I). Diaplectic quartz (i.e. anomalous shocked quartz) in the lunar regolith soils shows higher density-deviation (i.e. about 1%; Miura and Kato, 1991).

Pristine rocks on the lunar surface are completely destroyed by shock impacts due to meteorites, comets, solar wind and solar flares. Thus the present lunar surface is almost covered by thick layers of small fragments of lunar regolith soils and breccias. The lunar regolith soils collected by the Apollo missions consist of agglutinates, impact glass and ropy glassy materials, together with brecciated rocks, volcanic basaltic rocks and volcanic glasses (cf. Table I).

The agglutinates, which are small aggregates (<1 mm) bonded by vesicular, flow-banded glass, consist of comminuted lithic and mineral fragments, and very fine-grained, single iron domains (3 ~ 10 nm) bonded together by impacted glass.

The unique texture and composition of the agglutinates indicate that they are formed by bombardment of micrometeoroids and the solar wind onto the airless lunar surface.

The importance of unique lunar impact glass and agglutinates lies in the fact that hydrogen and helium from the solar wind have been stored within the lunar fine material. Hydrogen and helium can be more strongly fixed in the regular crystal structure frameworks of ilmenite and rutile oxide phases than in the loose and random frameworks of amorphous glass phases.

Thus, a rain of high-velocity meteorites and micrometeoroids onto the airless lunar surface produces the lunar fine-grained regoliths with agglutinates and the solar wind components.

4. Shock Metamorphism of Planetary Materials

In many impact craters, various shock metamorphosed rock types can be obtained that are considered to be similar to evolved igneous rocks produced by volcanic (cf. melt-sheet of impact crater) or plutonic (cf. central-peak of impact crater) processes. Table II shows typical processes of shock metamorphism on evolved and primordial planetary surfaces. On the evolved planets (including the Earth and Mars) and meteoroids, normal standard quartz (i.e. rock crystal) is changed to high-pressure silica minerals of coesite and stishovite by shock waves. The result is a relict phase of high-pressure silica coated by glass. Anomalous shocked quartz (with shock lamellae and high density) can be formed as a result of high-pressure processes (i.e. at impact craters and achondrites). This will crystallize as normal quartz by high-temperature metamorphism. On primordial planetary and lunar surfaces, plagioclase minerals are changed to diaplectic plagioclases and vaporized to silica minerals (with a minor content of plagioclase composition, ca. 3 wt.%) by strong shock waves, resulting in the formation of anomalous shocked quartz by recrystallization, followed by normal quartz by high-temperature metamorphism. By using the shock and high-temperature metamorphism, anomalous combinations of lunar rocks (including coarse-grained feldspar-quartz impact rocks) can be well explained as due to new evolutionary processes involving shock metamorphism as listed in Table II.

Acknowledgements

The present research is supported by Grant-in-Aid for Scientific Researches of Shock Waves Research and General Research (B) of the author, 1991 selected by the Japanese Ministry of Education, Science and Culture. The authors wish to thank Professors J. Henrard, S.V.M. Clube and S. Yabushita for revising the manuscript.

References

French, B.: 1968, *Shock Metamorphism of Natural Materials*, 1–17
Miura, Y.: 1989, *LPI Technical Report*, USA, 90–2, 52–53
Miura, Y.: 1991a, *Lunar and Planetary Science Conference XXI*, NASA-LPI, 905–908
Miura, Y.: 1991b, *Shock Waves, an International Journal* **1**, 35–41
Miura, Y. and Kato, T.: 1989, *Lunar and Planetary Science Conference XX*, NASA-LPI, 703–704
Miura, Y. and Kato, T.: 1991, *Symp. Antartic Meteorites*, NIPR, Tokyo, 18
Stöffler, D.: 1971, *Journal of Geophysical Research* **76**, 5474–5488
Tidman, D.A. and Krall, N.A.: 1971, *Shock Waves in Collisionless Plasmas*, Wiley-Interscience

EVIDENCE OF EARTH CATASTROPHE BY ANOMALOUS SHOCKED QUARTZ AT THE K/T BOUNDARY

YASUNORI MIURA, TOSHIO KATO and MASASHI IMAI
Faculty of Science, Yamaguchi University, Yoshida, Yamaguchi 753, Japan

Abstract. Anomalous shocked quartz with high density (less than 1 % of density-deviation) is considered to be a relict of ultra high-pressure at meteoritic impact. The shocked quartz grains can be found only in terrestrial and artificial impact craters, meteorites and the Cretaceous-Tertiary (K/T) boundary samples. Volcanic activity is considered to be started or accelerated by enormous impact event.

Key words: K/T boundary problem, shocked quartz, impact craters, density deviation, formation process.

1. Introduction

Geochemical data of the Cretaceous-Tertiary (K/T) boundary have been reported for anomalous contents of Pt-group elements including Ir (Alvarez *et al.*, 1980), with varying amounts and ratios of various isotopes and siderophile group and chalcophile elements (Kyte *et al.*, 1980). There are some problems as to the main cause of the K/T boundary event arising from sedimentary and tectonic mixing processes that cause the elementary concentration to vary depending on the location of water on the Earth. On the other hand , mineralogical data from the K/T boundary samples are important for discovering the origin of the K/T boundary event; for example, goyazite, pyrite, feldspar in spherules (Montanari *et al.*, 1983), shocked quartz grains (Bohor *et al.*, 1984; Miura, 1991a), and stishovite (McHone *et al.*, 1989). Among these mineralogical materials, worldwide distribution of shocked quartz with the characteristic features of shock lamellae and high density ($\rho/\rho_0 \sim +1\%$) is important evidence to the cause of the K/T boundary event and Earth catastrophe (Miura, 1991a,b).

The purposes of the present study are (1) to investigate shocked quartz grains with anomalous high density formed by artificial impact craters, compared with fine quartz from Un-zen volcanic ash, and (2) to clarify the circumstances of the Earth catastrophe due to meteoritic impacts at the K/T boundary and various terrestrial impact craters.

2. Experimental Procedure

The samples of shocked quartz used in this study are (1) from six K/T boundaries (European marine K/T sections at Stevns Klint in Denmark; Gams in Austria; Gubbio in Italy; African marine section at El Kef in Tunisia; Japanese marine section at Kawaruppu, Hokkaido; and a nonmarine section at Clear Creek North,

Celestial Mechanics and Dynamical Astronomy **54:** 249–253, 1992.

Colorado in U.S.A.), (2) from five impact craters (Quaternary km-size meteorite crater with sandstone target rocks at Barringer, U.S.A.; Tertiary 20 km-size impact crater with breccia at Lake Mistastin, Canada; Mesozoic 20 km-size impact crater with shattercones at Gosses Bluff, Australia; Paleozoic 20 km-size impact crater at Clearwater Lakes, Canada; Precambrian 100 km-size impact crater at Sudbury, Canada), and (3) from an artificial impact crater with target rock of Kohyama gabbroic anorthosite formed by a railgun apparatus with a velocity of about 2 km/sec.

Preliminary data (using an X-ray powdered diffractometer) were obtained in order to calculate the density for all the quartz samples. Single grains of shocked quartz were selected with a polarized optical microscope using an supersonic cutter. The final high precision data for calculating density were then obtained with the Rigaku automated single crystal four-circle X-ray diffractometer (AFC-5 FOS system) at the Yamaguchi University.

3. Results and Discussion

3.1. ANOMALOUS SHOCKED QUARTZ

Shocked quartz grains from marine and nonmarine K/T boundary samples, impact craters and meteorites are anomalous quartz with higher density (less than 1% of the density-deviation), undulose optical extinction, fine grain size (less than 1 mm in diameter), peak diffuseness of X-ray diffraction (about 0.3% deviation in the a-axis of Colorado K/T boundary sample SQ-20 compared with 0.04% in the quartz standard sample RC; cf. Table I), pseudo-monoclinic features characterized by structural discrepancies (i.e. about 0.3% of discrepancy between **a**- and **b**-axes in the sample SQ-20; the original uncorrected unit-cell dimensions from 25 measured peaks were **a**=4.904±0.014(Å), **b**= 4.919±0.016(Å), **c**= 5.385±0.011(Å)), and irregular shrinkage of atomic structure (about −1.6% deviation in the distance between oxygens in Si-O tetrahedra).

3.2. QUARTZ FROM IMPACT CRATERS, METEORITES, ARTIFICIAL CRATER AND VOLCANIC ASH

All values of density-deviation (ρ/ρ_0) of quartz listed in Table I are obtained from density (2.645 g/cm^3) of rock crystal from Hirase Mine, Japan, which was used as standard quartz in this study (i.e. sample RC). Representative values of the density-deviation from terrestrial impact craters are higher, from +0.2±0.1% in a shattercone of Gosses Bluff GB-3SC to +0.7±0.2% in the Barringer crater, as listed in Table I.

Anomalous quartz with the highest density is found in the extraterrestrial materials of Zagami meteorites; that is, +1.7±0.5% density-deviation (Miura and Kato, 1991; cf. Table I).

However, anomalous quartz with shock lamellae could not be found in volcanic, metamorphic or tectonic quartz. The density of quartz from volcanic ash has a

TABLE I
Calculated density data of quartz grains from impact craters, meteorites, artificial impact craters and volcanic ash. The numbers in parentheses are standard deviation refering to the last decimal place of the data

Sample	Vol. (Å^3)	ρ (g/cm^3)	ρ/ρ_0 %	Pressure, Temperature
Standard quartz				
Quartz(RC)	113.1(2)	2.645(3)	0.0	(standard)
Japanese K/T boundary				
JKT-75	112.7(2)	2.655(2)	+0.4(1)	P
Danish K/T boundary				
SK-23P	112.7(7)	2.656(16)	+0.4(5)	P
Italian K/T boundary				
GB-IJP	112.8(1)	2.652(2)	+0.3(1)	P
Austrian K/T boundary				
AG-3P	112.8(1)	2.653(1)	+0.3(1)	P
Tunisian K/T boundary				
ELK-8P	112.6(2)	2.658(4)	+0.5(1)	P
American K/T boundary				
SQ-2	112.2(2)	2.667(4)	+0.8(2)	P
Barringer meteorite crater				
B-3W	112.3(3)	2.664(6)	+0.7(2)	P
Lake Mistastin crater				
LM-2069	112.7(1)	2.656(1)	+0.4(1)	P
Gosses Bluff impact crater				
GB-3SC	112.9(1)	2.651(3)	+0.2(1)	P
Clearwater lakes craters				
DCW-16429	112.7(1)	2.656(1)	+0.4(1)	P
Sudbury impact breccia				
SB-1P	112.8(1)	2.653(1)	+0.3(1)	P
Meteorite				
Zagami-1	111.3(5)	2.689(12)	+1.5(5)	P
Artificial impact crater				
GAN-S-Ef	112.9(1)	2.651(2)	+0.2(1)	P
Volcanic ash				
UnzenLSS611	113.6(4)	2.635(7)	−0.4(3)	T

negative value (i.e. -0.4%) due to temperature-dependent formation (cf. Table I). The preliminary detailed data from fine quartz grains from volcanic ash indicate that anomalous density of quartz formed by volcanic eruptions could not be obtained by long exposure to the high temperature of pyroclastic lava flows.

3.3. Anomalous shocked quartz from the K/T boundary

Table I shows representative density data of shocked quartz from K/T boundary sections. Typical shocked quartz grains from the K/T boundary show higher density-deviations than that of standard rock crystal, from +0.3±0.1% in Italian marine section GB-IJP to +0.8±0.2% in the nonmarine Colorado CCN section SQ-2.

As similar anomalous shocked quartz from the K/T boundary can also be found in terrestrial impact craters, both on land and in the ocean, the anomalous shocked quartz with high density can be found at different horizons before and after the K/T boundary.

Meteoritic impacts with strong shock waves produce the remarkable effects of increased pressure (about 10 times by artificial experiments) within sea-water. Such effects prevailing throughout the world could well explain the serious damage to marine lives. However the extinction of nonmarine plants and animals such as the dinosaurs at the K/T boundary can be largely explained by the terrestrial environmental changes caused by lack of sunlight and changes in oxygen fugacity.

Anomalous shocked quartz formed by rapid release of high pressure shows a maximum density change of +0.8% in nonmarine K/T boundary and impact crater samples (which are calculated as about 41 GPa). Amorphous glassy silica materials formed at high shock pressure recrystallize to high-temperature type silica.

3.4. Formation processes of anomalous shocked quartz

There are two types of "shocked quartz": (1) shocked quartz formed by direct transformation from high-pressure silica, which is found in the terrestrial and Martian crusts and in meteoroids, and (2) fine shocked quartz crystallized by chemical change mainly from feldspar composition under vapour conditions of impact events such as found on the lunar and planetary surfaces. The preliminary data indicate that some silica minerals with about 2 wt.% of feldspar composition are relict minerals from feldspar impact melts.

The Earth catastrophe at the K/T boundary can be considered to be an example of a huge meteoritic impact followed by tectonic and volcanic activity.

Acknowledgements

The present research is partly supported by the Grant-in-Aid for Scientific Researches of the Japanese Ministry of Education and Science, 1991 of the senior author (03452071, 03238108). The authors wish to thank Professor J. Henrard, S.V.M. Clube and S. Yabushita for revising the manuscript.

References

Alvarez, L.W., Alvarez, W., Asaro, F. and Michel, H.V.: 1980, *Science* **208**, 1095–1108
Bohor, B.F., Modreski, P.J. and Foord, E.E.: 1987, *Science* **236**, 705–709

Kyte, F.T., Zhou, Z. and Wasson, J.T.: 1980, *Nature* **288**, 651–656
McHone, J.F., Niema, R.A., Lewis, C.F. and Yates, A.M.: 1989, *Science* **243**, 776–779
Miura, Y.: 1991a, *Shock Waves* **1**, 35–41
Miura, Y.: 1991b, *Lunar and Planetary Science XXII*, (LPI,Houston), 905–908
Miura, Y. and Kato, T.: 1991, *Symp. Antartic Meteorites*, NIPR, Tokyo, 18, 17–19
Montanari, A., Hay, R.L., Alvarez, W., Asaro, F., Michel, H.V., Alvarez L. and Smit, J.: 1983, *Geology* **11**, 668–671

HILL STABILITY OF THE PLANAR THREE-BODY PROBLEM: GENERAL AND RESTRICTED CASES

TAKESHI NAKAMURA
Rakuyo Girls' High School, Nakagyo-ku, Kyoto 604, Japan

and

JUNZO YOSHIDA
Department of Physics, Kyoto Sangyo University, Kita-ku, Kyoto 603, Japan

Abstract. The equations of motion of the planar three-body problem split into two parts, called an external part and an internal part. When the third mass approaches zero, the first part tends to the equations of the Kepler motion of the primaries and the second part to the equations of motion of the restricted problem.

We discuss the Hill stability from these equations of motion and the energy integral. In particular, the Jacobi integral for the circular restricted problem is seen as an infinitesimal-mass-order term of the Sundman function in this context.

Key words: Three-body problem, Hill stability.

1. Notations

Let $m_1,\ m_2,\ m_3$ be the masses of the three bodies, r_{ij} the mutual distances. We will use the usual Jacobi coordinates $\boldsymbol{r},\ \boldsymbol{\rho}$, where $\boldsymbol{r}$ is a vector from m_1 to m_2, and $\boldsymbol{\rho}$ from the center of mass of m_1 and m_2 to m_3. In what follows, all vectors are identified with complex numbers, thus $\boldsymbol{r},\ \boldsymbol{\rho}$ are denoted by

$$\boldsymbol{r} = r_1 + ir_2, \quad \boldsymbol{\rho} = \rho_1 + i\rho_2 , \tag{1}$$

where $i = \sqrt{-1}$. For any given complex number $\boldsymbol{a}$, its conjugate is denoted by $\bar{\boldsymbol{a}}$ and its absolute value by $a = |\boldsymbol{a}| = \sqrt{\boldsymbol{a}\bar{\boldsymbol{a}}}$.

Let

$$\begin{aligned} \alpha_1 &= \frac{m_1}{m_1+m_2}, \qquad \alpha_2 = \frac{m_2}{m_1+m_2}; \\ \mu_1 &= \frac{m_1 m_2}{m_1+m_2}, \qquad \mu_2 = \frac{m_3(m_1+m_2)}{m_1+m_2+m_3}. \end{aligned} \tag{2}$$

The equations of motion of the planar three-body problem are

$$\mu_1\ddot{\boldsymbol{r}} = grad_r U , \quad \mu_2\ddot{\boldsymbol{\rho}} = grad_\rho U , \tag{3}$$

where

$$grad_r = \partial/\partial r_1 + i\partial/\partial r_2 , \tag{4}$$

Celestial Mechanics and Dynamical Astronomy **54:** 255–260, 1992.

and

$$U = G\left(\frac{m_1m_2}{r_{12}} + \frac{m_2m_3}{r_{23}} + \frac{m_3m_1}{r_{31}}\right) , \tag{5}$$

where G is the gravitational constant.

The integrals of angular momentum and energy are

$$\begin{aligned} c &= \Im(\mu_1\bar{\boldsymbol{r}}\dot{\boldsymbol{r}} + \mu_2\bar{\boldsymbol{\rho}}\dot{\boldsymbol{\rho}}) \\ &= \mu_1(r_1\dot{r}_2 - r_2\dot{r}_1) + \mu_2(\rho_1\dot{\rho}_2 - \rho_2\dot{\rho}_1) , && (6) \\ h &= T - U , && (7) \end{aligned}$$

where the symbol $\Im$ stands for the imaginary part in parentheses, and

$$T = \frac{1}{2}(\mu_1|\dot{\boldsymbol{r}}|^2 + \mu_2|\dot{\boldsymbol{\rho}}|^2) . \tag{8}$$

The moment of inertia with respect to the center of mass of the three-body system is

$$R^2 = \mu_1 r^2 + \mu_2\rho^2 . \tag{9}$$

2. Transformation of Coordinates

A transformation of coordinates from $(\boldsymbol{r}, \boldsymbol{\rho}, \dot{\boldsymbol{r}}, \dot{\boldsymbol{\rho}})$ to a new set of coordinates $(\boldsymbol{x}, \theta, R, \boldsymbol{\xi}, \boldsymbol{\eta})$ is defined by

$$\begin{cases} \boldsymbol{x} &= \boldsymbol{\rho}/\boldsymbol{r} , \quad (\boldsymbol{r} \neq \mathrm{o}) \\ \theta &= \tan^{-1}(r_2/r_1) , \quad (\boldsymbol{r} = re^{i\theta}) \\ R &= \sqrt{\mu_1 r^2 + \mu_2\rho^2} , \\ \boldsymbol{\xi} &= \mu_1\bar{\boldsymbol{r}}\dot{\boldsymbol{r}} + \mu_2\bar{\boldsymbol{\rho}}\dot{\boldsymbol{\rho}} , \\ \boldsymbol{\eta} &= \boldsymbol{r}\dot{\boldsymbol{\rho}} - \boldsymbol{\rho}\dot{\boldsymbol{r}} , \end{cases} \tag{10}$$

where $\boldsymbol{x}$, $\boldsymbol{\xi}$ and $\boldsymbol{\eta}$ are complex numbers and θ and R are real numbers. The first coordinate $\boldsymbol{x}$ determines the shape of a triangle formed by m_1, m_2, m_3. The angle θ is the argument of $\boldsymbol{r} = re^{i\theta}$ and it determines the orientation of the triangle. The third coordinate R determines the scale of the triangle in its evolution. Note that the triangles discussed here include collinear ones, and especially colliding cases. The coordinate $\boldsymbol{\xi}$ expressed in terms of $(r_i, \dot{r}_i, \rho_i, \dot{\rho}_i)$, $(i = 1, 2)$, is also denoted by

$$\boldsymbol{\xi} = R\dot{R} + ic . \tag{11}$$

Therefore, the adoption of the coordinate $\boldsymbol{\xi}$ is equivalent to introducing real variables $\dot{R}$ and c as new coordinates. The last coordinate $\boldsymbol{\eta}$ is related to $\dot{\boldsymbol{x}}$ by

$$\dot{\boldsymbol{x}} = \frac{\boldsymbol{\eta}}{\boldsymbol{r}^2} . \tag{12}$$

The inverse transformation of (10) is given by

$$\begin{cases} \boldsymbol{r} &= \sigma^{-1} R e^{i\theta} \,, \\ \boldsymbol{\rho} &= \sigma^{-1} R \boldsymbol{x} e^{i\theta} \,, \\ \dot{\boldsymbol{r}} &= (\sigma R)^{-1} (\boldsymbol{\xi} - \mu_2 \bar{\boldsymbol{x}} \boldsymbol{\eta} e^{-2i\theta}) e^{i\theta} \,, \\ \dot{\boldsymbol{\rho}} &= (\sigma R)^{-1} (\boldsymbol{x} \boldsymbol{\xi} + \mu_1 \boldsymbol{\eta} e^{-2i\theta}) e^{i\theta} \,, \end{cases} \tag{13}$$

where

$$\sigma = \sqrt{\mu_1 + \mu_2 x^2} \,. \tag{14}$$

Finally, introducing

$$\boldsymbol{y} = \boldsymbol{\eta} e^{-2i\theta} \,, \tag{15}$$

we arrive at the final coordinates $(R, \theta, \boldsymbol{x}; \dot{R}, c, \boldsymbol{y})$.

3. Splitting of the Equations of Motion

PROPOSITION 1. *The equations of motion of the planar three-body problem split into the following two parts when expressed in terms of the coordinates* $(R, \theta, \boldsymbol{x}; \dot{R}, c, \boldsymbol{y})$*:*

1. *the external part with coordinates* $(R, \theta, \dot{R}, c)$*:*

$$\begin{cases} \dot{\theta} &= \dfrac{c}{R^2} - \dfrac{\mu_2}{R^2} \Im(\bar{\boldsymbol{x}} \boldsymbol{y}) \,, \\ \ddot{R} &= -\dfrac{V(\boldsymbol{x})}{R^2} + \dfrac{c^2}{R^3} + \mu_1 \mu_2 \dfrac{y^2}{R^3} \,, \\ \dot{c} &= 0 \,, \end{cases} \tag{16}$$

where

$$V(\boldsymbol{x}) = RU = G\sigma \left(m_1 m_2 + \frac{m_1 m_3}{|\boldsymbol{x} + \alpha_2|} + \frac{m_2 m_3}{|\boldsymbol{x} - \alpha_1|} \right) ; \tag{17}$$

2. *the internal part with coordinates* $(\boldsymbol{x}, \boldsymbol{y})$*;*

$$\begin{cases} \dot{\boldsymbol{x}} &= \dfrac{\sigma^2}{R^2} \boldsymbol{y} \,, \\ \dot{\boldsymbol{y}} + 2i\dot{\theta} \boldsymbol{y} &= \dfrac{\sigma^2}{\mu_1 \mu_2 R} grad_x V(\boldsymbol{x}) \,. \end{cases} \tag{18}$$

The energy integral is expressed in terms of the new coordinates as

$$h = \frac{1}{2R^2} (R^2 \dot{R}^2 + c^2 + \mu_1 \mu_2 y^2) - \frac{V(\boldsymbol{x})}{R} \,. \tag{19}$$

REMARK 1. *The coordinates $R, \theta, \boldsymbol{x}$ together with their conjugate momenta $P_R, P_\theta, \boldsymbol{P}$:*

$$P_R = \dot{R}\,, \quad P_\theta = c\,, \quad \frac{\sigma^2}{\mu_1\mu_2}\boldsymbol{P} = \boldsymbol{y} + \frac{ic}{\mu_1}\boldsymbol{x}\,,$$

form a set of canonical coordinates. The Hamiltonian of the problem is given by:

$$H = \frac{1}{2}\left\{P_R^2 + \frac{\sigma^2 P_\theta^2}{\mu_1 R^2} + \frac{\sigma^4}{\mu_1\mu_2 R^2}|\boldsymbol{P}|^2 + \frac{2\sigma^2 P_\theta}{\mu_1 R^2}(P_1 x_2 - P_2 x_1)\right\} - \frac{V(\boldsymbol{x})}{R}\,.$$

4. The Restricted Problem as Limit Case

Now, let us consider the limit case as $m_3 \to 0$ in the Proposition 1.

LEMMA 1. *If $m_3 \to 0$, then we have:*

1. $c \to \mu_1 c_0$, where c_0 is the angular momentum of the m_1, m_2-system:

$$c_0 = r_1\dot{r}_2 - r_2\dot{r}_1\,; \tag{20}$$

2. $h \to \mu_1 h_0$, where h_0 is the energy of the m_1, m_2-system:

$$h_0 = \frac{1}{2}(\dot{r}^2 + \frac{c_0^2}{r^2}) - \frac{G(m_1+m_2)}{r}\,; \tag{21}$$

3. $V(\boldsymbol{x}) = Gm_1m_2\sqrt{\mu_1} + m_3\sqrt{\mu_1}\tilde{V}(\boldsymbol{x}) + O(m_3^2)$, (Milani and Nobili, 1983) *where*

$$\tilde{V}(\boldsymbol{x}) = G(m_1+m_2)\left\{\frac{x^2}{2} + \frac{\alpha_1}{|\boldsymbol{x}+\alpha_2|} + \frac{\alpha_2}{|\boldsymbol{x}-\alpha_1|}\right\}\,. \tag{22}$$

PROPOSITION 2. *If $m_3 \to 0$, then*

1. the external part (16) tends to the Kepler motion of m_1 and m_2 in polar coordinates (r, θ) and with the angular momentum c_0:

$$\left\{\begin{array}{rcl} \dot{\theta} & = & c_0/r^2\,, \\ \ddot{r} & = & -\dfrac{G(m_1+m_2)}{r^2} + \dfrac{c_0^2}{r^3}\,, \\ \dot{c}_0 & = & 0\,; \end{array}\right. \tag{23}$$

2. the internal part (18) tends to

$$\left\{\begin{array}{rcl} c_0\dfrac{d\boldsymbol{x}}{d\theta} & = & \boldsymbol{y}\,, \\ \dfrac{c_0}{r^2}(\dfrac{d\boldsymbol{y}}{d\theta} + 2i\boldsymbol{y}) & = & \dfrac{1}{r}grad_x\tilde{V}(\boldsymbol{x})\,. \end{array}\right. \tag{24}$$

If the motion of m_1 and m_2 is elliptic, then the equations (24) become the equations of motion of the elliptic restricted three-body problem (Szebehely, 1967, §10.3.2). In particular, if $r = const.$ in (24), then the equations (24) become the equations of motion of the circular restricted three-body problem.

Similarly, the equations of motion of the parabolic or the hyperbolic restricted three-body problem are obtained.

5. Deduction of Hill Stability of the Restricted Problem

Here, we consider how the Hill stability of the restricted problem is deduced in our framework (Marchal and Bozis, 1982, Milani and Nobili, 1983).

5.1. HILL STABILITY OF THE GENERAL PROBLEM

Let

$$S(R,\dot{R};c,h) \equiv R\dot{R}^2 + \frac{c^2}{R} - 2hR \tag{25}$$

be the Sundman function. Using the energy integral (19), S is also denoted by:

$$S = 2V(\boldsymbol{x}) - \frac{\mu_1\mu_2 y^2}{R}\,. \tag{26}$$

If $h < 0, c \neq 0$, then S has a minimum value $2\sqrt{-2hc^2}$ at $\dot{R} = 0$ and $R = c/\sqrt{-2h}$. Thus

$$V(\boldsymbol{x}) \geq \sqrt{-2hc^2}\,. \tag{27}$$

From this we can discuss the Hill stability as is well known. For example, if $\sqrt{-2hc^2} > \max_{i=1,2,3} V(L_i)$, where $L_i, (i = 1,2,3)$ are the Lagrange collinear points, then the Hill region is separated into three disjoint components.

If $h < 0$, $c = 0$, it is open whether S has a minimum value or not.

In general, from (25) and the second equation of (16) and (19) we have

$$\frac{dS}{dt} = \frac{\mu_1\mu_2 y^2}{R^2}\dot{R}\,. \tag{28}$$

Let $\dot{R}_0$ and S_0 be the initial values of $\dot{R}$ and S at $t = 0$. It is known that if $h \geq 0$, and $\dot{R}_0 > 0$, then $\dot{R} > 0$ for all $t \geq 0$. Thus, from (28) the minimum value of S is S_0 for $t \geq 0$, and $2V(\boldsymbol{x}) \geq S_0$ for $t \geq 0$. Then, in the case $h \geq 0$, the condition of the Hill stability is:

$$\dot{R}_0 > 0, \quad and \quad S_0 > \max_{i=1,2,3} 2V(L_i)\,. \tag{29}$$

5.2. LIMIT CASE AS $m_3 \to 0$

From Lemma 1.3 and (26) we have for $m_3 \ll 1$

$$S = 2Gm_1m_2\sqrt{\mu_1} + m_3\sqrt{\mu_1}\left(2\tilde{V}(\boldsymbol{x}) - \frac{y^2}{r}\right) + O(m_3^2)\,. \tag{30}$$

PROPOSITION 3. *From (30) the formula (28) becomes*

$$\frac{d}{dt}\left(2\tilde{V}(\boldsymbol{x}) - \frac{y^2}{r}\right) = \frac{y^2}{r^2}\dot{r} \tag{31}$$

as $m_3 \to 0$, where $\tilde{V}(\boldsymbol{x})$ is defined by (22).

Case I $h < 0$, $c \neq 0$. Since from the Lemma 1, $h \to \mu_1 h_0$ and $c \to \mu_1 c_0$, the motion of m_1 and m_2 is elliptic or circular without collision. From (24) and (31), only in the case that $\dot{r} = 0$, i.e. in the circular motion, there appears the well-known Jacobi integral

$$J(\boldsymbol{x}, \boldsymbol{y}) \equiv 2\tilde{V}(\boldsymbol{x}) - \frac{y^2}{r} = const. \,, \tag{32}$$

where $r = const.$, and we can discuss the Hill stability for the circular restricted problem as usual. For example, if the Jacobi constant is greater than $2\tilde{V}(L_3)$, then the Hill region consists of three components, and the infinitesimal mass m_3 stays forever near m_1 or m_2, or else far apart from m_1 and m_2.

REMARK 2. *We notice from (30) that the function J is a term of $O(m_3)$ of the Sundman function up to a finite factor $\sqrt{\mu_1}$.*

For the elliptic restricted problem, it is open in our framework whether the equations of motion is Hill stable or not, since the sign of $\dot{r}$ is not fixed.

Case II $h \geq 0$. In this case the motion of m_1 and m_2 is parabolic or hyperbolic as $m_3 \to 0$. Let $t = 0$ be the time at which m_2 passes through the pericenter. Then we always have $\dot{r} > 0$ for all $t > 0$. Let J_0 be the initial value of J at $t = 0$, then from (31) we have

$$J > J_0 \,, \quad \text{for all} \quad t > 0 \,. \tag{33}$$

Thus,

$$2\tilde{V}(\boldsymbol{x}) \geq J_0 \,. \tag{34}$$

The condition of Hill stability is satisfied when $J_0 > 2\tilde{V}(L_3)$. From (34), we have the following:

PROPOSITION 4. *In the parabolic or hyperbolic restricted problem, if at $t = 0$*
(i) $\dot{r}_0 \geq 0$,
(ii) $y_0^2 < 2r_0(\tilde{V}(\boldsymbol{x}_0) - \tilde{V}(L_3))$,
are satisfied, then the infinitesimal mass m_3 is bounded near m_1 or m_2, or else far apart from m_1 and m_2 for all $t \geq 0$.

References

Marchal, C. and Bozis, G: 1982, 'Hill Stability and Distance Curves for the General Three-Body Problem', *Celest. Mech.*, **26**, 311–333.

Milani, A. and Nobili, A.M.: 1983, 'On Topological Stability in the General Three-Body Problem', *Celest. Mech*, **31**, 213–240.

Szebehely, V.: 1967, *Theory of Orbits*, Academic Press, New York.

INVISIBLE COMETS ON EVOLUTIONARY TRACK OF SHORT-PERIOD COMETS

TSUKO NAKAMURA
National Astronomical Observatory, Mitaka, Tokyo, Japan

and

MAKOTO YOSHIKAWA
Communications Research Laboratory, Kashima Space Research Center, 893-1 Hirai, Kashima, Ibaraki 314, Japan

Abstract. We systematically surveyed the orbits of short-period (SP) comets that show a large change of perihelion distance (q) between 1–2 AU (visible comets) and 4–5 AU (invisible comets) during 4400 years. The data are taken from Cosmo-DICE (Nakamura and Yoshikawa 1991a), which is a long-term orbital evolution project for SP comets. Recognizing that q is the most critical element for observability of comets, an invisibility factor (f), defined as the ratio of unobservable time span to observable span during 4400 years, is calculated for each of the large-q-change comets. A detection limit for each comet is obtained from the heliocentric distance at discovery and/or the absolute magnitude at recent apparitions. A mean f value for 35 SP comets with $2.9 \leq J$ (J is the Tisserand's invariant) is found to be 19.8. This implies that for each visible SP comet of this J-range, at every epoch of time, there exist about 20 invisible comets near the capture orbits by Jupiter, under the assumptions of steady-state flux and ergodicity for the SP-comet population.

Key words: Short period comets, observability of comets.

1. Introduction

It is well known that the statistics of observed comets is more or less affected by observational selection. Such a typical example is the distribution of perihelion distance (q) of observed SP comets as shown in Figure 1.

Although we see very few comets with q larger than 3 AU in the histogram, this is obviously due to observational selection caused by the low activity of sublimation of cometary nuclei at large heliocentric distances. As mentioned below, there are good reasons to believe that there exist a considerable number of invisible comets also in the orbits with q of 4 - 5 AU. The aim of this paper is therefore to estimate the number of such unobservable comets in the capture orbits by Jupiter, from the discovery circumstances of SP comets and their long-term orbital behavior.

2. Orbital Behavior of Comets Captured by Jupiter

Everhart (1972) first demonstrated numerically, using a simplified dynamical model, that long-period (LP) comets in the orbits with low inclination ($i \leq 9^o$) and 4 AU $\leq q \leq$ 6 AU (he called these orbits the capture region) can be most efficiently captured into SP comets by Jovian perturbation. We have integrated the orbits of all the observed SP comets (about 160) for 3400 years and 1000 years

Celestial Mechanics and Dynamical Astronomy **54:** 261–266, 1992.

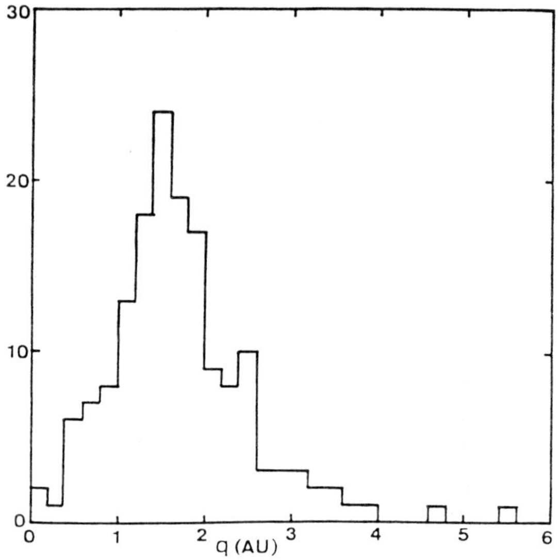

Fig. 1. Distribution of perihelion distance for 160 SP comets.

TABLE I
Invisibility factors for SP comets of large q-change

Group	J-range	N_1	N_2	n		$\langle f \rangle$
A	$2.9 \leq J < 3.1$	50	44	39	(all)	17.9
				35	($f > 2$)	19.8
B	$2.8 \leq J < 2.9$	21	13	13	(all)	13.4
				10	($f > 2$)	17.3
C	$2.7 \leq J < 2.8$	33	21	15	(all)	6.5
D	$2.5 \leq J < 2.7$	20	7	5	(all)	1.3
E	$J < 2.5$	35	1	1		1.3

toward the past and the future respectively by taking account of all the planets (Nakamura and Yoshikawa, 1991a). We have confirmed Everhart's view (1972) of cometary capture for *actual* SP comets (Nakamura and Yoshikawa, 1991b). The following is a brief summary of their orbital behavior.

Since the heliocentric distance or equivalently q is the most sensitive factor for observability conditions of comets, we classify in Table 1 the orbital behavior of q for SP comets by the range of their value of Tisserand's invariant (J). This is because the J value of a comet should be kept nearly constant during its orbital evolution; in fact, the variation of J during 4400 years stays within 0.05 - 0.1 (in the units of the orbital radius of Jupiter) for most of the comets in question.

N_1 in Table 1 is the total number of SP comets belonging to each group classified by the range of J. N_2 is the number that experienced the large-q change between

1-2 AU (visible) and 3.5-5 AU (invisible) during 4400 years' orbital evolution. It is apparent that most comets with $2.9 \leq J$ and the majority with $2.8 \leq J \leq 2.9$ suffer from large variation of q; these comets correspond to those in the capture region pointed out by Everhart (1972). Average ranges of semi-major axis, q, and aphelion distance before capture are 6–10 AU, 4.5–5.5 AU, and 8–15 AU respectively. In section 3, using these 86 comets of large-q-change, their observability is investigated.

3. Detection Limit

The heliocentric distance or q as a detection limit of comets is strongly dependent on the level of detection technology. Detection systems of high sensitivity can discover comets at remote distances. Hence, for the purpose of estimating the number of invisible comets, we have to specify the detection level first. Here we try to calculate the number of invisible comets on the basis of modern discovery statistics, that is, the discovery data for 64 new SP comets from 1964 through 1989 (Kosai and Nakamura, 1991). Concretely, we use two criteria as follows.

3.1. Detection Limit for SP Comets Discovered after 1964

For these comets, we regard the heliocentric distance (r_{dis}) at their discovery as a detection limit. That r_{dis} is a good approximation for the true detection limit during 1964-1989 can be indicated by the fact that most of the new SP comets are discovered near their perihelion. Actually a least-squares-fitted relation between r_{dis} and q for these comets can be expressed by $q = r_{\mathrm{dis}} - 0.2$ (AU).

3.2. Detection Limit for SP Comets Discovered before 1964

It is not appropriate to use their r_{dis} for these comets, because it is very probable that the r_{dis} for comets discovered at old times (namely before 1964) will be systematically smaller than that after 1964. Therefore, in order to get a detection limit which conforms the one after 1964, we instead use the mean apparent magnitude $< m >$ of the 64 new SP comets after 1964 and the absolute magnitude H_{10} for each comet at recent apparitions. Then we have the following equation to be solved for d as the detection limit:

$$< m > - H_{10} = 5\log(d - 0.8) + 10\log(d), \qquad (1)$$

where a statistical relation between geocentric distance (Δ) and the r at discovery, $\Delta = r - 0.8$ (AU) obtained by a least-squares fitting for the 64 comets, is incorporated in equation (1). The mean apparent magnitude at discovery for the 64 comets is found to be $< m > = 15.0 \pm 2.3$. It is noted that this value is brighter by a few magnitudes than those for recovery observations.

When we think about the detection limit for such a long time interval as 4400 years, we must take into account of the effects of secular change of H_{10} on the

$r_{\rm dis}$. Unfortunately, the current knowledge on the secular change of H_{10} (ΔH_{10}) is fairly uncertain. If the data from Vsekhsvyatskii (1964) are taken at its face value, an average ΔH_{10} per apparition amounts to 0.30 mag/rev (Hughes and Daniels, 1983). However, this value is unrealistically too large, because the difference of the mean H_{10}s for LP comets and SP comets is 3.5 mag or so (e.g., Nakamura, 1979). Therefore, we tentatively adopt here 0.002 mag/rev as ΔH_{10}, which is recommended by Hughes and Daniels (1983). The correction to $r_{\rm dis}$ due to ΔH_{10} during 4400 years is generally less than 0.2-0.3 AU for the average SP comets.

Figure 2 shows an example of the line of detection limit of q ($r_{\rm dis}$ is converted to q using the relation $q = r_{\rm dis} - 0.2$) for comet P/Schorr, plotted on its orbital evolution of q. The bold line segments represent the visible intervals of time for this comet. We express the total sum of visible intervals by $T_{\rm vis}$ (yr). Next we define an invisibility factor (f) by the equation:

$$f = (4400yr - T_{\rm vis})/T_{\rm vis} \; . \tag{2}$$

Then the f for P/Schorr can be calculated as 5.7 in reference to Figure 2. This value implies that this comet stays about six times longer in unobservable orbits than in visible orbits during 4400 years.

4. Number of Invisible Comets in the Capture Region

The f factor can thus be calculated for each of the large-q-change SP comets, as far as the recent $r_{\rm dis}$ or H_{10} is available. The analyzed number of comets for each J-group is given in Table 1 as the n-column. The mean f-value averaged over the n is also calculated in the $< f >$-column. We see from Table 1 that most of A-group comets and majority of B-group comets spend only about 200-300 years in observable orbits during 4400 years.

Now we assume that this population of comets is in steady state. Since the observed time span of the orbital data of comets is much shorter than their dynamical lifetime, the steady state seems to be the only acceptable assumption. Another assumption we adopt is an ergodicity hypothesis, implying that the time-average is equal to the sample-average. If our integration time (4400 years) is much longer than the time scale for the substantial orbital change to take place, this assumption will be justified. We see that this is actually the case, at least for A- and B-group comets. Hence, given the above assumptions being correct, we can conclude that there exist about 20 invisible comets per observed SP comet with q = 1–2 AU and $2.8–2.9 \leq J$ in the orbits of the Jovian capture region.

Joss (1973) first pointed out that there is a large imbalance of the steady-state fluxes between LP and SP comets in their orbital evolution. Although some parameters adopted by Joss now need substantial revision, this flux discrepancy is still an unsettled problem and has motivated researchers to look for new LP comet sources like the Kuiper belt (e.g., Ip and Fernandez 1991). In relation to such a

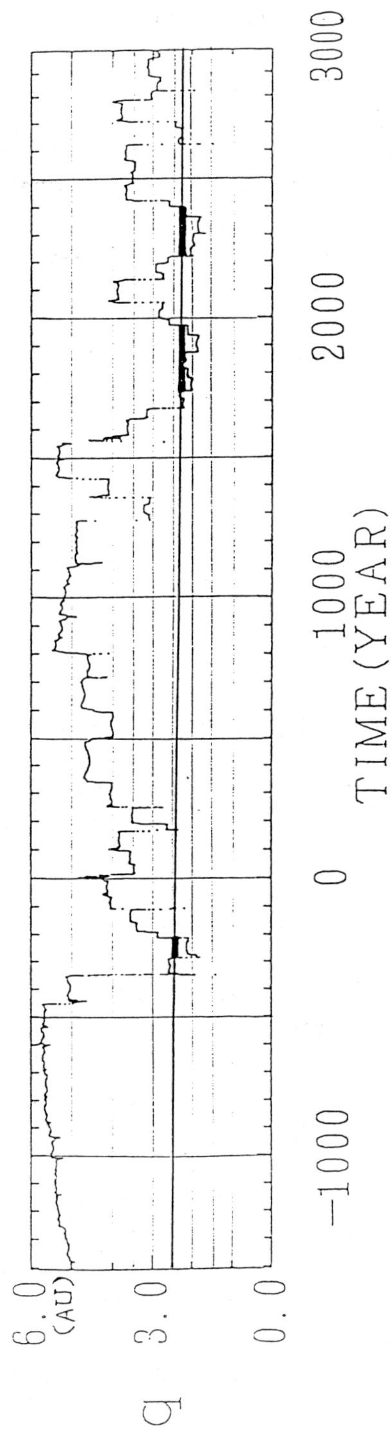

Fig. 2. Orbital evolution of q for P/Schorr and its detection limit. Bold line segments correspond to visible time spans.

background, we expect that our obtained f-value will help to make the dynamical conversion factor from LP to SP comets more quantitatively accurate.

References

Everhart, E.: 1972, *Astrophys. Letters* **10**, 131–135
Hughes, D.W. and Daniels, P.A.: 1983, *Icarus* **53**, 444–452
Ip, W.H. and Fernandez, J.A.: 1991, *Icarus* **92**, 185–193
Joss, P.C.: 1973, *Astron. and Astrophys.* **25**, 271–273
Kosai, H. and Nakamura, T.: 1991, *Publ. Nat. Astron. Obs. Japan* **2**, 63–101
Nakamura, T.: 1979, *Publ. Astron. Soc. Japan* **31**, 815–820
Nakamura, T. and Yoshikawa, M.: 1991a, *Publ. Nat. Astron. Obs. Japan* **2**, 293–383
Nakamura, T. and Yoshikawa, M.: 1991b, Proceeding of ACM91 (Arizona).
Vsekhsvyatskii, S.K.: 1964, *Physical characteristics of comets*, Israel translation program, Jerusalem

THE EVOLUTION OF GROUPS OF VERY LIGHT STARS IN THE GALAXY

IRINA V. PETROVSKAYA
St. Petersbourg University, Astronomical Observatory,
Bibliotechnaya pl., 2, St. Petersbourg—Petrodvorets, 198904, CIS

Abstract. The evolution of the velocity distribution function of a group of zero mass bodies in the gravitational field of a stellar cluster or a galactic disk is considered. We use the model of a purely discontinuous random process.

Key words: Stellar cluster, galaxy, velocity distribution.

We consider the encounters of members of a system of gravitating bodies as a purely discontinuous random process. The system may be a stellar cluster or a galaxy or a cluster of galaxies or some other systems. The motion of the test bodies changes by jumps and when the energy reaches a critical value, the test body leaves the system (the absorbing screen).

This purely discontinuous random process is described by Kolmogorov-Feller's equations (Petrovskaya, 1969a). We consider here the second equation describing the evolution of the velocity distribution function ($\phi(t,v)$) of the stellar group,

$$\frac{\partial\phi(t,v)}{\partial t} = \int_0^{v_\ell} \phi(t,z)\Phi(z,v)\,dz - \phi(t,v)\int_0^{\infty} \Phi(z,v)\,dz\;, \tag{1}$$

where $\phi(0,v) = \phi_0(v)$ is the initial velocity distribution function and v_ℓ is the escape velocity (Petrovskaya, 1969b). When the velocity of the test star becomes larger or equal to v_ℓ, the star escapes from the system.

We consider here the simplest case of variation of the velocity module through encounters with field stars and do not take into account the regular potential of the system as in (Petrovskaya, 1969a).

Equation (1) can be solved by the Fourier method (Petrovskaya, 1969b) or by the method of integral transformations (Petrovskaya, 1983). The kernel of Equation (1), $\Phi(v, v+\Delta v)$, is the density of probability that, during the encounter of the test body with the field star, the velocity of the test body changes from v to $v+\Delta v$. For zero-mass stars, this function was obtained by Agekyan (1959) for the three-dimensional case and in this work for the two-dimensional case.

We consider a set of zero-mass bodies with initial velocity distribution $\phi_0(v)$. In order to investigate the time evolution of the velocity distribution, we represent

Celestial Mechanics and Dynamical Astronomy **54:** 267–269, 1992.

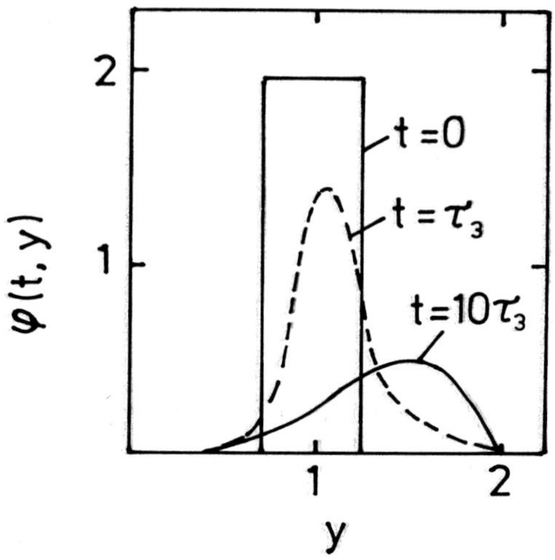

Fig. 1. The velocity distribution function of zero-mass stars in a stellar cluster at different times. The normalized velocity y is equal to $v/\bar{v}$ where $\bar{v}$ is the mean square velocity of the field stars.

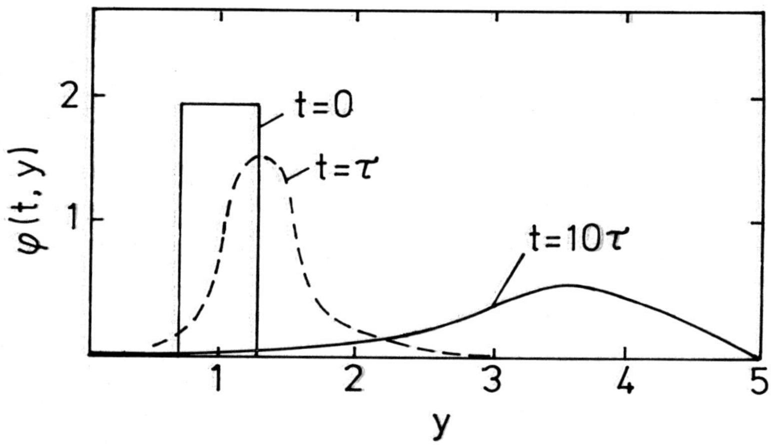

Fig. 2. The evolution of the velocity distribution of zero-mass stars in a flat stellar system.

the solution of Equation (1) as a series,

$$\phi(t,v) = \sum_j C_j \exp^{-\mu_j t} \phi_j(v) \,, \tag{2}$$

where the coefficients C_j are determined by the initial conditions. Substituting (2) into (1), we find that μ_j and $\phi_j(v)$ are the eigenvalues and the eigenfunctions of the integral operator,

$$\mu_j \phi_j(v) = -\int_0^{v_\ell} \phi_j(z)\Phi(z,v)\,dz + \phi_j(v)\int_0^{\infty} \Phi(v,z)\,dz \,. \tag{3}$$

The first eigenfunction determines the stationary state to which the velocity distribution approaches when $t \to \infty$. The rate of dissipation (μ_1) and the energy loss (E_1) are (when $t \to \infty$)

$$\mu_1 = \frac{\int_0^{v_\ell} \phi_1(v) \int_{v_\ell}^{\infty} \Phi(v,z)\, dz\, dv}{\int_0^{v_\ell} \phi_1(v)\, dv} \; ; \tag{4}$$

$$E_1 = \frac{\int_0^{v_\ell} \phi_1(v) \int_{v_\ell}^{\infty} (z^2 - v_\ell^2)\Phi(v,z)\, dz\, dv}{\int_0^{v_\ell} v^2\phi_1(v)\, dv} \, . \tag{5}$$

The results of the calculations are presented in Figure 1 for a three dimensional stellar cluster (Kaliberda and Petrovskaya, 1970) and in Figure 2 for a planar system (Ashurov and Petrovskaya, 1992). In both cases the initial velocities of the zero-mass stars are uniformly distributed in the interval $0.75 \leq v/\bar{v} \leq 1.25$ (where $\bar{v}$ is the mean square velocity of the field stars) and the velocity distribution varies as a result of the encounters with the field stars. The escape velocity is taken as $v_\ell = 2\bar{v}$ for a stellar cluster. Taking into account the rotational velocity of 220 km s^{-1} in the vicinity of the Sun, the escape velocity is taken in the case of a galaxy as $v_\ell = \bar{v}(380 - 220)/30 \approx 5\bar{v}$.

The stationary velocity distribution is reached after a time of $10\tau_3$ for the stellar cluster field and of $10\tau_2$ for the planar system. We found $\tau_3 \approx 10^7$ yr for a galactic open cluster, $\tau_3 \approx 10^9$ yr for a globular cluster and $\tau_2 \approx 3\ 10^8$ yr for the galactic disk. From (5) we find the relative loss of zero-mass stars (dN/N) and the relative loss of energy (dE/E) to be $dN/N = 0.08,\ dE/E = 0.22$ per 10^7 yr for the open cluster, $dN/N = 0.77\ 10^{-3},\ dE/E = 0.20\ 10^{-2}$ for the globular cluster and $dN/N = 0.44\ 10^{-7},\ dE/E = 0.12\ 10^{-5}$ per 10^9 yr for the galactic disk.

The estimated time for reaching a stationary velocity distribution in the case of a galactic disk is much lower than the estimate of Jeans and Chandrasekhar (Chandrasekhar, 1942). This comes from the fact that we have considered zero mass stars in the stellar field of a perfectly flat disk while the estimation of Jeans and Chandrasekhar are found for finite mass stars and a disk of finite thickness.

References

Agekyan, T.A.: 1959, *Sov. Astron. J.* **36**, 41
Agekyan, T.A.: 1961, *Astron. J.* **38**, 1055
Ashurov, A.E. and Petrovskaya, I.V.: 1992, in preparation
Petrovskaya, I.V.: 1969a, *Sov. Astron. J.* **46**, 824
Petrovskaya, I.V.: 1969b, *Sov. Astron. J.* **46**, 1220
Petrovskaya, I.V.: 1983, *Vestnik Leningr. Univ.* **13**, 94
Petrovskaya, I.V., Chumak, Z.N. and Chumak, O.V.: 1984, *Sov. Astron. J.* **61**, 467

NEAR-PARABOLIC COMETARY FLUX IN THE OUTER SOLAR SYSTEM

T. TSUJII
National Aerospace Laboratory, Chofu, Tokyo 182, Japan

Abstract. The flux of near-parabolic comets in the outer-planetary region is estimated on the presumption that the major planets and the galactic tide control the dynamics of comets. It is found that the flux of the Oort cloud comets (semi-major axis > 20000 AU) is similar to the case of a strong comet shower derived on the presumption that the galactic tidal force were not operative. On the other hand, the flux of comets with semi-major axes ≤ 20000 AU is found to be an increasing function of q (perihelion distance) until q reaches 20 AU, while for $a \simeq 45000$ AU it is a rapidly increasing function for $q \leq 12$ AU. In other words, for comets of the inner extension of the Oort cloud the planetary perturbation acts as a strong barrier for them to penetrate into the inner planetary region.

Key words: Comet, Oort's cloud.

1. Introduction

Recent years have seen active discussions on the possible role of comet shower on the mass-extinction of fauna and crater formation on the Earth. Some authors (Hills, 1981; Heisler and Tremaine, 1986) argue that the solar system is surrounded by a dense core of comets which is the inner extension of the Oort cloud comets; occasional penetration of this core by a passing star causes a large number of comets to be scattered into the inner Solar System, thereby hitting the Earth. According to these suggestions, the presently observed comets with long periods could be remnants of a recent disturbance of the inner core.

On the other hand, it has been shown (Byl, 1986; Heisler and Tremaine, 1986, Yabushita, 1989; Matese and Whitman, 1989) that the galactic tidal force is more efficient than stellar encounters in changing the perihelion distance of NP comets, apart from deep stellar penetrations of the as yet hypothetical inner cometary cloud. This means that one can calculate the cometary flux of NP comets in the outer solar system on the presumption that the current solar system is not suffering from a comet shower. The purpose of this paper is to obtain flux distributions caused by the galactic tide, which are in the steady state, and compare them with distributions due to the occasional comet shower.

2. Method

The method of calculating the cometary flux is based on the dynamical evolution of LP comets as described in Yabushita and Tsujii (1989). Comets dynamically evolve under the perturbations of the galactic tide as well as perturbations of

Celestial Mechanics and Dynamical Astronomy **54:** 271–274, 1992.

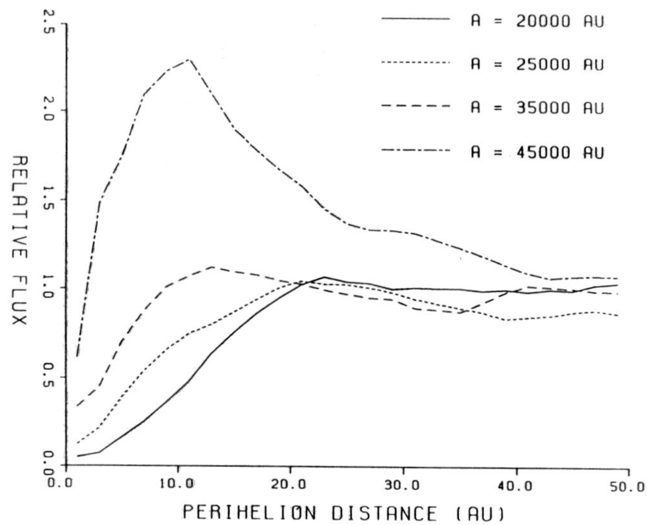

Fig. 1. Relative flux of nearly parabolic comets is plotted against perihelion distance q. Initial a = 20000, 25000, 35000, 45000 AU.

the four major planets. Equations of motion for a comet are regularized by the $K - S$ transformation (Stiefel and Scheifele, 1971) and are integrated by a fourth-order Runge-Kutta method. We calculate the flux of the NP comets of the Oort cloud and that of its inner extension. The initial values of the semi-major axis are 25000, 30000, 35000, 40000, 45000 AU (Oort cloud), and 15000, 20000 AU (inner extension of the Oort cloud). Initial perihelia ($q_{\rm initial}$) are first chosen such that they are well outside the planetary region. According to the magnitude of the galactic tide, values of $q_{\rm initial}$ are different (Table I). In order to save computational time, a geometrical factor $F = \sin 2\omega \sin^2 i$ is chosen to be positive initially. Perihelion distances of those comets are decreasing first. Except for this restriction, orbits are randomly generated relative to the galactic plane using a Monte Carlo method.

Calculations are started when comets are at aphelia and comets are followed until (1) they are ejected from the Solar System, (2) or the perihelion distance begins to increase and becomes more than 10 AU greater than the initial value of perihelion distance, (3) or a becomes less than 10000 AU, (4) or the number of revolution exceeds 20 (for the Oort cloud comets), or 50 (for comets of inner extension), which are listed in Table I as $N_{\rm max}$. As comets which revolve many times are suffering from stellar perturbations significantly, they would escape from the outer Solar System. The cometary flux for a given interval of perihelion distance, q can then be obtained by calculating the sum $\sum_i a_i^{-3/2}$ for the given interval, where the subscript i stands for the i-th comet.

TABLE I
The ratio f of the flux in the outer planetary region to the one in the inner planetary region

a_{initial} (AU)	q_{initial} (AU)	$N_{\max}$	No. of runs	Total No. of rev.	f
15000	70	50	371	8475	98.3
20000	70	50	852	10007	13.8
25000	100	20	2698	17303	4.9
30000	100	20	6546	26124	2.1
35000	150	20	13005	40054	2.3
40000	200	20	15326	40145	2.3
45000	250	20	15865	35362	0.8

3. Flux of Nearly Parabolic Comets

The calculated relative flux distributions are shown in Figure 1. The level flux is obtained by averaging values of flux at large perihelion distances. It is apparent from Figure 1 that the larger the semi-major axis becomes the deeper comets penetrate into the inner planetary region. A parameter which specifies the variation in the flux with q is the factor f which is the ratio of the flux in the outer planetary region to the one in $0 < q < 5$ AU. The numerical value of the factor f, number of runs, and how many orbits have been computed to obtain the flux distribution are summarized in Table I.

Now we wish to compare the obtained flux distribution with that calculated on the presumption that the galactic tide were not operating, using the factor f. The flux in the absence of the galactic tide is reproduced in Yabushita in this volume. During a strong shower, the flux in the inner planetary region rises to a significant fraction of the one at large perihelion distances. One finds that the flux in $0 < q < 5$ AU is 25% of that at large q values during a strong shower, while it is some 3% during a moderate shower. when this is converted to the factor f, $f \simeq 4$ (strong shower) and $f \simeq 30$ (moderate shower), and $f \geq 100$ (between showers). Then if showers are characterized by a factor f, the values of f given in Table I correspond to very strong showers if $a \geq 25000$ AU, to moderate shower if $a \simeq 20000$ AU, and to between showers if $a \simeq 15000$ AU. There is a different feature in the case where the semi-major axis is 45000 AU. Although one can infer from Figure 1 that the flux distribution in the strongest comet showers is independent of q (in this case, $f = 1$), the flux of comets with $a = 45000$ AU in the planetary region exceeds the level flux. In other words, passing stars do not increase the flux of comets with $a = 45000$ AU.

4. Conclusions

1. For comets with $a \leq 20000$ AU the planetary perturbation acts as a strong barrier for them to penetrate into the inner planetary region. If a strong comet shower occurs, the flux of these comets would increase considerably.

2. The galactic tide is very efficient to make comets penetrate into the inner planetary region when $a > 20000$ AU. Therefore, even during a comet shower, the flux distribution would be similar to the one obtained here.

3. Since the flux in the planetary region is more than that at larger perihelion distances for comets with $a \geq 45000$ AU, any comet shower will hardly contribute to increasing the flux in the inner solar system.

Acknowledgements

The author is grateful to S. Yabushita for his constructive comments on the present paper.

References

Byl,J.: 1986. *Earth, Moon and Planets*, **36**, 263.
Heisler, J. & Tremaine, S.: 1986. *Icarus*, **65**,13.
Hills, J.G.: 1981. Astron. J., **86**, 1730.
Matese, J.J. & Whitmann, P.G.: 1989. *Icarus*, **82**, 389.
Stiefel, E.L. & Scheifele, G.: 1971. *Linear and Regular Celestial Mechanics*, Springer, Berlin.
Weissmann, P.R.: 1986. *Bull. Amer. Astron. Soc.*, **18**,840.
Yabushita, S.: 1989. *Astron. J.*, **97**, 262.
Yabushita, S. & Tsujii, T.: 1990. *Mon. Not. R. Astr. Soc.*, **244**, 633.

ORGANIC MODELS OF INTERSTELLAR GRAINS

K. WADA
Department of Nuclear Engineering, Kyoto University, Kyoto 606, Japan

T. INAGAKI
Department of Physics, Osaka Kyoiku University, Tennoji, Osaka 543, Japan

and

S. YABUSHITA
Department of Applied Mathematics and Physics, Kyoto University, Kyoto 606, Japan

Abstract. A number of experimental studies have been carried out to verify the claim that interstellar grains largely consist of organic material, including biological cells. Our spectroscopic studies on biological cells and organic extracts from carbonaceous compounds have failed to identify the well-known 2200 Å interstellar extinction peak with the organic material.

Key words: Interstellar grains, organic material.

Our interest in the organic models of interstellar (IS) grains was inspired by a paper by Hoyle *et al.* (1985) where they claimed to have found a 2200 Å peak in the spectra of E. coli and of yeast. They argued that the well-known peak at $\lambda = 2200$ Å in the interstellar extinction curve could thus be interpreted in terms of biological cells. The objective of the present note is to describe the results obtained on the organic models of IS grains. From the description of the experiment in Hoyle *et al.* (1985), it seemed an easy task to verify the claim and an exactly similar experiment was done, by using a spectrometer in the visible and UV region (Yabushita and Wada, 1985). No peak at $\lambda = 2200$ Å has been found.

Later, Yabushita *et al.* (1986) carried out a more detailed investigation, adopting the vacuum UV technique to measure absorbances of E. coli, yeast and of the spore of Bacillus subtilis. The reason for adopting the vacuum UV technique is that a spurious result is sometimes got toward the end of the wavelength for which an adopted spectrometer can be used. Although a peak has been found in the absorbance in the UV region of E. coli and of yeast, the peak is centered at 1900 Å, instead of at 2200 Å (Figure 1).

It is conceivable that the peak position may be shifted when grains are supposed to have finite radii, instead of a thin film. Optical constants of the E. coli and of yeast have been determined using a standard procedure of spectroscopy, and Mie calculations have been done to see if the peak positions could be shifted to 2200 Å. No matching with observation has thereby been achieved (Yabushita *et al.*, 1987a).

These situations led us to cast doubt on the work of Sakata *et al.* (1977) where it was first claimed that organic extracts from a carbonaceous chondrite (Murchison)

Celestial Mechanics and Dynamical Astronomy **54:** 275–277, 1992.

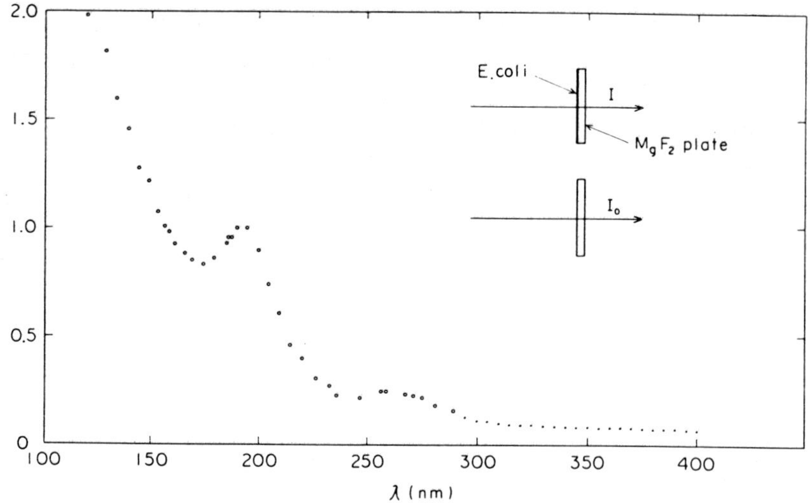

Fig. 1. Absorbance of a dry film of E. coli cast on a MgF_2 plate is plotted against wavelength. The ordinate is $\log I_0/I$. A circle o denotes a measurement with air discharge source while a dot represents a measurement using Xe lamp. Note a peak at $\lambda = 190nm$.

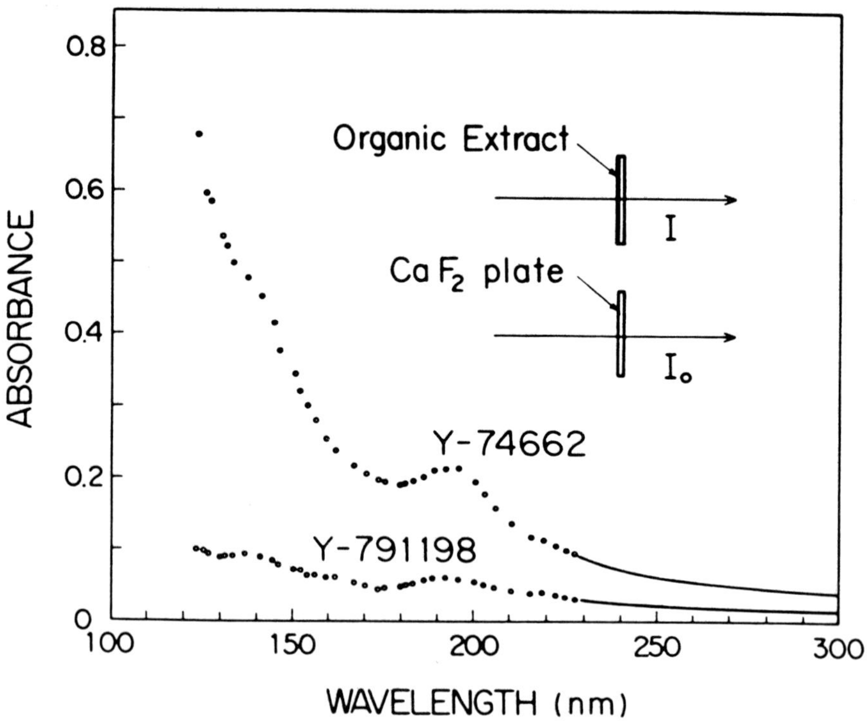

Fig. 2. Absorbance of the extract from carbonaceous chondrite Yamato-74662 and 791198 in a form of thin solid film cast on a calcium fluoride plate plotted as a function of wavelength. The absorbance is given by $\log_{10}(I_0/I)$, where I and I_0 were the light intensities transmitted through the plate with and without the sample film. Reference disc had a thin film of hexafluoroisopropanol which was allowed to evaporate.

son) exhibits an absorption peak at $\lambda = 2200$ Å. The National Institute of Polar Research, Japan, kindly provided us with Yamato carbonaceous chondrite, and Yabushita *et al.* (1987b) measured absorbance of organic extract from the chondrites, which is reproduced here in Figure 2. A peak is found at $\lambda = 1900$ Å, but no significant peak is seen near $\lambda = 2200$ Å.

It appears therefore that the well-known IS extinction peak at $\lambda = 2200$ Å cannot be ascribed to organic material from biological cells. It seems safe to conclude that organics such as biological cells or organic extract from carbonaceus chondrites are not the careers of the well-known IS peak at $\lambda = 2200$ Å.

References

Hoyle, F., Wickramasinghe, N. C. and Al-Mufti, S.: 1985, *Astrophys. Space Sci.*, **111**, 65.

Sakata, A., Nakagawa, N., Iguchi, T., Isobe, S., Morimoto, M., Hoyle, F. and Wickramasinghe, N. C.: 1977, *Nature*, **266**, 241.

Yabushita, S. and Wada, K.: 1985, *Astrophys. Space Sci.*, **110**, 405.

Yabushita, S., Wada, K., Takai, K., Inagaki, T., Young, D. and Arakawa, E.T.: 1986, *Astrophys. Space Sci.*, **124**, 377.

Yabushita, S., Inagaki, T., Kawabe, T. and Wada, K.: 1987a, *Astrophys. Space Sci.*, **132**, 409.

Yabushita, S., Inagaki, T., Arakawa, E. T. and Wada, K.: 1987b, *Mon. Not. R. astr. Soc.*, **229**, 459.

NUMERICAL INVESTIGATION ON THE ORBITAL EVOLUTION OF 1:2 RESONANT ASTEROIDS BETWEEN JUPITER AND SATURN

NORIAKI WATANABE
Chiba University of Commerce, Ichikawa, Chiba 272, Japan

Abstract. According to some investigations (Lecar and Franklin,1973; Franklin *et al.*, 1989; Soper *et al.*, 1990) asteroids cannot remain for a long time between Jupiter and Saturn. But as it is well known there is a near 5:2 commensurability between Jupiter and Saturn. So there might be a possibility that asteroids between Jupiter and Saturn could be trapped in a resonant relation.

In order to investigate this possibility, the changes of orbital elements of an asteroid whose initial value of semi-major axis corresponds to that of a 1:2 resonant orbit were investigated by means of a double precision Cowell method. The integration routine was kindly supplied by Dr Yoshikawa.

We considered first a planar restricted problem of three bodies, Sun-Jupiter-Asteroid, then a four body problem, Sun-Jupiter-Asteroid-Saturn. When integrating the equations of motion, short periodic terms were not eliminated and in the second test the interactions between Jupiter and Saturn were retained. Whether a close approach occured or not was not investigated. In every case $a_J = 5.20$, $a_S = 9.54$ and $a = 8.26$ were adopted as initial values of the semi-major axis of Jupiter, Saturn and Asteroid respectively.

Key words: Asteroids, resonance, restricted problem, N-body problem.

1. The Restricted Problem

We investigated the circular as well as the elliptic version of the restricted problem. In both cases $e = 0.0$ was adopted as an initial value of the eccentricity of the asteroid. In the circular case integrations were executed over 15000 and 30000 years by taking as an integration step 15-day and 30-day respectively. The results show that both the semi-major axis and the eccentricity oscillate quite regularly and this implies that the 1:2 resonant orbit is stable in the circular restricted case (Figure 1). In the elliptic case $e_J = 0.048$ was taken as a value of the eccentricity of Jupiter and integrations were performed over 30000 and 45000 years by taking as an integration step 30-day and 45-day respectively. From the results we see that the semi-major axis and the eccentricity change periodically and that the maximum of the aphelion distance exceeds a_S (Figure 2).

Celestial Mechanics and Dynamical Astronomy **54:** 279–286, 1992.

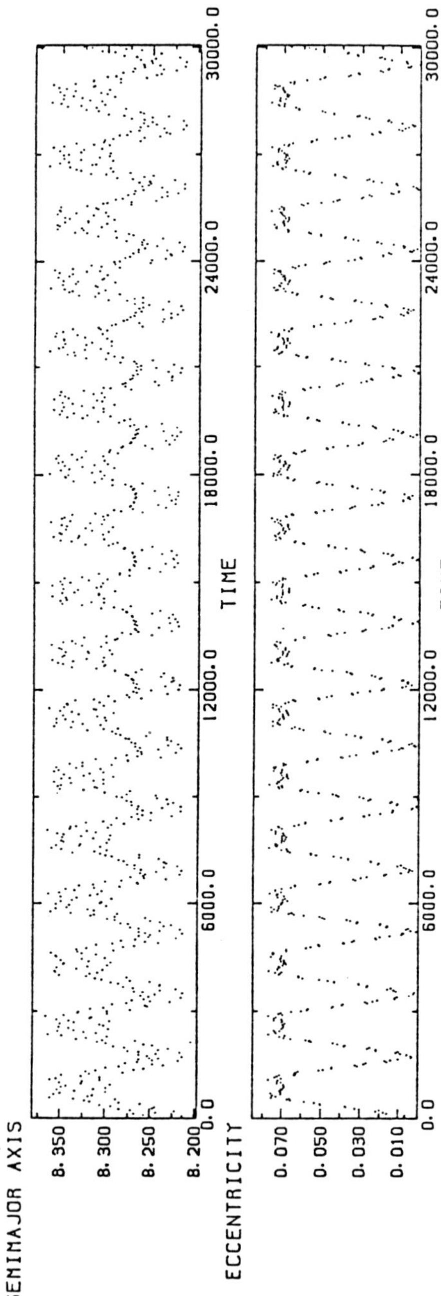

Fig. 1. Circular restricted problem. Integration step = 30 days.

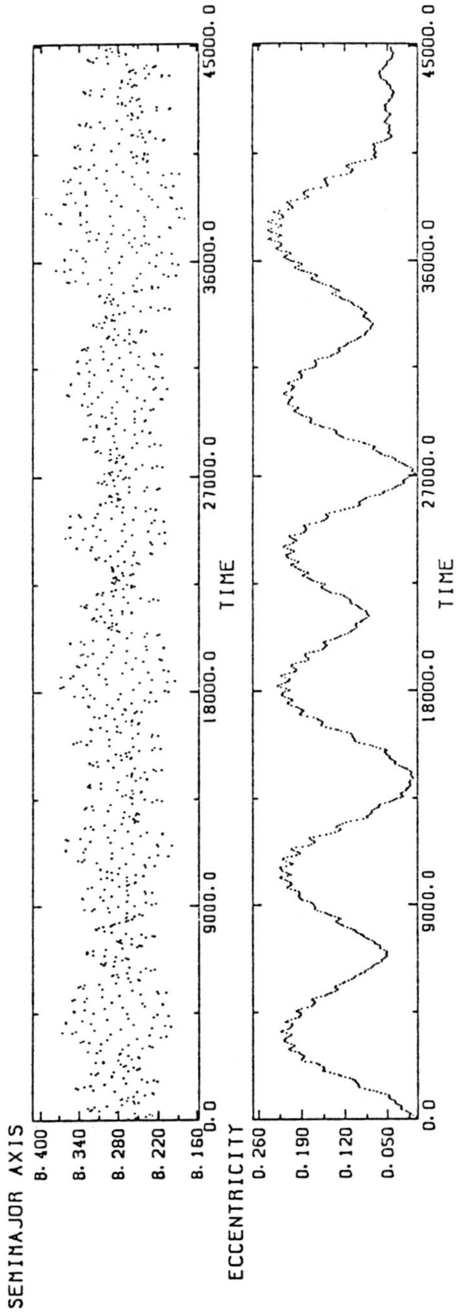

Fig. 2. Elliptic restricted problem. Integration step = 45 days.

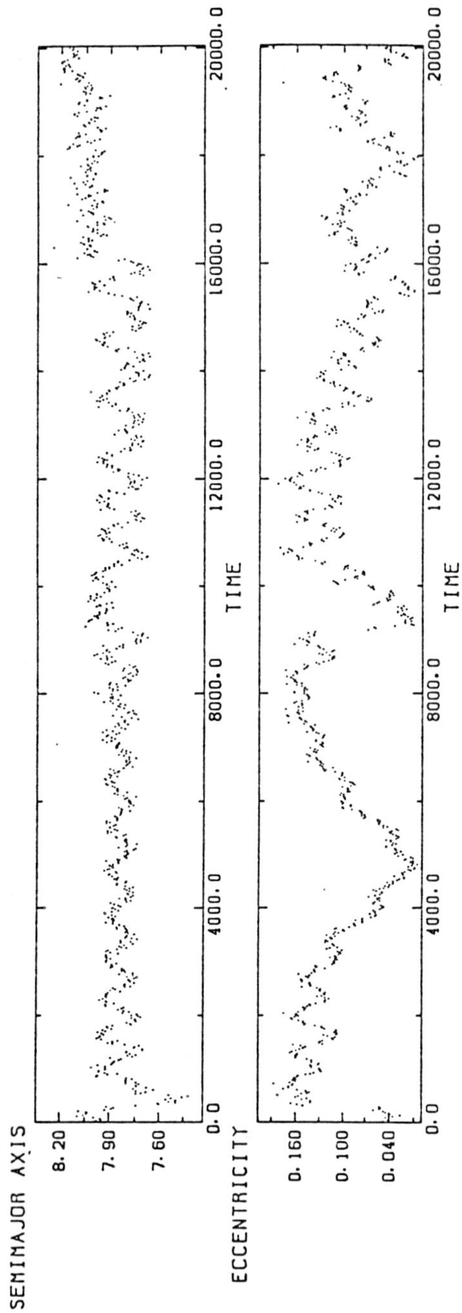

Fig. 3. Four-body problem. Integration step = 20 days.

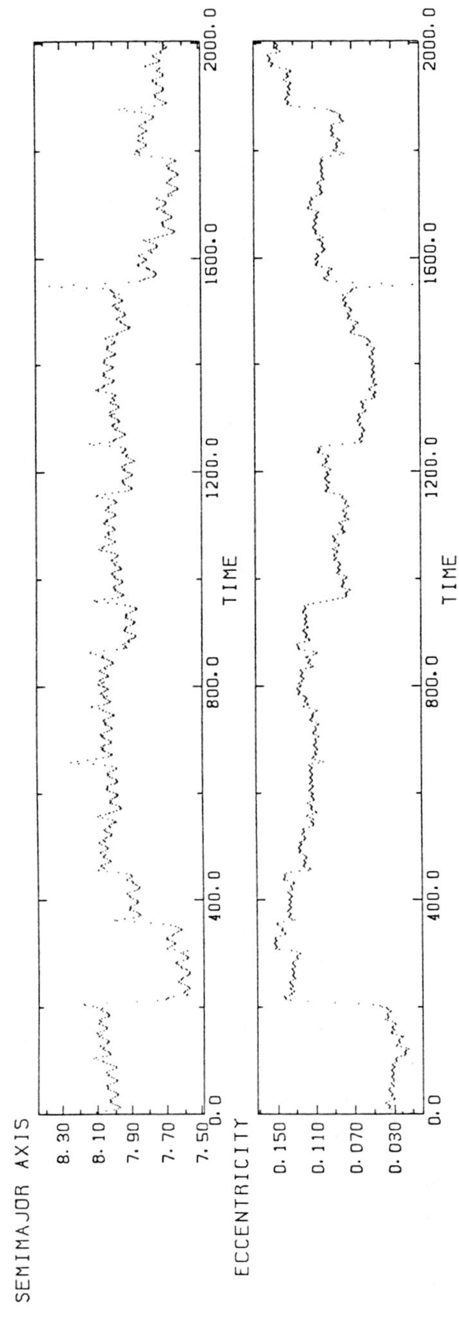

Fig. 4. Four-body problem, $e = 0.01$. The orbit remains trapped in the 2:1 resonance.

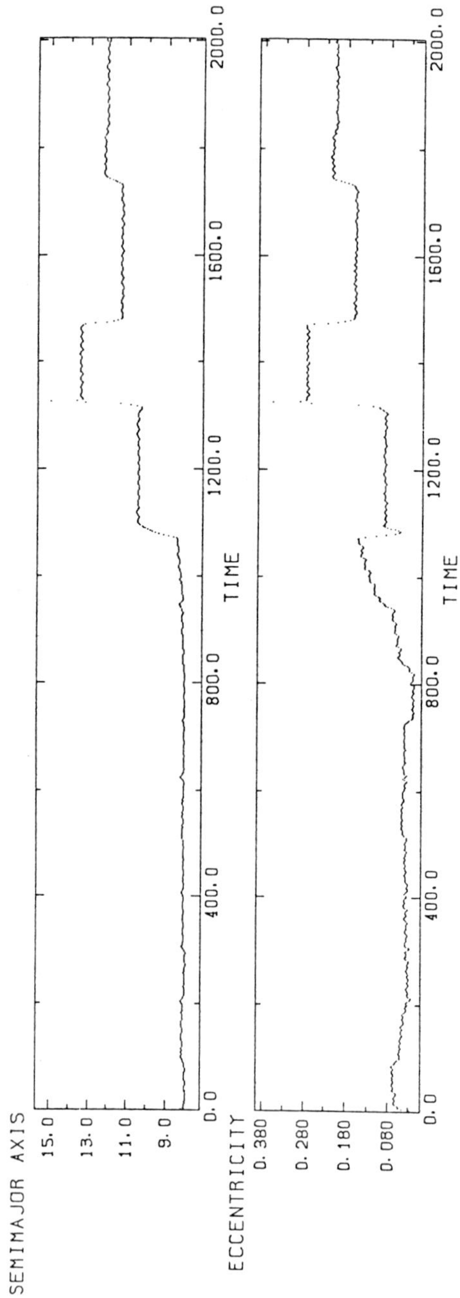

Fig. 5. Four-body problem, $e = 0.01$, $\omega_S = 180°$, $f_S = 0°$, $\omega = 0°$, $f = 180°$. The semi-major axis becomes larger than the one of Saturn.

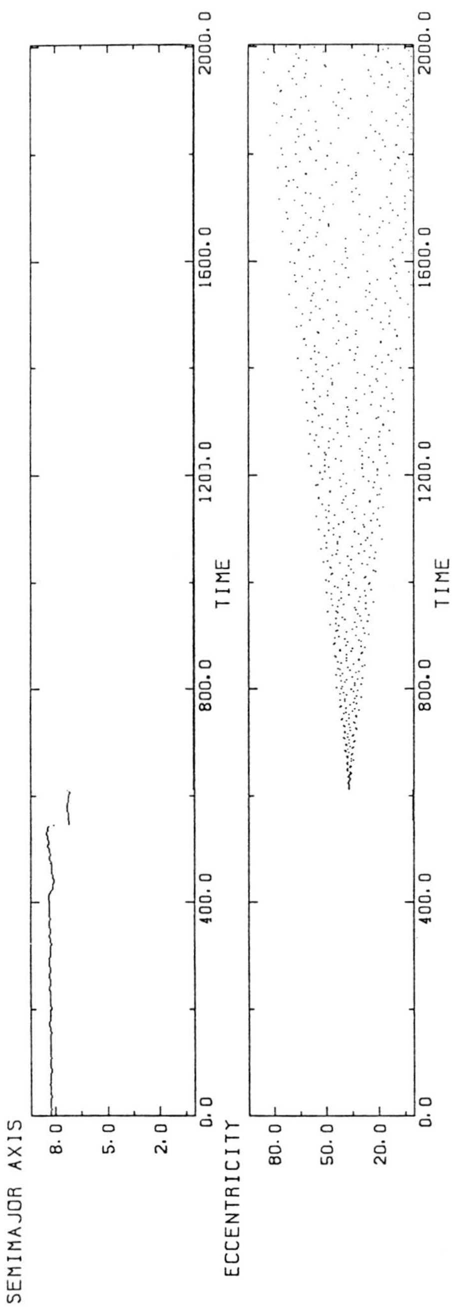

Fig. 6. Four-body problem, $e = 0.01$, $\omega_S = 0°$, $f_S = 180°$, $\omega = 0°$, $f = 0°$. The numerical integration becomes unstable.

2. The Four-Body Problem

In the 4 body problem the following values were adopted as initial values of the eccentricity, the argument of perihelion and the true anomaly of Jupiter, Saturn and Asteroid respectively: $e_J = 0.048$, $\omega_J = 0°$, $f_J = 0°$; $e_S = 0.056$, $\omega_S = 0°/180°$, $f_S = 0°/180°$; $e = 0.0/0.01/0.05$, $\omega = 0°/180°$, $f = 0°/180°$.

In the case of $e = 0.0$, integrations were carried out over 5000, 10000, 20000 and 30000 years by taking as an integration step 5, 10, 20 and 30-day and adopting $\omega_S = f_S = \omega = f = 0°$. In the first three cases the orbit seems stable (Figure 3). In the last case, the integration became unstable after about 27000 years. This may be due to a close approach.

When integrations were executed in the case of $e = 0.01$ by taking as an integration step 5-day or a longer one, there were many cases where the numerical integrations became unstable. So 1-day was finally adopted as an integration step and integrations were carried out over 2000 years in the case of $e = 0.01, 0.05$. The results falls into three classes.

In a first class, to which belongs only one case with $\omega_S = f_S = \omega = f = 0°$, the semi-major axis stays near that of the 1:2 resonant orbit during the full integration period (Figure 4). For a second class of orbits, the semi-major axes deviate from those of the initial 1:2 resonant values and in almost all cases end up with a value larger than the one of Saturn (Figure 5). For a third class, the numerical integration becomes unstable (Figure 6).

References

Lecar, M. and Franklin, F.A.: 1973, *Icarus*, **20**, 422–436
Franklin, F.A., Lecar, M. and Soper, P.: 1989, *Icarus*, **79**, 223–227
Soper, P., Franklin, F.A. and Lecar, M.: 1990, *Icarus*, **87**, 265–284

NUMERICAL INVESTIGATION OF MOTIONS OF RESONANT ASTEROIDS IN THE THREE-DIMENSIONAL SPACE

MAKOTO YOSHIKAWA
Communications Research Laboratory, Kashima Space Research Center, 893-1 Hirai, Kashima, Ibaraki 314, Japan

Abstract. Motions of asteroids in mean motion resonances with Jupiter are studied in three-dimensional space. Orbital changes of fictitious asteroids in the Kirkwood gaps are calculated by numerical integrations for $10^5 - 10^6$ years. The main results are as follows: (1) There are various motions of resonant asteroids, and some of them are very complicated and chaotic and others are regular. (2) The eccentricity of some asteroids becomes very large, and the variation of the inclination is large while the eccentricity is large. (3) In the 3:1 resonance, there is a long periodic change in the variation of the inclination, when $(\dot{\varpi} : \dot{\Omega})$ is a simple ratio (ϖ: longitude of perihelion, Ω: longitude of node). (4) In the 7:3 resonance, the variation of the inclination of some resonant asteroids is so large that prograde motion becomes retrograde. Some asteroids in the 7:3 resonance can collide with the Sun as well as with the inner planets.

Key words: Asteroid, Kirkwood gap, resonance.

1. Introduction

It is well known that in the distribution of asteroids there are clear gaps at the regions where the ratios of mean motions of asteroids and Jupiter are 3:1, 5:2, 7:3, and 2:1. These gaps are called Kirkwood gaps, and their origin is still not clear. One of the possible scenarios of gap formation is that the eccentricity of asteroids becomes so large that they collide with planets. This scenario was first proposed by Wisdom (1982). Until now there are many studies concerning the Kirkwood gaps (see, for example, the references shown in Yoshikawa, 1990, 1991), but most studies treat planar motions. As Yoshikawa (1991) shows, in the resonances which correspond to the Kirkwood gaps, there are some regions where the eccentricity of asteroids does not become large enough to cross the orbits of planets. In order to solve this difficulty, we should take account of the effects of Saturn and the inclination of asteroids. For example, Wisdom (1987) and Yoshikawa (1989, 1990) show that the variation of eccentricity of asteroids becomes large and chaotic when the effects of Saturn and other planets are considered. However Yoshikawa (1989) also shows that there are some cases where the variation of eccentricity is small even if the effect of Saturn is taken into account. As for the effect of the inclination of asteroids, Wisdom (1987) shows that the variations of orbital elements of resonant asteroids become very complicated and large, when the motions of asteroids are studied in the three-dimensional space. In this study, the effect of inclination of asteroids is

Celestial Mechanics and Dynamical Astronomy **54:** 287–290, 1992.

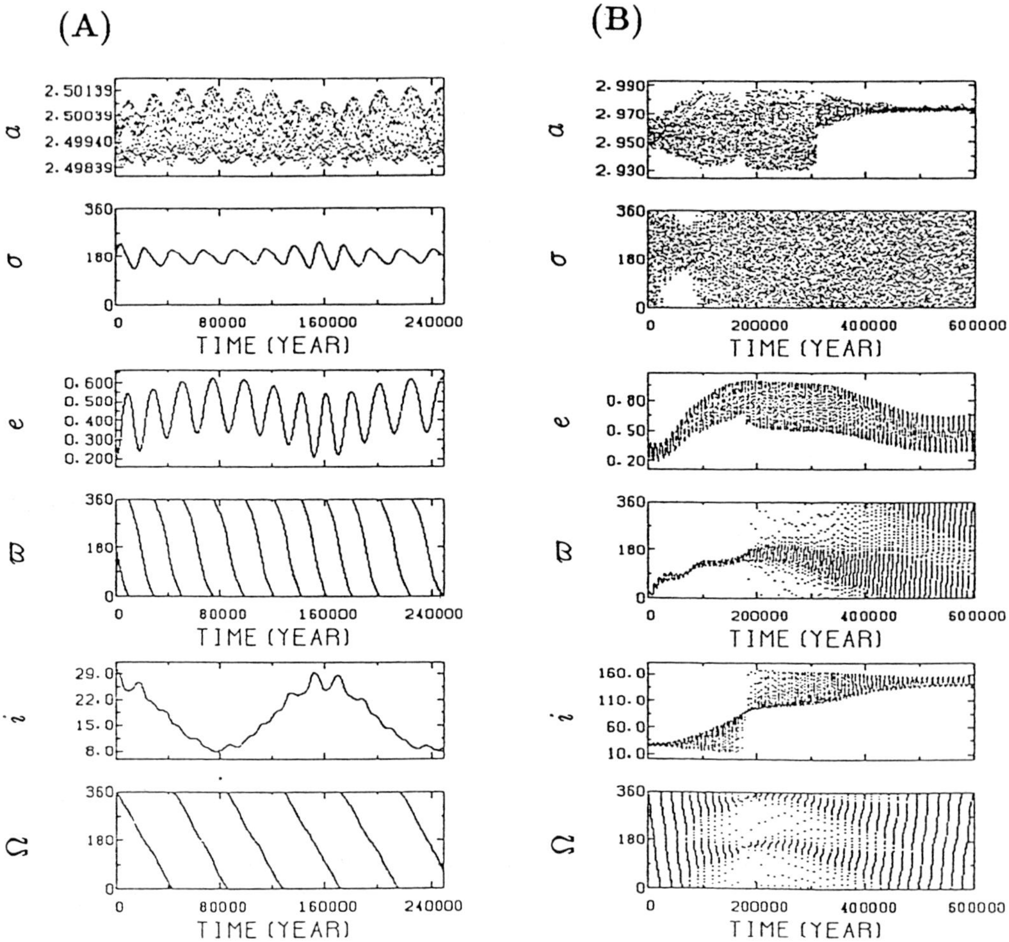

Fig. 1. The results for the 3:1 resonance (A) and the 7:3 resonance (B).

investigated by numerical integrations for the Kirkwood gaps. Since the parameter regions are too large to survey full resonance regions, we choose about 70 sets of initial values of orbital elements of asteroids for each resonance. The qualitative feature are summarized in this paper.

2. Analysis

The model that we treat here is the elliptical restricted three body problem, that is, Sun - Jupiter - Asteroid, where asteroids are in one of the resonances of 3:1, 5:2, 7:3, and 2:1. The orbital evolutions of asteroids are obtained by numerical integrations of Cowell method and extrapolation method. The initial values of asteroids are chosen as in Table I.

In this table, the notations, a, σ, i, Ω, e, ϖ indicate semimajor axis, critical

TABLE I
Initial conditions of asteroids

Elements	Initial values
a	resonance center
σ	resonance center
i	5°, 10°, 15°, 20°, 25°, 30°
Ω	0°, 180°, (90°)
e	0.1, 0.2, 0.3, (0.4)
ϖ	0°, 180°

argument, inclination, longitude of node, eccentricity, and longitude of perihelion, respectively. The definition of the critical argument is $\sigma = (p+q)\lambda_J - p\lambda - q\varpi$ for the $(p+q):p$ resonance, where λ is the mean longitude. The subscript J indicates the elements of Jupiter. The values of a_J and e_J are 5.2026 AU and 0.0485, and i_J, Ω_J, and ϖ_J are all 0. The calculations are carried out for all the combinations of the values without parentheses, and the values with parentheses are used only for several cases.

3. Results

Asteroids in mean motion resonances show various motions even in the planar case. In the three-dimensional case, the motions become much more complicated. The following two features are commonly seen in the four resonances: (1) Depending on the initial values, chaotic and regular motions appear. (2) The variation of i becomes large, when the value of e is large ($e \sim 0.8$). The period of this variation is 2ω, where ω is argument of perihelion.

Some special features are obtained in the motions of 3:1 and 7:3 resonances. In some cases of the 3:1 resonance, the inclination changes largely ($\Delta i \sim 30°$) with long period ($\sim 10^5$ years), when $\dot{\varpi} : \dot{\Omega}$ is simple numbers, such as, 5:2, 9:4, and 2:1. One of the results is shown in Figure 1(A). Such motion occurs in rather wide parameter region of the 3:1 resonance. Similar motion is reported in Yoshikawa (1989).

As for the 7:3 resonance, very large change of e and i takes place, when the initial inclination is relatively high ($\sim 30°$). In this case, the eccentricity becomes almost 1, and the inclination becomes as large as 165° (Figure 1(B)). This means that prograde motion becomes retrograde motion. However such a transition does not actually occur, because asteroids collide with the Sun before the inclination exceeds 90°. This large change occurs not due to close encounters of asteroids to planets but due to the effect of resonance.

The motions of asteroids in the 2:1 resonance are generally regular, and the eccentricity does not become large enough to cross the orbits of planets. Therefore the origin of the gap at the 2:1 resonance still cannot be explained.

Many detailed results will be reported in separate papers.

4. Summary

When we consider the resonant motions of asteroids in the three-dimensional space, various motions are seen, and the eccentricity is very likely to become large. The results of this study generally support the hypothesis of Wisdom (1982). However we should extend the research much further to learn the real nature of the resonant motions of asteroids, because some resonant asteroids seem to be very stable.

References

Wisdom, J.: 1982, 'The origin of the Kirkwood gaps: A mapping for asteroidal motion near the 3/1 commensurability', *Astron. J.* **87**, 577–593

Wisdom, J.: 1987, 'Urey Prize Lecture: Chaotic dynamics in the solar system', *Icarus* **72**, 241–275

Yoshikawa, M.: 1989, 'A survey of the motions of asteroids in the commensurabilities with Jupiter', *Astron. Astrophys.* **213**, 436–458

Yoshikawa, M.: 1990, 'Motions of asteroids at the Kirkwood gaps I. On the 3:1 resonance with Jupiter', *Icarus* **87**, 78–102

Yoshikawa, M.: 1991, 'Motions of asteroids at the Kirkwood gaps II. On the 5:2, 7:3, and 2:1 resonances with Jupiter', *Icarus* **92**, 94–117

CANDIDATES FOR FAMILIES IN APOLLO-AMOR TYPE ASTEROIDS

MANABU YUASA
Research Institute for Science and Technology, Kinki University, Higashi-Osaka, 577 Japan

Abstract. The asteroids whose perihelion distances ($\bar{q}$) are smaller than 0.983 AU and aphelion distances ($\bar{Q}$) are larger than 1.017 AU are called Apollo type objects. Similarly Amor type objects are defined by the conditions: 0.987 AU $< \bar{q} <$ 1.382 AU and $\bar{Q} >$ 1.666 AU. Both types are peculiar asteroids and have common dynamical behaviors. So, we regard these two type asteroids as one group and try to search for families among them.

Key words: Asteroids, Apollo-Amor, families of asteroids.

1. Conditions for Asteroid Families

From the theory of secular perturbations of asteroids, we can adopt the following three conditions to judge whether some asteroids constitute a family or not:

(1) Their semi-major axes (a) are almost the same;

(2) If we plot these asteroids on the v, u plane ($v = e \cos \varpi$ and $u = e \sin \varpi$, where e is the eccentricity and ϖ is the longitude of perihelion), the plotted points are distributed on a circle;

(3) If we plot these asteroids on the q, p plane ($q = \sin I \cos \Omega$ and $p = \sin I \sin \Omega$, where I is the inclination and Ω is the longitude of node), the plotted points are distributed on a circle.

To investigate the conditions (1), (2) and (3), we arrange the 111 Apollo-Amor type asteroids in order of semi-major axis. Then we plot them on the v, u and the q, p plane.

2. Results

In our previous investigation (Yuasa, 1985), we found amongst 65 Apollo-Amor type objects, two groups of asteroids, which were candidates for forming a family. For convenience, we call them the Eros family and the Amor family respectively. Figures 1 and 2 show the distribution of the elements of these two candidates on the v, u and the q, p plane as we obtained them in our previous investigation. In the present investigation, the same method is applied for an increased number (111) of Apollo-Amor type asteroids. The results are as follows:

(a) We do not find any other new candidates for families.

(b) The Eros family seems to be almost disappearing.

(c) The Amor family is still existing.

Figures 3 and 4 show the new distribution of points on the v, u and the q, p plane in the same region of semi-major axis as in the previous investigation.

Celestial Mechanics and Dynamical Astronomy **54:** 291–293, 1992.

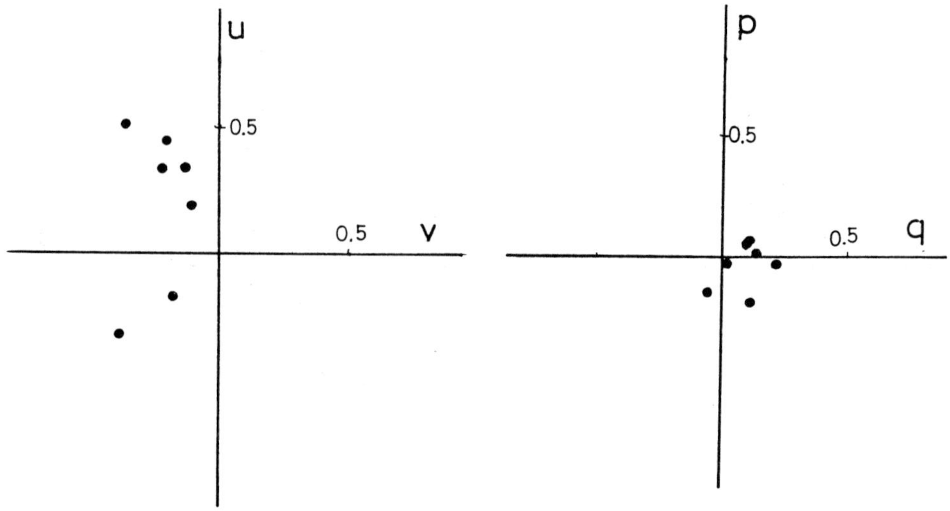

Fig. 1. Eros family.

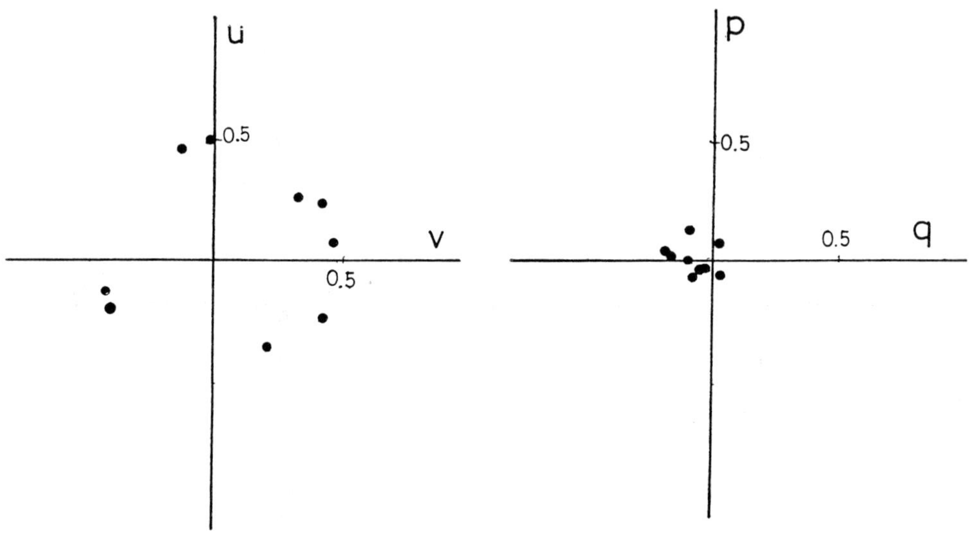

Fig. 2. Amor family.

As a concluding remark, let us point out that, in both our investigations, the number of objects considered is small and the osculating elements are used for simplicity. In order to clarify the situation, more detailed methods will have to be applied to an increased number of Apollo-Amor type objects, if possible by using mean orbital elements. We plan to do this in the near future.

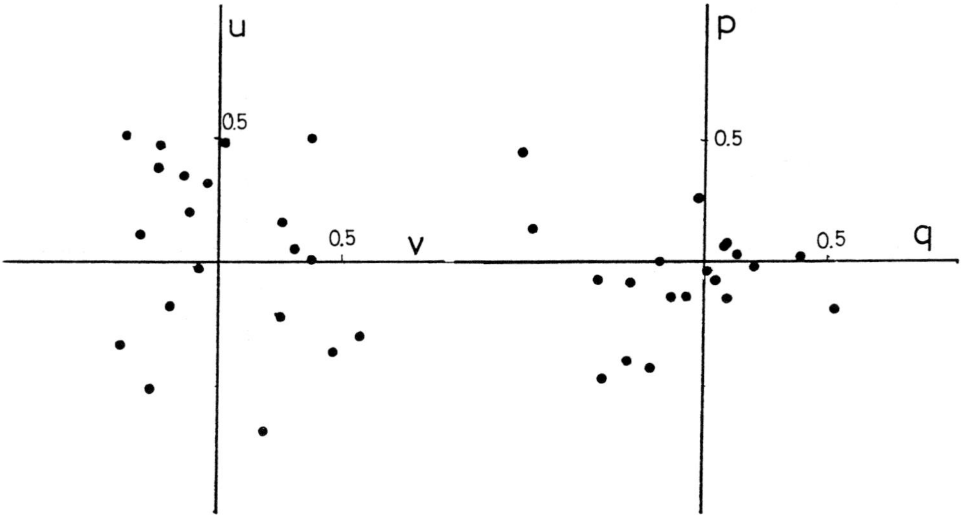

Fig. 3. Eros family (is almost disappearing).

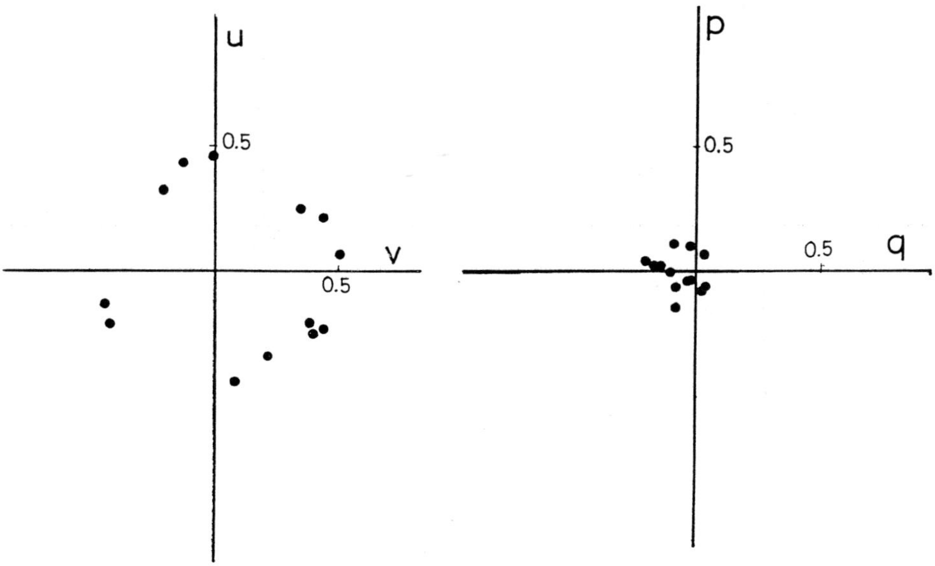

Fig. 4. Amor family (is still existing).

Reference

Yuasa M.: 1985, *Proc. of the 18th Symp. on Celestial Mechanics*, (in Japanese), 50–57.